∃L

✓ $\exists v \mathcal{A}$
$\mathcal{A}[c/v]$

(Here c must be a constant
new to the branch.)

∃R

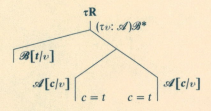

$\forall v \mathcal{A}$ ✓
$\mathcal{A}[c/v]$

(Here c must be a constant
new to the branch.)

∀L

∗ $\forall v \mathcal{A}$
$\mathcal{A}[t/v]$

=L

$\ell = \ell'$
$\mathcal{A}$
$\mathcal{A}[\ell/\ell']$
(or $\mathcal{A}[\ell'/\ell]$)

$\ell = \ell'$
$\mathcal{A}$
$\mathcal{A}[\ell/\ell']$
(or $\mathcal{A}[\ell'/\ell]$)

=R

$\ell = \ell$
Cl

τL

✓ $(\tau v\!:\mathcal{A})\mathcal{B}$
$\mathcal{B}[c/v]$
$\forall v(v = c \leftrightarrow \mathcal{A})$

(Here c must be a constant
new to the branch.)

τR

$(\tau v\!:\mathcal{A})\mathcal{B}^{*}$

$\mathcal{B}[t/v]$

$\mathcal{A}[c/v]$ — $\mathcal{A}[c/v]$
$c = t$ — $c = t$

(Here c must be a constant new to the
branch, but t may be any closed term
other than c.)

DEDUCTION

DEDUCTION
INTRODUCTORY SYMBOLIC LOGIC

DANIEL BONEVAC
UNIVERSITY OF TEXAS AT AUSTIN

MAYFIELD PUBLISHING COMPANY
PALO ALTO, CALIFORNIA

Library of Congress Catalog Card Number: 86-062995
International Standard Book Number: 0-87484-772-9

Manufactured in the United States of America
10 9 8 7 6 5 4 3 2 1

Mayfield Publishing Company
285 Hamilton Avenue
Palo Alto, California 94301

Sponsoring editor: James Bull
Manuscript editor: Antonio Padial
Managing editor: Pat Herbst
Production editor: Jan deProsse
Art director: Cynthia Bassett
Designer: Joseph di Chiarro
Production manager: Cathy Willkie
Compositor: Syntax International
Printer and binder: Maple-Vail

Contents

PREFACE

This book is a comprehensive introduction to symbolic logic. Logic has played a central role in the training of educated people since its origin in the writings of Aristotle. In the nineteenth century, mathematical techniques reshaped the traditional, Aristotelian conception of logic. The result, symbolic logic, has dominated the study of reasoning in the twentieth century. My aim in this book is to introduce readers to twentieth-century logic. I survey both the well-established core of modern logic, consisting of sentential and quantificational logic, and its less-settled but exciting periphery, extensions of sentential and quantificational logic. Such extensions—modal, conditional, and deontic logic—have sparked much research and philosophical excitement since the 1950s.

Logic occupies an important position in contemporary university curricula for much the same reason it occupied such a position in ancient academies, medieval centers of learning, and Enlightenment universities: Its object of study, reasoning, is fundamental to all intellectual activity and to most other human endeavors. In this book, I focus on reasoning in natural, spoken language; the problems at the ends of sections illustrate applications of logical techniques to reasoning in a wide variety of contexts. The intended audience is undergraduate students with little or no previous training in logic.

This book differs from virtually all other introductory symbolic logic texts by containing extensive discussions of modal, counterfactual, and deontic logics. Part III (Chapters 9–12) extends the logical systems of previous chapters to incorporate symbols representing *must, can, would, might, ought, may,* and related words and phrases. The resulting methods are accessible to beginning students; extending sentential logic to include modal connectives, for example, is less complicated than extending it to include quantifiers. The theories in Part III are nevertheless serious logical theories, not oversimplifications.

Extensions of classical logic play an important role in an introductory presentation of symbolic logic. First, they demonstrate that logic is not cut-and-dried but instead filled with activity, excitement, and controversy. Near the end of the eighteenth century, Immanuel Kant wrote that logic was a closed and completed subject, to which nothing significant had been contributed since the time of Aristotle and to which nothing significant remained to be contributed. Too many contemporary students receive a similar impression from their introductory courses, except that in today's world Russell and Whitehead occupy the venerated position that Aristotle held in Kant's time. In Part III of this book I try to show that logic is open-ended; that research on understanding reasoning in natural language is active and growing; and that many critical issues, even in areas already being investigated, remain to be settled.

Second, modal, counterfactual, and deontic logics can be used to analyze notions crucial to much natural-language reasoning. Modal and counterfactual systems offer approaches to conditionals—*if . . . then* sentences—that may be seen as competing with or complementing a truth-functional account. Modal and counterfactual concepts are fundamental to all reasoning concerning possibility or necessary connections; deontic concepts are central to moral and practical reasoning and, thus, to deliberation.

Third, a good introduction to logic should cover not only the essentials of logical theory and the application of theory to actual bits of reasoning but also communicate a mode of thinking. It should help students understand what it is to think like a logician and how logicians construct and evaluate theories. In Part III, accordingly, I present various logical systems extending classical logic and also discuss their logical and philosophical motivations and commitments. Part III encourages students to think about how to evaluate existing theories, how to choose among competing theories, and how to construct new theories.

ORGANIZATION

The first two parts of this book outline classical symbolic logic. Part I presents the central ideas of logic and develops a system of sentential logic. Part II develops quantificational logic, including identity, function symbols, and descriptions. Part III extends classical symbolic logic to modal notions, counterfactual conditionals, and deontic concepts.

Part I begins with a chapter on fundamental logical concepts: validity, implication, equivalence, contradictoriness, and the like. It defines these concepts rigorously while focusing on their usefulness in understanding natural-language inference. Chapter 2 presents the language of sentential logic, its relationship to natural language, and the method of truth tables. Chapter 3 concerns semantic tableaux, an efficient and straightforward technique for testing for validity, implication, and so on. Each connective has two rules: One tells us what we can infer if a formula of a certain form is true; the other tells us what to infer if a formula of that form is false. Chapter 4 develops a method of natural deduction for sentential logic. Most connectives similarly have two rules: One introduces formulas of a certain form into a deduction; the other shows how to exploit the presence of such formulas to deduce others. Additionally, there are methods of indirect and conditional proof and a variety of derivable rules that make deductions more efficient.

Part II presents quantificational logic. Chapter 5 introduces the language of quantificational logic and discusses the symbolization of natural-language sentences in detail. Chapter 6 extends the semantic tableau technique to predicate logic while also presenting the semantics of quantifiers. Chapter 7 introduces rules for quantifiers into the natural deduction method, and Chapter 8 develops ways of incorporating identity, function symbols, and definite descriptions into the quantificational language. This chapter presents two approaches to definite descriptions: In one approach, they are treated as singular terms; in the other, as quantifierlike expressions.

Part III extends classical logic to include several nontruth-functional connectives. Chapter 9 presents sentential modal logic, concentrating chiefly on S5 but also addressing M (or T) and S4. It develops both tableau and deduction techniques. Chapter 10 develops natural deduction systems for counterfactuals. One system corresponds to David Lewis's theory of counterfactuals; another corresponds to Robert Stalnaker's theory. The basic system exploits the commonalities in these approaches. Chapter 11 presents a system of deontic logic, with both semantic tableau rules and natural deduction rules. Finally, Chapter 12 concerns quantified modal logic, developing both classical and "free" tableau systems and deduction systems.

PROBLEMS

To understand a logical system, the student must apply its methods to particular problems. This book incorporates many problems of varying difficulty. I have marked particularly challenging exercises with an asterisk (*). Some problems have no clear-cut correct answers; they are intended to prompt discussion or essay answers; I have marked those with two asterisks (**). Solutions to roughly 20% of the problems appear in the back of the book; the symbol ▶ indicates that the solution to a problem appears in the back.

ADDITIONAL FEATURES

Several features of this book have proven especially useful in the classroom. First, the book presents new semantic and proof-theoretic logical techniques that are both simple and powerful. Based on E. W. Beth's semantic tableaux, my semantic method bears much similarity to Richard Jeffrey's truth trees. The technique is very easy to teach and to learn. I strive to make my natural deduction system as straightforward and natural as possible. The pattern of rules is easy to understand; most connectives have introduction and exploitation rules. The system's form greatly simplifies deductive rules and strategies. Neither the tableau system nor the deduction system uses free variables; instantiation always involves a constant or closed function term. Indeed, in this book, formulas never contain free variables; quantifiers on the same variable never overlap in scope. The need for distinguishing free from bound variables never arises.

Second, this book takes natural language more seriously than many logic texts do. In talking about the translation of English sentences into quantification theory, for example, I do more than offer examples and suggestions; I outline strategies that almost add up to an algorithm for a certain fragment of English. This explicit treatment of translation has several advantages. Students learn to translate into logical notation faster and more successfully when given explicit instructions for performing the task. Students also learn a great deal more about the limitations of the logical systems they study when they see what does not translate into the notation at all, or what translates only with some loss of meaning. Most importantly, an explicit treatment of translation can help students to become more careful readers and writers. For example, few students understand the difference between a restrictive and appositive relative clause. But the distinction makes a difference for translation; sentences can have different truth conditions, depending on whether a relative clause is restrictive or appositive. By considering these differences, students derive an increased understanding of various linguistic mechanisms and are thus better able to use their language to express themselves clearly and precisely.

ACKNOWLEDGMENTS

I am grateful to Jim Bull of Mayfield Publishing Company for his frequent advice and encouragement during my work on the manuscript and to the reviewers whose reactions did much to improve the book: Wayne Davis, Georgetown University; Ruth Manor, San Jose State University; Richard Parker, California State University, Chico; and Bangs Tapscott, University of Utah. I owe thanks to Antonio Padial for his editing and to Jan de Prosse for her work in the production of the book.

I am deeply indebted to Gerald Massey; the conception of introductory symbolic logic that this book manifests is, in many respects, his. Additionally, he formulated the version of Beth's semantic tableaux used in this book, read earlier versions of several chapters, and offered some much-needed encouragement. I also owe a great deal to my other logic teachers: Ermanno Bencivenga, Carl Posy, Charles Parsons, Kenneth Manders, and Nuel Belnap.

I am very grateful to my colleagues Nicholas Asher and Hans Kamp, who used earlier versions of the manuscript to teach introductory logic courses and suggested many improvements. I would also like to thank Maria Slowiaczek, Richard Larson, and Thomas Seung, who read early versions of various chapters.

I owe much to the students and teaching assistants at the University of Texas who used versions of this book and helped to improve it, especially

Owen Goldin, John Watkins, Clarence Bonnen, William Boon, Christopher Colvin, Craig Hansen, and Mark Bauder. I am grateful to the Center for Cognitive Science of the University of Texas at Austin for its support, and, in particular, to Adrienne Diehr and Kim Krull, without whose help the manuscript would have been completed much later. Finally, I want to thank Sheila Asher, Roy Flukinger, Martha Preston, and, above all, my family—my wife Beverly and Jessica, Gwen, Sarah Jane, and Leela—for their support and their friendship.

NOTES ON COURSE ORGANIZATION

To give instructors flexibility, I include far more than a semester's worth of material in the book. Many courses may be carved from it. The discussion of classical logic is independent of the treatment of its extensions; instructors can easily restrict attention to Parts I and II. Presentations of the tableau and deduction methods are also independent so that instructors can use the book to teach only one technique if they choose. Furthermore, the order is flexible: Topics from Chapters 9–11 may be discussed before the presentation of quantification theory in Chapters 5–8. The table below helps instructors plan courses and guides readers interested in particular topics. The following table indicates what each section of the book presupposes.

Sections	Presuppositions
1	None
2.1–2.5	1
2.6–2.7	2.1–2.5, 1
3	2.1–2.5, 1
4	2.1–2.5, 1
5	2.1–2.5, 1
6.1–6.2	5, 3, 2.1–2.5, 1
6.3	5, 2.1–2.5, 1
6.4	6.1–6.3, 5, 2.1–2.5, 1
7	5, 4, 2.1–2.5, 1
8.1	5, 2.1–2.5, 1
8.2	8.1, 6.1–6.2, 5, 3, 2.1–2.5, 1
8.3	8.1, 7, 5, 4, 2.1–2.5, 1
8.4–8.5 (t)	8.2, 8.1, 6.1–6.2, 5, 3, 2.1–2.5, 1
8.4–8.5 (d)	8.3, 8.1, 7, 5, 4, 2.1–2.5, 1
9.1–9.3	2.1–2.5, 1
9.4–9.6	9.1–9.3, 3, 2.1–2.5, 1
9.7	9.1–9.3, 4, 2.1–2.5, 1
9.8 (t)	9.1–9.6, 3, 2.1–2.5, 1
9.8 (d)	9.7, 9.1–9.3, 4, 2.1–2.5, 1

Sections	Presuppositions
10.1	9.1–9.3, 2.1–2.5, 1
10.2–10.4	10.1, 9.7, 9.1–9.3, 4, 2.1–2.5, 1
11.1	9.1–9.3, 2.1–2.5, 1
11.2	11.1, 9.1–9.6, 3, 2.1–2.5, 1
11.3	11.1, 9.7, 9.1–9.3, 4, 2.1–2.5, 1
11.4	11.1, 9.1–9.3, 2.1–2.5, 1
12.1 (t)	9.1–9.6, 6.1–6.2, 5, 3, 2.1–2.5, 1
12.1 (d)	9.7, 9.1–9.3, 7, 5, 4, 2.1–2.5, 1
12.2 (t)	12.1 (t), 9.1–9.6, 6.1–6.2, 5, 3, 2.1–2.5, 1
12.2 (d)	12.1 (d), 9.7, 9.1–9.3, 7, 5, 4, 2.1–2.5, 1
12.3 (t)	12.1 (t), 9.1–9.6, 8.5 (t), 8.1–8.2, 6.1–6.2, 5, 3, 2.1–2.5, 1
12.3 (d)	12.1 (d), 9.7, 9.1–9.3, 8.5, 8.3, 8.1 (d), 7, 5, 4, 2.1–2.5, 1
12.4 (t)	12.1–12.2 (t), 11.1–11.2, 9.1–9.6, 6.1–6.2, 5, 3, 2.1–2.5, 1
12.4 (d)	12.1–12.2 (d), 11.1, 11.3, 9.7, 9.1–9.3, 7, 5, 4, 2.1–2.5, 1

(Note: (d) indicates a section's discussion of deduction methods, and (t) indicates a section's discussion of tableau methods, in those sections that discuss both.)

DEDUCTION

PART

I

CLASSICAL SENTENTIAL LOGIC

1

Basic
Concepts
of
Logic

lerk: Mr. McClory moves to postpone for ten days further consideration of whether sufficient grounds exist for the House of Representatives to exercise constitutional power of impeachment unless by 12 noon, eastern daylight time, on Saturday, July 27, 1974, the president fails to give his unequivocal assurance to produce forthwith all taped conversations subpoenaed by the committee which are to be made available to the district court pursuant to court order in United States v. Mitchell. . . .

Mr. Latta: . . . I just want to call [Mr. McClory's] attention before we vote to the wording of his motion. You move to postpone for ten days unless the president fails to give his assurance to produce the tapes. So, if he fails tomorrow, we get ten days. If he complies, we do not. The way you have it drafted I would suggest that you correct your motion to say that you get ten days providing the president gives his unequivocal assurance to produce the tapes by tomorrow noon.

Mr. McClory: I think the motion is correctly worded; it has been thoughtfully drafted.

Mr. Latta: I would suggest you rethink it. . . .

Mr. Mann: Mr. Chairman, I think it is important that the committee vote on a resolution that properly expresses the intent of the gentleman from Illinois [Mr. McClory], and if he will examine his motion he will find that the words *fail to* need to be stricken. . . .

Mr. McClory: If the gentleman will yield, the motion is correctly worded. It provides for a postponement for ten days unless the president fails tomorrow to give his assurance, so there is no postponement for ten days if the president fails to give his assurance, just one day. I think it is correctly drafted. I have had it drafted by counsel, and I was misled originally, too, but it is correctly drafted. There is a ten-day postponement unless the president fails to give assurance. If he fails to give it, there is only a twenty-four-hour or there is only a twenty-three-and-a-half-hour day.

House Judiciary Committee, July 26, 1974

Logic is the study of correct reasoning. Logic pertains to virtually all subjects, since people can reason about anything they can think about. Politics, the arts, literature, business, the sciences, and everyday problems are all subjects open to reasoning. As the House debate quoted above suggests, sometimes the reasoning is good; sometimes, not so good. People use logic to tell the difference.

Using logic, we can evaluate bits of reasoning as proper or improper, good or bad. Logic is not the study of how people do reason, but how they should reason. We might put this point differently by saying that logic does not describe real reasoning, with its errors, omissions, and oversights; it prescribes methods for justifying reasoning, that is, for showing that a given bit of reasoning is proper. Logic thus describes an ideal that actual reasoning strives for but often fails to reach.

Logic begins with the study of how people use language. Before we can develop a system of logic, it's necessary to understand how people actually reason. To eliminate the errors that creep into people's performance, we need to examine people's considered judgments about the correctness or incorrectness of inferences. No matter what mental processes people go through to achieve the right result, they try to follow rules for putting sentences together to form proper bits of reasoning. Logic is not merely a description of reasoning; logicians examine people's evaluations of bits of reasoning to say what the rules of correct reasoning are. Logic describes not the process of reasoning but the rules of correct reasoning.

The logical systems discussed in this book are designed primarily to explicate the meanings of certain expressions or constructions in natural languages such as English, Chinese, Swahili, and German. Empirical facts about language are obviously highly relevant to any evaluation of these systems. The systems succeed only insofar as they really do account for the kinds of reasoning they are trying to explain. The chief aim of the book is to help readers understand and evaluate natural-language arguments. The first and second parts develop a framework for analyzing reasoning; in virtually all approaches to logic, such a framework is the starting point. The third part applies the framework to some interesting concepts and constructions important to reasoning in natural language.

1.1 ARGUMENTS

Arguments are reasoning in language. Frequently, we think of arguments as heated debates, disagreements, or disputes. Sometimes, however, we speak of a politician arguing for the passage of a bill, a lawyer arguing a case, or a moviegoer arguing that *North by Northwest* is better than *The 39 Steps*. In this latter sense, an argument starts with some assertions and tries to justify a particular thesis. In this sense, one argues *for* the thesis.

Many arguments in natural language are complicated. A lawyer arguing for the innocence of a client, for instance, offers many specific arguments in presenting the case. The lawyer may argue that a piece of evidence is inadmissible, that the results of a lab test are ambiguous, that the client could not have reached the scene of the crime by the time it was committed, and so on. All these smaller arguments form part of the larger argument for the client's innocence.

We can divide natural-language arguments, then, into two groups: *extended* arguments, which contain other arguments, and *simple* arguments, which do not. Extended arguments may have several conclusions. Such arguments may consist of several simple arguments in sequence. They may contain other extended arguments. And they may consist of a list of premises, followed by several conclusions stated at once.

Mathematical proofs are extended arguments. A mathematician often begins a proof by stating some assumptions. The mathematician then draws out consequences of the assumptions, perhaps making other assumptions along the way. Finally, the proof ends with a conclusion, the theorem it proves. A mathematical proof is thus a series of simple arguments.

We'll begin by focusing on simple arguments. A simple argument, like an extended argument, starts with some assertions justifying a thesis. The initial assertions of the argument are its *premises;* the thesis the argument tries to justify is its *conclusion.* We'll be so often concerned with simple arguments that we'll drop the adjective *simple* and speak simply of *arguments.* (Later, when we examine proofs, we'll just call them *proofs.*)

> **DEFINITION.** An *argument* is a finite sequence of sentences, called *premises,* together with another sentence, the *conclusion,* which the premises are taken to support.

An argument in ordinary language or in mathematics is a string or sequence of sentences. The sentences that make up the argument are in a particular order, whether the argument is spoken, written, or encoded in a computer language. For our purposes in this text, the order of the premises will make no difference. So, we generally won't worry about order of presentation. But we will require that the string of premises be finite. No one has the patience to listen to an argument that runs on forever.

Arguments consist of sentences. In this text, we'll be interested only in sentences that can be true or false. Many ordinary sentences, including almost every one in this book, fall into this category. They say something about the way the world is, and they might be correct or incorrect in so describing it. But commands, for example, are different: 'Shut the door' can be appropriate or inappropriate, irritating or conciliatory, friendly or hostile, but it cannot be true or false. Questions, too, are neither true nor false: Consider 'What time is it?' or 'What is the capital of Zaire?' Interjections—'Ouch!,' 'All right!,' 'Alas!,' and most curses—are likewise neither true nor false.

A sentence is true or false in a particular context: as used on a particular occasion, by a particular speaker, to a particular audience, in a given circumstance, as part of a given discourse. Without all this contextual information, we can't begin to say whether a sentence such as 'I love you' is true or false. The truth value of the sentence clearly depends on who I and you are, when the sentence is spoken, and so on. Sentences have truth values only relative to a certain context of use.

Nevertheless, very little in the following pages will involve context directly. So, we'll generally speak of sentences as having truth values, trusting ourselves to remember that these values are relative to context.

An argument, according to our definition, contains one sentence that is its conclusion. This is an idealization: In natural language, a conclusion may be a clause in a sentence; it may be spread across several sentences; or it may be left unstated. Premises, too, may be clauses in sentences, spread across several sentences, or left unstated.

The definition does not specify how to pick out the conclusion of an argument. In English, certain words or phrases typically signal the conclusion of an argument, while others signal premises.

Conclusion Indicators
---

therefore, thus, hence,
so, consequently, it follows that,
in conclusion, as a
result, then, must

Premise Indicators
---

because, for, since

All these words and phrases have other uses; they do not always indicate premises or conclusions. But these words and phrases can, and often do, serve as indicators because they can attest to relations of support among the sentences of an argument. 'Since Fred forgot to go to the interview, he won't get the job' presents a simple argument in a single English sentence. The word *since* indicates that we should take 'Fred forgot to go to the interview' as a premise, supporting the conclusion 'he won't get the job.' (In the sentence 'Since Fred forgot to go to the interview, he's been depressed,' however, *since* indicates a temporal relationship.) Similarly, 'Jane's business must be doing well; she drives a Mercedes' constitutes an argument. The auxiliary verb *must* marks 'Jane's business is doing well' as the conclusion, supported by the evidence in 'she drives a Mercedes.'

Premise indicators often signal not only that one or more sentences are premises but also that a certain sentence is a conclusion. *Since,* for example, exhibits a relation of support between the sentences it links. The occurrence of *since* in 'Since Fred forgot to go to the interview, he won't get the job' points out not only that the sentence immediately following it is a premise but also that the sentence 'he won't get the job' is a conclusion. Similarly, the occurrence of *for* in 'Northern Indiana Public Service will not pay its usual dividend this quarter, for the court refused to allow expenditures on its now-canceled nuclear project into the rate base' indicates both that 'the court refused to allow expenditures on its now-canceled nuclear project into the rate base' is a premise and that 'Northern Indiana Public Service will not pay its usual dividend this quarter' is a conclusion.

Indicators provide important clues to the structure of arguments. Often, however, no explicit indicators appear. Sometimes the conclusion isn't even stated. In such cases, we must consider the point of the argument. What is the author trying to establish? In some arguments, the conclusion is quite clear, but other arguments are hard to analyze.

Consider some examples:

> Suppose we argued that what was true was true for us, that two assertions met on no common ground, so that neither was "really true" or "really false." This position went further than skepticism and declared the belief in error itself to be erroneous. Royce called this view that of the total relativity of truth, and he had an argument against it. If the statement "There is error" is true, there is error; if it is false, then there is, ipso facto, error. He could only conclude that error existed; to deny its existence was contradictory.
>
> *Bruce Kuklick,*
> The Rise of American Philosophy

> If it were permitted to reason consistently in religious matters, it is clear that we all ought to become Jews, because Jesus Christ was born a Jew, lived a Jew, and died a Jew, and because he said that he was accomplishing and fulfilling the Jewish religion.
>
> *Voltaire*

These constitute extended arguments. The conclusion of Royce's smaller argument is plainly 'error exists'; the words *conclude that* make this obvious. Royce then uses this conclusion to argue that the view of the total relativity of truth is false. The sentence 'error exists' thus functions both as the conclusion of one argument and as a premise of another, all within the same extended argument.

Voltaire seems to be arguing for the conclusion 'we all ought to become Jews.' Here the key word is *because,* which indicates that the rest of the argument is a list of premises. Of course, Voltaire, a satirist, is really aiming not at this conclusion but at another. Everything he says is supposed to follow from the hypothetical 'if it were permitted to reason consistently in religious matters.' Like Royce, he is offering an argument within an extended argument. The conclusion of the larger argument is not stated. Nevertheless, it's easy to see that Voltaire is trying to establish that it is not permitted to reason consistently in religious matters. The conclusion of the smaller argument— 'we all ought to become Jews'—is an observation that few Christians would willingly accept, even though, according to Voltaire, their own doctrine commits them to it.

A final example is a mathematical proof: the traditional proof that the square root of 2 is irrational.

[Suppose] for the sake of argument that $\sqrt{2}$ is rational, i.e., that there are two integers, say m and n, which are mutually prime and which are such that $m/n = \sqrt{2}$ or $m^2 = 2n^2$. From this it follows that m^2 must be even and with it m, since a square number cannot have any prime factor which is not also a factor of the number of which it is the square. But if m is even, n must be odd according to our initial supposition that they are mutually prime. Assuming that $m = 2k$, we can infer that $2n^2 = 4k^2$, or $n^2 = 2k^2$; and from this it can be shown by a repetition of the reasoning used above that n must be even. Our hypothesis, therefore, entails incompatible consequences, and so it must be false.

W. Kneale and M. Kneale,
The Development of Logic

Like almost any proof, this one is an extended argument; in fact, it is a series of simple arguments. The proof begins with the assumption that $\sqrt{2}$ is rational. The first simple argument concludes that m^2 must be even; very quickly follows another simple argument concluding that m must also be even. The third simple argument concludes that n is odd. The fourth concludes that $2n^2 = 4k^2$; the fifth, that $n^2 = 2k^2$; the sixth, that n is even. Finally, the proof ends with a seventh simple argument that the hypothesis that the square root of 2 is rational is false.

When we write an argument "officially," in what we'll call *standard form,* we'll list the premises in the order in which they are given, and then list the conclusion. So, in our official representations, conclusions will always come last. This isn't true in natural language, as Voltaire's argument shows; conclusions may appear at the beginning, in the middle, or at the end of arguments, if they are stated at all. In addition, we'll preface the conclusion with the symbol $\therefore$, which means "therefore."

To see how these representations work, let's write Royce's smaller argument in standard form:

If the statement 'There is error' is true, there is error.
If the statement 'There is error' is false, there is error.
$\therefore$ There is error.

Royce's larger argument, then, is:

> Error exists.
> The view of the total relativity of truth holds that the belief in error
> is erroneous.
> ∴ The view of the total relativity of truth is false.

We can similarly express Voltaire's two arguments in standard form:

> Jesus Christ was born a Jew, lived a Jew, and died a Jew.
> Jesus Christ said he was accomplishing and fulfilling the Jewish
> religion.
> ∴ If it were permitted to reason consistently in religious matters, it
> is clear that we all ought to become Jews.

> If it were permitted to reason consistently in religious matters, it is
> clear that we all ought to become Jews.
> (It's not clear to religious Christians that we all ought to become
> Jews.)
> ∴ It is not permitted to reason consistently in religious matters.

Finally, we can express the proof of the irrationality of $\sqrt{2}$ as a series
of simple arguments:

> $\sqrt{2}$ is rational, i.e., there are two integers, say m and n, that are
> mutually prime and that are such that $m/n = \sqrt{2}$ or $m^2 = 2n^2$.
> ∴ m^2 is even.

> m^2 is even.
> A square number cannot have any prime factor that is not also a
> factor of the number of which it is the square.
> ∴ m is even.

> m is even.
> m and n are mutually prime.
> ∴ n is odd.

> $m = 2k$
> ∴ $2n^2 = 4k^2$

> $2n^2 = 4k^2$
> ∴ $n^2 = 2k^2$

> $n^2 = 2k^2$
> (A repetition of the reasoning used above.)
> ∴ n is even.

The hypothesis that $\sqrt{2}$ is rational entails incompatible consequences.
∴ The hypothesis that the square root of 2 is rational is false.

Problems

Write each of the following arguments in standard form. If there are several arguments in a passage, write each separately.

1. The Bears did well this year, and so they'll probably do well again next year.

2. John must have left already; his books are gone.

3. Few contemporary novels deal explicitly with political themes. The study of contemporary literature is therefore largely independent of the study of political culture.

4. Mary dislikes Pat. Consequently, it's unlikely that they'll work on the project together.

▶ 5. Most criminals believe that their chances of being caught and punished are small; thus, the perceived costs of a life of crime are low.

6. The building will generate large tax write-offs. As a result, it will be a good investment even if it yields little direct profit.

7. No one has ever constructed a convincing case that Bacon or someone else wrote the plays we generally attribute to Shakespeare. Shakespeare, then, almost certainly wrote the plays we attribute to him.

8. There are no centaurs, for centaurs are mythical creatures, and no mythical creatures really exist.

9. Nobody will ever find an easy way to get rich, because people have been looking for centuries, and nobody's ever found one yet.

▶ 10. Swedish is an Indo-European language, but Finnish isn't. Hence Finnish is more difficult for English-speakers to learn than Swedish.

11. Many people are easily shocked by unusual or threatening events. No one who is thunderstruck can think clearly. It follows that the emotions can obstruct reason.

12. Since happiness consists in peace of mind, and since durable peace of mind depends on the confidence we have in the future, and since that confidence is based on the science we should have of the nature of God and the soul, it follows that science is necessary for true happiness (Leibniz).

13. In Europe, pupils devote time during each school day to calisthenics. American schools rarely offer a daily calisthenic program. Tests prove that our children are weaker, slower, and more short-winded than European children. We must conclude that our children can be made fit only if they participate in school calisthenics on a daily basis (LSAT test, 1980).

14. Marks found on the dented case indicated that it had been loaded in and extracted from a weapon at least three times (26H449). In addition, it had "three sets of marks on the base" that were not found on the others or on any of the numerous test cartridges obtained from Oswald's rifle (26H449). A ballistics expert testified that these anomalous marks were possibly caused by a "dry firing" run—that is, by inserting the empty cartridge case in the breech while practicing with the rifle (3H510). Of all the various marks discovered on this case, only one set links it to Oswald's rifle, and this set was identified as having come from the magazine follower. Yet the magazine follower marks only the last cartridge in a clip, a position that must have been occupied on November 22 not by the dented case but by the live round subsequently found in the chamber. Thus, unlike the other two cases that bear marks from the chamber and bolt of Oswald's rifle, the only mark borne by the dented case, linking it to Oswald's rifle, could not have been incurred on November 22 (Josiah Thompson).*

▶ 15. The earth receives radiant heat from the sun and loses heat to outer space by its own radiative emissions. The energy received undergoes many transformations. But in the long run no appreciable fraction of this energy is stored on the earth, and there is no persistent trend toward higher or lower temperatures (LSAT test, 1980).*

16. To describe an equilateral triangle on a given finite straight line: Let *AB* be the given straight line; it is required to describe an equilateral triangle on *AB*. From the center, *A*, at the distance *AB*, describe the circle *BCD*. From the center, *B*, at the distance *BA*, describe the circle *ACE*. From the point *C*, at which the circles cut one another, draw the straight lines *CA* and *CB* to the points *A* and *B*. *ABC* shall be an equilateral triangle. Because the point *A* is the center of the circle *BCD*, *AC* is equal to *AB*. And because the point *B* is the center of the circle *ACE*, *BC* is equal to *BA*. But it has been shown that *CA* is equal to *AB*; therefore *CA* and *CB* are each of them equal to *AB*. But things which are equal to the same thing are equal to one another. Therefore *CA* is equal to *CB*. Therefore, *CA*, *AB*, *BC* are equal to one another. Wherefore the triangle *ABC* is equilateral, and it is described on a given straight line *AB* (Euclid).*

17. One may well ask, "How can you advocate breaking some laws and obeying others?" The answer is found in the fact that there are two types of laws: There are *just* laws and there are *unjust* laws. I would be the first to advocate obeying just laws. One has not only a legal but a moral responsibility to obey just laws. Conversely, one has a moral responsibility to disobey unjust laws. I would agree with Saint Augustine that "An unjust law is no law at all."

Now what is the difference between the two? How does one determine whether a law is just or unjust? A just law is a man-made code that squares with the moral law or the law of God. An unjust law is a code that is out of harmony with the moral law. To put it in the terms of Saint Thomas Aquinas, an unjust law is a human law that is not rooted in eternal and natural law. Any law that uplifts human personality is just. Any law that degrades human personality is unjust.

All segregation statutes are unjust because segregation distorts the soul and damages the personality . . . (Martin Luther King, Jr.).

18. Theorem (mean-value theorem). Let $a, b \in \mathcal{R}$, $a < b$, and let f be a continuous real-valued function on $[a, b]$ that is differentiable on (a, b). Then there exists a number $c \in (a, b)$ such that $f(b) - f(a) = (b - a)f'(c)$.

Proof: Define a new function F: $[a, b] \to \mathcal{R}$ by

$$F(x) = f(x) - f(a) - \frac{f(b) - f(a)}{b - a} \cdot (x - a)$$

for all $x \in [a, b]$. (Geometrically $F(x)$ is the vertical distance between the graph of f over $[a, b]$ and the line segment through the end points of this graph.) Then F is continuous on $[a, b]$, differentiable on (a, b), and $F(a) = F(b) = 0$. By Rolle's theorem, there exists a $c \in (a, b)$ such that $F'(c) = 0$. Thus,

$$F'(c) = f'(c) - \frac{f(b) - f(a)}{b - a} = 0$$

proving the result (M. Rosenlicht).*

19. If the premise, that independence means complete absence of subconscious bias, were carried to its logical conclusion, no one could be found independent in any absolute sense. One is bound to be influenced by his environment, and everyone has some subconscious bias. On this premise it could be argued that the accountant who frequently lunches with a client, or plays golf or bridge with him, or serves with him on a board of trustees of a church or school, should not be considered independent in certifying the financial statements of that client. . . . [This would lead] to the conclusion that certified public accountants who act as independent auditors for a company must avoid all other relationships with that company, leaving the accounting work to be performed by other CPAs, and accepting accounting work only from clients who retain other independent auditors.

This conclusion seems almost fantastically inconsistent with settled practice and the temper of the community generally. There would be serious economic loss in depriving the businessman who needs accounting service of the knowledge that the independent auditor gains about the business, particularly its accounting aspects, in the course of his audit (William Carey).*

20. I should like to suggest that neither composition nor literature is an intellectual field in its own right. Literary study obviously connects with a number of genuine intellectual fields like history and philosophy. Composition, too, has disciplinary connections with linguistics and psychology. But neither literary study nor composition is an intellectual discipline. Both are primarily cultural subjects with cultural missions of unparalleled importance. To the extent that we evade those missions under the banner of some neutral formalism or disciplinary pretense, we are neglecting our primary educational responsibilities and are also making an empirical mistake. This clearly implies that we should return to an integrated conception of "English" based on the pattern of the old literature-and-composition course originated by Blair and followed traditionally in the schools and colleges (E. D. Hirsch, Jr.).*

1.2 VALIDITY

Some arguments are good; others aren't. According to the definition of argument given above, any string of sentences counts as an argument if it's possible to single out one sentence as the conclusion, purportedly supported by the others. What distinguishes good from bad arguments? What makes a good argument good? What makes a bad argument unsuccessful?

People typically demand many things of an argument. Throughout this book, we'll focus primarily on one of them. A good argument should link its premises to its conclusion in the right way. There should be some special connection between the premises and the conclusion.

To see what this special connection is, consider an argument that has true premises and a true conclusion, but is nevertheless bad:

Harrisburg is the capital of Pennsylvania.
Richmond is the capital of Virginia.
∴ Austin is the capital of Texas.

What's wrong with this argument? Basically, the facts cited in the premises have nothing to do with the truth or falsehood of the conclusion. Texas could move its capital—to, say, Del Rio—while Harrisburg and Richmond remained the capitals of their respective states. That is, the conclusion of this argument could turn out to be false, even when the premises were true. The truth of the premises does nothing to guarantee the truth of the conclusion. This is the mark of a *deductively invalid* argument: Its premises could all be true in a circumstance in which its conclusion is false.

In a *deductively valid* argument, the truth of the premises guarantees the truth of the conclusion. If the premises are all true, then the conclusion has to be true. Consider, for example, this argument:

> Paris is, and has always been, the capital of France.
> Ed has never visited Paris.
> ∴ Ed has never visited the capital of France.

In any circumstance in which the premises of this argument are true, the conclusion must be true as well. It's impossible to conceive of a state of affairs in which Ed has visited the French capital without visiting Paris, if Paris is, and has always been, the capital of France. To put this differently, the only way to imagine Ed visiting the French capital without visiting Paris is to imagine the case that Paris is not the capital of France. In a deductively valid argument, the truth of the premises guarantees the truth of the conclusion. Or, to say the same thing, if the conclusion of a deductively valid argument is false, at least one premise must also be false.

> **DEFINITION.** An argument is *deductively valid* if and only if it's impossible for its premises all to be true while its conclusion is false.

It's possible, then, for a deductively valid argument to have true premises and a true conclusion, (at least some) false premises and a false conclusion, or false premises and a true conclusion. But no deductively valid argument has true premises and a false conclusion.

Some Deductively Valid Arguments

True Premises, True Conclusion	False Premises, False Conclusion	False Premises, True Conclusion
Daniel is human. All humans are mortal. ∴ Daniel is mortal.	Daniel is a dog. All dogs eat mice. ∴ Daniel eats mice.	Daniel is a dog. All dogs sleep. ∴ Daniel sleeps.

Each of these arguments is deductively valid: In each case, there is no possible circumstance in which the premises are all true but the conclusion is false. How could it be true that Daniel is a dog, and true that all dogs eat mice, but false that Daniel eats mice? Whether the premises and conclusion are actually true or false makes little difference to the validity of the argument. What matters is that if the premises are true the conclusion cannot be false.

Thus, not every argument with true premises and a true conclusion is deductively valid, as the argument concerning state capitals shows. Similarly, many arguments with false premises and a true conclusion are deductively invalid. The same is true for arguments with false premises and a false conclusion. So, although valid arguments can have any of these three combinations of truth and falsity, not every argument with those combinations is valid. We've said that an argument is deductively invalid if it's possible for the premises to be true while the conclusion is false. Similarly, an argument is deductively valid if and only if its conclusion has to be true if its premises are all true.[1]

Some deductively invalid arguments nevertheless have some legitimate force in reasoning. Although the truth of the premises of such an argument does not guarantee the truth of its conclusion, it does make the truth of the conclusion probable. Consider for example, this argument:

Most cats like salmon better than beef.
Gwen is a cat.
∴ Gwen likes salmon better than beef.

It's possible for the premises to be true while the conclusion is false. Gwen may be atypical; she may prefer beef to salmon. So the argument is deductively invalid. Nevertheless, the premises lend some support to the conclusion. Given just the information in the argument itself, the conclusion is more likely to be true than false. Arguments such as this are called *inductively valid*. They are extremely important in both scientific and everyday reasoning. Evaluating them, however, requires developing theories of probability and statistics. In this book, therefore, we'll restrict our attention to deductive validity and invalidity.

When we imagine a circumstance in which some sentences are true and others are false, we normally imagine no more than we must. Above, for example, we imagined a case in which Texas moved its capital, but Pennsylvania and Virginia didn't. That case was all we imagined, or, apparently, needed to imagine, to convince ourselves that the argument was invalid. But that case did not even approximate a complete description of an entire world. We said nothing about what happened to Montana, or Alaska, or Afghanistan, or the pennant hopes of the Mets, or the price of pork bellies on the Chicago Board of Trade. The case we've described, therefore, isn't very determinate. The world could be many different ways and still fit our description. So, it might be more correct to say that we imagined, not a single case, but a kind of case in which the premises are all true and the conclusion is false. Many circumstances might fit the description we gave.

So far we've examined only one criterion for success in arguments: We want arguments to be valid. Deductively valid arguments always preserve truth; if they begin with true premises, they carry us to true conclusions. But we might demand other things of arguments as well.

A *sound* argument has true premises and is valid. Furthermore, since, in any valid argument, the truth of the premises guarantees the truth of the conclusion, it also has a true conclusion.

DEFINITION. An argument is *sound* if and only if (a) it is valid and (b) all its premises are true.

Most of this book focuses, not on soundness, but on validity. Logicians have always concentrated on validity.[2] This focus is easy to understand.

First, validity is obviously a crucial component of soundness. We can't evaluate whether an argument is sound without first determining whether it's valid.

Second, evaluating soundness requires judging the actual truth or falsehood of premises. Logicians could do this only if they knew physics, history, psychology, financial theory, and whatever else might be relevant to an argument. In short, logicians could evaluate soundness only if they dominated all knowledge.

Third, although we usually want to argue from true premises, many useful arguments start from false ones. Some argue that a certain sentence is false by using it as a premise to reach an outrageous or absurd conclusion. Others adopt a premise purely as a hypothesis, to see what would follow if it were true. Aristotle first realized how important such arguments are; he characterized them as having dialectical, rather than demonstrative, premises. As we shall see later, these forms of argument are much more common and useful than most people imagine. A simple example occurs at the beginning of this chapter in the proof of the irrationality of $\sqrt{2}$. The proof starts with the assumption that $\sqrt{2}$ is rational, and a contradiction is deduced from it. The point of this argument is precisely to show that the assumption that $\sqrt{2}$ is rational is false. The success of the argument, therefore, depends solely on validity, not on soundness.

We might demand other things of arguments. Successful arguments generally lead us from premises based on good evidence to a relevant conclusion that follows from those premises. The definition of validity, however, mentions neither evidence nor relevance. A variety of valid arguments seem, from an intuitive point of view, peculiar. Some violate our ordinary notion of evidence. Suppose that the earth will be invaded by little green men in 2025, but that we possess no evidence now to support this. Then the argument

The earth will be invaded by little green men in 2025.
∴ The earth will be invaded by little green men in 2025.

is sound. Given our supposition, the premise is true, and the conclusion is surely true whenever the premise is, since they are the same sentence. But this argument won't convince anyone that we ought to be building defenses; it doesn't establish its conclusion in the usual, evidence-related sense of *establish*. Some arguments also count as sound even though they violate our usual notion of relevance. The argument

Coffee ice cream is more popular than chocolate in Rhode Island.
∴ Cats are cats.

is sound, since the premise is true, and the conclusion can never be false while the premise is true, simply because the conclusion can never be false. Yet this argument, too, seems bizarre. The premise is irrelevant to the conclusion. Thus, validity constitutes a part, but only a part, of our intuitive notion of success in argumentation.

Nevertheless, this book concentrates on validity. Considerations of evidence and relevance, though important, are extremely complex. Evidence raises the issue of inductive validity. Relevance has been treated in a variety of ways; none are simple, and none command general acceptance as a full theory of relevance.[3]

Problems

Evaluate these arguments as valid or invalid. If the argument is invalid, describe a circumstance in which the premises would be true but the conclusion would be false.

1. John and Mary came to the party. Hence, Mary came to the party.

2. Larry got angry and stormed out of the room. Consequently, Larry stormed out of the room.

3. If Susan's lawyer objects, she will not sign the contract. Susan's lawyer will object. Therefore, she will not sign the contract.

4. If Frank takes the job in Cleveland, he'll make a lot of money on the sale of his house. Frank won't take the job in Cleveland. It follows that Frank won't make a lot of money on the sale of his house.

▶ 5. If Strawberry hits 30 home runs, the Mets will be contenders. The Mets will be contenders. So, Strawberry will hit 30 home runs.

6. If Lynn testifies against the mobsters, she'll endanger her life. So, she won't testify against them, since she won't put her own life in danger.

7. Max is mayor of either Abilene or Anarene. Max isn't mayor of Abilene. Hence, Max must be mayor of Anarene.

8. Pamela played Shelley for the tournament trophy. Consequently, Pamela played either Shelley or Tracy for the trophy.

9. Henry doesn't know anyone. So, Henry doesn't know Kim.

▶ 10. Rocky has beaten everyone he's faced. Thus, Rocky has beaten Mad Moe, if he's faced him.

11. Since Jim applied, and all who have been accepted have scores over 1300, either Jim has been accepted, or his scores weren't over 1300.

12. Since Jim applied, and all who have been accepted have scores over 1300, either Jim hasn't been accepted, or his scores were over 1300.

13. Everyone who has thought about the political tensions of the Middle East realizes that they're complicated. Deborah doesn't realize that these political tensions are complicated. So, she mustn't have thought about them.

14. Everyone who admires Frost also admires Dickinson. Some people who normally hate poetry admire Frost. Therefore, some people who normally hate poetry admire Dickinson.

▶ 15. Some politicians are demagogues, but no demagogues are good leaders. Hence, some politicians are not good leaders.

16. All scientists have a deep interest in the workings of nature. All who devote their lives to the study of the physical world have a deep interest in the workings of nature. Consequently, all scientists devote their lives to the study of the physical world.

17. Some modern art shows the strong influence of primitivism. No modern art is primarily representational. Thus, some art that exhibits the influence of primitivism is not primarily representational.

18. Most Americans like baseball. Anyone who likes baseball likes sports. So, most Americans like sports.

19. Most medieval theories of motion were, in essence, Aristotelian. No theory of motion that includes a concept corresponding to inertia is essentially Aristotelian. It follows that most medieval theories of motion included no concept corresponding to inertia.

▶ 20. The patient will surely die unless we operate. We will operate. Therefore, the patient will not die.

21. Jerry will take the job unless we match the salary offer. Since we won't match the offer, Jerry will take the job.

22. The launch will be delayed unless the weather clears. So, if the weather clears, the launch won't be delayed.

23. The meeting will take place only if both parties agree on the agenda. So, if the parties don't agree on the agenda, the meeting will not take place.

24. Marilyn will finish the brief on time only if she gets an extension on the Morley case. Therefore, if Marilyn gets an extension on the Morley case, she will finish the brief on time.

▶ 25. If Jack understands how important this sale is, he'll devote most of the next two weeks to securing it. It follows that Jack won't devote most of the next two weeks to securing this sale unless he understands how important it is.

26. The boss won't understand what you're trying to say unless you put it in the bluntest possible terms. Consequently, if you don't put what you're trying to say in the bluntest terms possible, the boss won't understand it.

27. Either the city will raise electric rates, or it will raise taxes. Thus, if the city does not raise electric rates, it will raise taxes.

28. Nancy will not marry Alex unless he signs a prenuptial agreement. So, if Alex signs a prenuptial agreement, Nancy will marry him.

29. This album will sell only if it contains at least one hit song. Hence, unless it contains a hit song, this album will not sell.

▶ **30.** John is watching Mary run through the park. So, Mary must be running through the park.

31. Few students fully appreciate the value of an education while they are in school. Only those who fully appreciate the value of their education while they are in school devote themselves to their studies as much as they ought to. Therefore, most students don't devote themselves to their studies as much as they ought to.*

32. Corporate taxes result in higher prices for consumer goods, increases in interest rates, reduced employment at lower wages, and reduced levels of savings and investment, depending on whether corporations pass along the cost of taxation to the consumer, borrow to replace these funds, take steps to reduce labor costs, or reduce the return they offer to shareholders. Consequently, corporate taxes should be repealed.*

33. Most Americans who travel in Europe know no language other than English. All Americans who travel in Europe are affluent. Thus, most affluent Americans know no language other than English.*

34. The President didn't know that several of his subordinates had started "the company within the company," a small, highly secret group within the CIA. All of the President's subordinates belong to the President's political party. It follows that the President didn't know that several people of his own political party started a secret group within the CIA.*

▶ **35.** Few mathematics students take courses in logic. All accounting majors take courses in logic. So, few accounting majors are students of mathematics.*

36. Almost all Asian nations have socialist or statist or otherwise centralized economies. All our allies in Eastern Asia are, of course, Asian nations. So, most of our Eastern Asian allies have centralized economies.*

37. By 1988, the capital of Israel will be either Tel Aviv or Jerusalem. Thus, if in 1988 the Israeli capital is not Jerusalem, it will be Tel Aviv.**

38. Dogs are animals. Dogs bark. John owns a dog. So, John owns an animal that barks.**

39. Terry's mother gave her permission to go to the movies or to the park. Thus, Terry's mother gave her permission to go to the park.**

40. The collapse of the Austro-Hungarian Empire at the end of World War I caused the fragmentation and political divisions that led, ultimately, to an easy Soviet takeover of most of Eastern Europe at the end of World War II. Thus, if the end of World War I had not witnessed the collapse of Austria-Hungary, the Soviets would have found it more difficult to take over most of Eastern Europe after World War II.**

1.3 IMPLICATION AND EQUIVALENCE

A concept closely related to validity is *implication*. The verb *imply* has various uses in English. Here we'll discuss just one, highly specialized use. We can express the idea that an argument is valid by saying that its conclusion *follows from* its premises. Equivalently, we can say that its premises *imply* or *entail* its conclusion. At least part of what we mean, in either case, is that the truth of the premises guarantees the truth of the conclusion. If the premises are true, the conclusion has to be true, too. Implication, then, is very similar to validity. But validity is a property of arguments; implication is a relation between sentences and sets of sentences.

A set of sentences implies a given sentence just in case the truth of that sentence is guaranteed by the truth of all the members of the set.[4]

> **DEFINITION.** A set of sentences, *S*, *implies* a sentence, *A*, if and only if it's impossible for every member of *S* to be true while *A* is false.

It should be clear from this definition that, if an argument is valid, the set consisting of its premises implies its conclusion.

We can also speak of a single sentence implying another sentence.

> **DEFINITION.** A sentence, *A*, *implies* a sentence, *B*, if and only if it's impossible for *A* to be true while *B* is false.

One sentence implies another, that is, just in case the truth of the former guarantees the truth of the latter. In every circumstance in which the first is true, the second must be true as well.

Consider these two pairs of sentences.

(1) a. Mary likes Chinese food, but Bill hates it.
  b. Mary likes Chinese food.

(2) a. Susan is going to spend her summer in either Palo Alto or Pittsburgh.
  b. Susan is going to spend her summer in Pittsburgh.

Sentence (1)a implies (1)b. It's impossible to conceive of a situation in which it's true that Mary likes Chinese food, but Bill hates it, and false that Mary likes Chinese food. In such a circumstance, Mary would have to like and not like Chinese food; the sentence 'Mary likes Chinese food' would have to be both true and false at the same time. There are no such circumstances. No sentence can be both true and false at the same time. So the truth of (1)a guarantees the truth of (1)b. Does the truth of (2)a similarly guarantee the truth of (2)b? Obviously, the answer is no. Imagine a world in which Susan is going to spend her summer in Palo Alto, never setting foot outside California. In this situation, (2)a is true, but (2)b is false. So (2)a does not imply (2)b.

A sentence, A, implies a sentence, B, just in case B is true in all those possible circumstances in which A is true. B implies A, of course, just in case A is true in all those cases in which B is true. If A implies B and B implies A, then A and B must be true in exactly the same circumstances. In such a case, we say that A and B are *equivalent*.

> **DEFINITION.** A sentence, A, is *equivalent* to a sentence, B, if and only if it's impossible for A and B to disagree in truth value.

If A and B are equivalent, then they must be true in the same circumstances, and false in the same circumstances. There could be no situation in which one would be true while the other would be false. Thus, equivalence amounts to implication in both directions. A is equivalent to B just in case A implies B and B implies A. But the equivalence of A and B does not mean that A and B are *synonymous,* that is, have the same meaning.

To make this more concrete, consider four more pairs of sentences:

(3) a. No apples are oranges.
    b. No oranges are apples.

(4) a. All apples are fruits.
    b. All fruits are apples.

(5) a. The senator is neither worried nor angry about the investigation.
    b. The senator is not worried about the investigation; he is not angry about the investigation.

(6) a. Professor Pinsk saw that no one left.
    b. Professor Pinsk saw no one leave.

The sentences in (3) are equivalent. Any circumstance in which no apples are oranges is one in which no oranges are apples, and vice versa. Both sentences say that nothing is both an orange and an apple. In (4), however, the sentences are obviously not equivalent. All apples are fruits, so (4)a is true. But not all fruits are apples, so (4)b is false. The real world is thus a case in which these sentences disagree in truth value.

Similarly, the sentences in (5) are equivalent; they are true in exactly the same circumstances. If the senator is neither worried nor angry, then he is not worried, and he is not angry. Conversely, if he is not worried, and is not angry, then he is neither worried nor angry. The sentences in (6), however, are not equivalent. Imagine a case where a student left without being observed by the professor. In such a case, it could well be true that Professor Pinsk saw no one leave; it would nonetheless be false that the professor saw *that* no one left, since, in fact, someone did leave.

Problems

Consider the sentences in each pair: Are they equivalent? If not, does either sentence imply the other?

1. (a) Vivian and Beth both majored in English in college. (b) Beth majored in English in college, and so did Vivian.

2. (a) Pittsburgh will face Dallas or New York in the championship game. (b) Either Pittsburgh will face Dallas in the championship game, or Pittsburgh will face New York.

3. (a) Pluto or Uranus is now directly aligned with Neptune. (b) Pluto and Uranus are now directly aligned with Neptune.

4. (a) Columbia and Universal cannot both be the year's most successful studio. (b) Neither Columbia nor Universal is the year's most successful studio.

▶ 5. (a) Aunt Alice will not come to the wedding, and neither will Uncle Harry. (b) Not both Uncle Harry and Aunt Alice will come to the wedding.

6. (a) Either the Babylonians or the Assyrians employed the *lex talionis*. (b) If the Assyrians employed the *lex talionis,* the Babylonians didn't.

7. (a) If pay-per-view television catches on, cable companies will make huge profits. (b) If pay-per-view TV doesn't catch on, cable companies will not make huge profits.

8. (a) Universities will continue to grow only if they find new markets for their services. (b) If universities do not find new markets for their services, they will not continue to grow.

9. (a) If Elizabeth did not sign this letter, then her assistant did. (b) If Elizabeth had not signed this letter, her assistant would have.

▶ 10. (a) If Caesar had not crossed the Rubicon, he would never have become consul. (b) If Caesar had become consul, he would have crossed the Rubicon.

11. (a) No high-paying job is easy; (b) No easy job is high-paying.

12. (a) Some small law firms have specialists in municipal bonds. (b) Some law firms who have specialists in municipal bonds are small.

13. (a) All corporations primarily in the metals business are looking to diversify. (b) All corporations looking to diversify are primarily in the metals business.

14. (a) Several cities with populations over 700,000 have no baseball franchises. (b) Several cities without baseball franchises have populations over 700,000.

▶ **15.** (a) Anybody who can speak effectively can find a job in sales. (b) Anybody who can find a job in sales can speak effectively.

16. (a) Either the Federal Reserve Board or foreign investments will increase the supply of capital. (b) If the Federal Reserve Board doesn't increase the supply of capital, foreign investments will.

17. (a) Meg thinks that Oswald did not shoot Kennedy. (b) Meg realizes that Oswald did not shoot Kennedy.

18. (a) It wasn't necessary for things to turn out as they did. (b) Things could have turned out differently.

19. (a) Many films that make a lot of money are tailored to the teenage audience. (b) Many films tailored to the teenage audience make a lot of money.**

▶ **20.** (a) It's not true that Donna will come to the party but won't enjoy herself. (b) If Donna comes to the party, she'll enjoy herself.**

21. (a) Good wine isn't inexpensive. (b) Inexpensive wine isn't good.**

22. (a) Even Ralph found your comments offensive. (b) Ralph found your comments offensive.**

23. (a) We have some excellent redfish today, if you would like some. (b) We have some excellent redfish today.**

24. (a) If historians revise their analysis of the impact of refugees from the Weimar Republic on American intellectual history, they will revise their entire conception of that history. (b) Historians will revise their analysis of the impact of refugees from the Weimar Republic on American intellectual history only if they revise their entire conception of that history.**

Consider the statement: 'The patient will die unless we operate immediately.' What follows from this, together with the information listed?

▶ **25.** The patient will die.

26. The patient will not die.

27. We will operate immediately.

28. We won't operate immediately.

Consider the statement: 'If a fetus is a person, it has a right to life.' Which of the following sentences follow from this? Which imply it?

29. A fetus is a person.

▶ **30.** If a fetus has a right to life, then it's a person.

31. A fetus has a right to life only if it's a person.

32. A fetus is a person only if it has a right to life.

33. If a fetus isn't a person, it doesn't have a right to life.

34. If a fetus doesn't have a right to life, it isn't a person.

▶ **35.** A fetus has a right to life.

36. A fetus isn't a person only if it doesn't have a right to life.

37. A fetus doesn't have a right to life only if it isn't a person.

38. A fetus doesn't have a right to life unless it's a person.

39. A fetus isn't a person unless it has a right to life.

▶ **40.** A fetus is a person unless it doesn't have a right to life.

41. A fetus has a right to life unless it isn't a person.

In a quotation at the beginning of this chapter, Mr. McClory introduces a motion to this effect:

> There will be a ten-day postponement unless the president fails to give his assurance to produce the White House tapes.

Throughout the subsequent debate, congressmen paraphrase this motion in ways that they believe better express Mr. McClory's intentions, and Mr. McClory also expresses his motion in other terms that he takes to be equivalent to his original. Which of these paraphrases are in fact equivalent to the original motion?

42. If the president fails to give assurance, there is a ten-day postponement. (Congressman Latta proposes this as equivalent.)*

43. If the president gives his assurance, there is no postponement. (Congressman Latta proposes this as equivalent.)*

44. There is a ten-day postponement provided that the president gives his assurance. (Congressman Latta suggests this as a nonequivalent revision.)*

▶ **45.** There is a ten-day postponement unless the president gives his assurance. (Congressman Mann suggests this as a nonequivalent revision.)*

46. There is no postponement if the president fails to give his assurance. (Congressman McClory proposes this as equivalent.)*

A *factive* verb has special properties. For example, if a factive verb is substituted in

Mary ⟨VERB⟩s that Jane left.
∴ Jane left.

the resulting argument is valid. The verb *know,* for example, is factive; 'Mary knows that Jane left; so, Jane left' is valid. You can't know what isn't so. Which of these verbs are factive?

47. believe	**48.** realize	**49.** hope
▶ **50.** remember	**51.** doubt	**52.** deny
53. fear	**54.** be surprised	▶ **55.** prove
56. be certain	**57.** decide	**58.** mean
59. conclude	▶ **60.** understand	**61.** see
62. hear	**63.** be angry	**64.** say
▶ **65.** sense	**66.** suppose	**67.** imply
68. establish	**69.** determine	▶ **70.** be unaware

1.4 LOGICAL PROPERTIES OF SENTENCES

We use logic primarily to analyze the connections between sentences. Nevertheless, we can also classify individual sentences. The following sentences, depending on what the facts are, could be either true or false. It's possible to conceive of cases in which they would be true and other cases in which they would be false.

(7) The snow is falling all over Ireland.
(8) The earth is the third planet from the sun.

Such sentences are *contingent.*

> **DEFINITION.** A sentence is *contingent* if and only if it's possible for it to be true and possible for it to be false.

Contingent sentences could be true, given the right set of circumstances. Of course, they could also be false, depending on the facts of the situation. They assert, in effect, that the real circumstance is among those in which they are true.

Some sentences, in contrast, cannot help being true. It's simply impossible for them to be false. They are true in every possible circumstance. Such sentences are *valid,* or *logically true:*

> **DEFINITION.** A sentence is *valid* (or *logically true*) if and only if it's impossible for it to be false.

In the realm of the sentential logic of Chapters 2, 3, and 4, valid sentences are often called *tautologies.*

If you doubt that there are any sentences that cannot be false, no matter what the facts may be, then try to imagine circumstances in which these sentences are false.

(9) A rose is a rose.
(10) Wherever you go, there you are.
(11) Either some of my friends are crazy, or none of them are.

These sentences are true in every possible world. They also seem to say very little. But not all valid sentences are so straightforward and unsurprising. (12), for example, is logically true:

(12) If nobody likes a quitter, and Sam likes Jeanne, then Jeanne didn't quit the team.

This sentence doesn't seem as trivial as (9)–(11). Valid sentences can be useful. They often set up the structure of an argument, as when a mathematician begins a proof by saying, "The number n is either prime or not prime. If it is prime . . . "

Some sentences could never be true. They are false, regardless of the facts. These sentences are *contradictory* (or *contradictions*).

> **DEFINITION.** A sentence is *contradictory* if and only if it's impossible for it to be true.

Here are some examples of contradictions:

(13) Fred is both Mexican and not Mexican.
(14) Nobody's seen the trouble I've seen.
(15) The natural number n is both odd and even.

In no conceivable circumstance could these sentences be literally true. Try, for example, to imagine a situation in which Fred is both Mexican and not Mexican at the same time. Whatever Fred's nationality, it seems he's either Mexican or not Mexican but not both. Similarly, n must be either odd or even, but it can't be both. And, since I've seen the trouble I've seen, somebody (namely, me) has indeed seen the trouble I've seen. Contradictions tend to signal that we should interpret some terms generously, since we assume that our colleagues in communication are trying to say something that could be true. Hearing (14), then, we tend to read *nobody* as *nobody else,* interpreting the sentence as 'Nobody else has seen the trouble I've seen.' Contradictions, too, may fulfill important functions in arguments. (15) played a role in the proof that $\sqrt{2}$ is irrational.

Nevertheless, contradictions can be disruptive, and it's useful to have a term for sentences that, whether valid or contingent, at least are not contradictory. Such noncontradictory sentences are *satisfiable.*

> **DEFINITION.** A sentence is *satisfiable* if and only if it's not contradictory.

Obviously, a sentence is satisfiable just in case it's either contingent or valid. That is, it must be possible for the sentence to be true. Since every sentence is either valid, contingent, or contradictory, we can divide sentences into three groups, as shown in this table:

	Sentences	
Valid	Contingent	Contradictory
True in every circumstance	True in some circumstances, false in others	False in every circumstance
	(Satisfiable)	

Problems

Classify these sentences as valid, contradictory, or contingent.

  1. I am who I am.

  2. All dogs are dogs.

  3. Some dogs are not dogs.

  4. Some cars are red.

▶ **5.** All red automobiles are automobiles.

  6. All red automobiles are red.

  7. Every German car is a car.

  8. Every German car is a German.

  9. I know what I know.

▶ **10.** Some people are friendly, and others aren't.

  11. Some people are friendly and not friendly.

  12. Some people aren't friendly, but everybody is friendly.

  13. There are many trees in Yosemite National Park.

  14. Every student studies.

▶ **15.** Nobody loves everybody.

  16. Everyone who loves everyone loves every loser.

  17. Everyone who loves every loser loves everyone.

18. Everyone who drives a Mercedes drives a car.

19. Not everyone who drives a car drives a Mercedes.

▶ 20. If what you say is true, then it's false.*

21. Nobody can defeat everyone without being defeated at least once.*

22. Today is the first day of the rest of your life.*

23. No batter ever made a hit with the bat on his shoulder (John McGraw).*

24. You are what you eat (Ludwig Feuerbach).*

▶ 25. Everything is what it is, and not another thing (Joseph Butler).*

26. The business of America is business (Calvin Coolidge).*

27. There are two kinds of people in the world: those who divide the world into two kinds of people, and those who don't (H. L. Mencken).*

28. I am never less alone than when I am alone, nor less at leisure than when I am at leisure (Scipio Africanus).

29. There comes a time to put principle aside and do what's right (Michigan legislator).*

▶ 30. I don't know what the previous speaker said, but I agree with him (Texas legislator).*

31. this poem is the reader and the reader this poem (Ishmael Reed).**

32. I exist.*

33. Some dogs are dogs.*

34. No dogs are dogs.*

▶ 35. Most dogs are dogs.*

36. Many dogs are dogs.*

37. All former congressmen are congressmen.*

38. Some fake diamonds are diamonds.*

39. John likes baseball but hates all sports.*

▶ 40. Alice taught Sarah some chemistry, but Sarah learned no chemistry from Alice.*

41. Ned rented an apartment, but nobody rented an apartment to him.*

42. If Carlotta is eating, then Carlotta is eating something.*

43. If Hector is laughing, then Hector is laughing at something.*

44. The round square is round.**

45. Water is H_2O.**

46. Cats are animals.**

47. No cats are dogs.**

48. Being bad is bad.**

49. It's foolish to be foolish.**

50. $2 + 2 = 4$.**

51. Nothing can be in two places at once.**

52. Nothing can be both black and red all over.**

53. Suppose that a sentence, A, implies another sentence, B. What can we conclude about B if A is (a) valid? (b) contingent? (c) satisfiable? (d) contradictory?*

54. Suppose that a sentence, A, implies another sentence, B. What can we conclude about A if B is (a) valid? (b) contingent? (c) satisfiable? (d) contradictory?*

The fourteenth-century logician Pseudo-Scot (so-called because his writings, for many years, were attributed to John Duns Scotus) raised several objections to definitions of validity such as those of the last two sections. The following two arguments, he thought, showed that there was a problem with saying that an argument is valid if and only if it's impossible for its premises to be true while its conclusion is false. Do these arguments really pose a problem for such definitions of validity? Explain.

▶ **55.** Every sentence is affirmative.
∴ No sentence is negative.**
(Note: Assume that no sentence can be both affirmative and negative. The argument seems valid. But, Pseudo-Scot argued, even though the premise could be true, the conclusion can't be; it refutes itself, since it is negative.)

56. God exists.
∴ This argument is not valid.**
(Note: Pseudo-Scot assumes that the premise is necessarily true. Any necessary truth would serve here in place of 'God exists.' So, Pseudo-Scot argued, if the argument is valid, the conclusion must be true; but then the argument isn't valid, contradicting the hypothesis that it is valid. So suppose the argument is not valid. Then the conclusion is true; in fact, it must be necessarily true. In that case, we have an invalid argument in which it cannot happen that the premise is true while the conclusion is false, because the conclusion can never be false.)

1.5 Sets of Sentences

A sentence is satisfiable just in case it is not contradictory; that is, just in case it can be true. Any true sentence, obviously, is satisfiable. But false sentences can also be satisfiable, so long as they are true in some other possible circumstance.

We can speak of sets of sentences, too, as satisfiable or contradictory. It's easy to think of sets of sentences that, in some sense, contain contradictions, even though each sentence in the set is itself satisfiable:

(16) a. Beer and sauerkraut are very good together.
    b. Beer and sauerkraut aren't very good together.

(17) a. Many of my friends belong to the Flat Earth Society.
    b. Nobody in the Flat Earth Society believes in modern science.
    c. All my friends believe in modern science.

The sentences in (16), like those in (17), are not themselves contradictions. Taken individually, each could be true. Taken together, however, they describe an impossible situation. Though each could be true, the sentences in (16) or (17) couldn't be true together. In such cases, the *set* of sentences is contradictory, whether or not any individual sentence in the set is contradictory.

> **Definition.** A set of sentences is *contradictory* if and only if it's impossible for all its members to be true. A set is *satisfiable* otherwise.

If a set is contradictory, we can also say that its members are *mutually inconsistent,* and that any member *contradicts,* or *is inconsistent with,* the set containing all the rest. If the set is satisfiable, then its members are *mutually consistent,* and each member is *consistent* or *compatible* with the set containing all the rest. Two sentences contradict each other just in case the set containing just the two of them is contradictory. If a set is satisfiable, then all its subsets are satisfiable: Each member is consistent or compatible with each other member of the set.

From a logical point of view, contradictory sets of sentences can be described in two ways. First, the sentences in the set can't all be true in the same circumstance. Second, the set implies a contradiction. Although a contradictory set of sentences might not contain a contradiction, it must imply one. The sentences in (17), for example, together imply 'Although many of my friends don't believe in modern science, all my friends do believe in modern science.' This is an outright contradiction.

To see that these two characterizations come to the same thing, recall that a set of sentences, S, implies a sentence, A, just in case it's impossible for every sentence in S to be true while A is false. Contradictions, of course, are always false. When *A* is a contradiction, then *S* implies *A* if and only if it's impossible for every sentence in *S* to be true. Therefore, a set of sentences implies a contradiction just in case the set is itself contradictory. Or, to put it another way, satisfiability is freedom from contradiction.

Satisfiability is important: Sets of sentences that are not satisfiable can't be true. They must contain at least one false sentence, no matter what the facts might be. A satisfiable set may also contain false sentences, but at least there is a possibility that all the sentences it contains are true.

Satisfiability has great significance in legal contexts. A lawyer may try to trap an opposing witness in a contradiction. The lawyer, in most cases, cannot alone provide any direct testimony relevant to the case but may introduce witnesses to dispute what the opposing witness says. If the opposing witness falls into a contradiction, however, then the witness must be saying something false, regardless of the facts of the case.

Even more fundamentally, people often use arguments to disprove someone else's contention. To refute an assertion, we have to recognize when we have shown something that contradicts that assertion. So the notion of refutation depends on the notion of contradiction.

Finally, the concept of satisfiability has been very important in modern mathematics. Around the turn of the century, several mathematicians and logicians deduced contradictions from mathematical theories in use at the time. Ever since, mathematicians have sought, whenever possible, proofs that theories are satisfiable. This has led to some of the most important developments in twentieth-century logic, mathematics, and computer science.

Problems

Evaluate these sets of sentences as satisfiable or contradictory:

1. The yard isn't white unless it's snowing. It's not snowing. But the yard is white.

2. If a student's GPA is very high, he or she will get into a good graduate school. Frank's GPA is not very high. Nevertheless, he'll get into a good graduate school.

3. John is a good guitarist. John is also an accountant, but not a good one.

4. Everyone who can cook a good chicken kung pao knows the value of hot peppers. Some who know the value of hot peppers don't themselves like hot food. Anybody who can cook a good chicken kung pao likes hot food.

▶ 5. If Marsha takes a job with a state commission, she'll gain much experience in new areas, although she won't get to travel. If she takes a job with a private company, she'll get to travel, and she'll be paid well, although she won't gain much experience outside her area. Marsha won't be paid well, but she will get to travel.

6. If the court's decision here is consistent with the decision in Yick Wo v. Hopkins, it will hold that statistical arguments alone can suffice to establish discrimination. If, however, it is compatible with the decision in several recent cases, it will hold that establishing discrimination requires something beyond purely statistical argumentation.

7. I like this painting, even though I don't think it's very good. I like everything that Elmer likes, and Elmer likes every painting that's good.

8. Many Indo-European languages are descended from Latin. All languages descended from Latin developed a word for *yes* from the Latin *sic* (meaning *thus*). But few Indo-European languages have a word for *yes* developed from *sic*.

9. No drugs are approved for use without careful screening. Careful screening takes years. A few drugs in great demand, however, are approved for use in less time.

▶ 10. Few communist parties in Europe seek to identify themselves with the Soviet party. Parties seeking to identify themselves with the Soviets have a difficult time becoming part of coalition governments. Almost all European communist parties find it difficult, however, to become part of coalition governments.

11. Stocks of companies with high debt-equity ratios are fairly risky. If a stock is fairly risky, it must reward investors with better-than-average returns, or they will eschew the risk. But many stocks that fail to reward investors with better-than-average returns are those of companies with high debt-equity ratios.*

12. People have a right to life. Fetuses are not people. If something has a right to life, it's wrong to kill it. Abortion is the killing of a fetus. Abortion is wrong.*

13. Few contemporary composers write anything that could reasonably be called twelve-tone compositions. If so, then atonal music is defunct. But atonal principles of composition still exert some influence on contemporary composers. And nothing that still exerts influence is defunct.*

14. Many football stars never graduate from the colleges where they first become famous. Most of these colleges insist that almost all their football players receive degrees. These schools are telling the truth.*

▶ 15. Most actresses begin their careers as successful models. Every woman who begins her career as a successful model is very glamorous. Nevertheless, few actresses are very glamorous.*

16. Many well-known American novels deal with the character of a specific region of the country. Every well-known American novel, of course, portrays a certain conception of America itself. Nonetheless, many novels

that portray a conception of America do not deal with any specific region of the country.*

17. My barber in town shaves every man in town who doesn't shave himself. Furthermore, my barber doesn't shave anyone in town who does shave himself.*

18. If 3 were an even number, then 4 would be odd. But, if 3 were even, then 6 would be even, and if 6 were even, then 4 would be even too.**

19. If God exists, it's surely true that He exists necessarily. It's possible that God exists. It's also possible that He doesn't exist.**

▶ 20. Because you promised, you ought to take your brother to the zoo. But, since you also have duties as club treasurer, you have an obligation to go to the club meeting. And you can't do both.**

True or false? Explain.

21. If a set of sentences is satisfiable, no member of that set implies a contradictory sentence.

22. If no member of a set implies a contradictory sentence, that set is satisfiable.

23. Some satisfiable sets of sentences imply contradictions.

24. Some satisfiable sets of sentences imply no contingent sentences.

▶ 25. Every contradictory set of sentences implies every contradiction.

26. No satisfiable sets of sentences imply every sentence.

27. Some contradictory sets of sentences imply every sentence.

28. If *A* implies *B*, then the set consisting of *A* and *B* together is satisfiable.

29. If the set consisting of just *A* together with *B* is contradictory, then *A* implies that *B* is false.

▶ 30. Any argument with a contradictory set of premises is valid.

31. Arguments with satisfiable sets of premises have satisfiable conclusions.

32. Every satisfiable set of sentences contains at least one true sentence.

33. Every contradictory set of sentences contains at least one false sentence.

34. Any set consisting of all valid sentences is satisfiable.

▶ 35. Any set consisting of all contingent sentences is satisfiable.

The Englishman William of Ockham (1280–1348), perhaps the most influential philosopher and logician of the fourteenth century, recorded eleven rules of logic in a chapter of his *Summa Totius Logicae*. Ten of these use concepts we've already developed. Say whether each is true, given the definitions of this chapter, and explain why.

36. The false never follows from the true.

37. The true may follow from the false.

38. Whatever follows from the conclusion of a valid argument follows from its premises.

39. The conclusion of a valid argument follows from anything that implies the argument's premises.

▶ **40.** Whatever is consistent with the premises of a valid argument is also consistent with the argument's conclusion.

41. Whatever is inconsistent with the conclusion of a valid argument is also inconsistent with the argument's premises.

42. The contingent does not follow from the valid.

43. The contradictory does not follow from the satisfiable.

44. Anything whatsoever follows from the contradictory.

▶ **45.** The valid follows from anything whatsoever.

Notes

[1] These links between deductive validity and truth or falsehood were first recognized explicitly by the Greek philosopher Aristotle (384–322 B.C.), the father of logic.

[2] Indeed, although Aristotle and some earlier thinkers talked of valid arguments, a term for soundness was introduced only later, by logicians of the Stoic school, which thrived in Greece from the third century B.C. to the second century A.D.

[3] Relevance logic builds relevance into the concept of validity. Alan Ross Anderson and Nuel D. Belnap, Jr., in *Entailment: The Logic of Relevance and Necessity*, Volume 1 (Princeton, N.J.: Princeton University Press, 1975), argue in favor of such a strategy and develop several systems of relevance logic. For a very different approach to relevance, see H. P. Grice, "Logic and Conversation," in P. Cole and J. L. Morgan (eds.), *Syntax and Semantics 3: Speech Acts* (New York: Academic Press, 1975): 45–58.

[4] Throughout this book, 'just in case' will be used as a synonym for 'if and only if.'

2

SENTENCES

entential logic is the study of sentences as basic units and the examination of sentence relationships as they pertain to reasoning. This chapter focuses on only a portion of that logic. We'll develop a theory only of the connections between sentences that are *truth-functional*. The logical character of these connections is rather clear-cut. The truth values of compound sentences formed with certain connecting words and phrases depend entirely on the truth values of the smaller sentences they connect. The result of forming these connections is thus completely predictable on the basis of the truth values of the smaller sentences alone.

2.1 SENTENCE CONNECTIVES

The word *sentence* has at least two distinct uses relevant to logic. The most straightforward is this: A sentence is any grammatical string of words that ends with a period, exclamation point, or question mark. When Churchill stressed the tremendous practical value of being able to write an English sentence, he probably had this sense of the word in mind. Similarly, a computer program or writer counting the number of sentences in a paragraph is counting the strings between the above punctuation marks or, perhaps, just counting those marks themselves.

Linguists, however, use *sentence* in a slightly different and more complex way. Consider this string of words.

(1) Fred acted rather crudely, but his enthusiasm was contagious.

How many sentences does this string contain? In the sense of *sentence* discussed above, the answer is clearly one. In another sense, however, the answer might be three. The entire string does constitute a sentence. Within it, furthermore, are two other sentences: 'Fred acted rather crudely' and 'his enthusiasm was contagious.' If we are interested in how to form sentences, then we might naturally conjecture that the string is the result of combining these two shorter sentences. Grammarians, accordingly, call such sentences *compound* sentences. The two shorter sentences in (1) are its *components*. The word *but,* which links these sentences together to form a compound, is a *sentence connective.* Throughout this chapter, we'll use the term *sentence* in this second sense, in which one sentence may be said to be a part or component of another.

Sentences that appear within other sentences are often called *embedded* or *subordinate clauses.* Sentences may be embedded in others in various ways. *Complemented* verbs, such as *believe;* temporal conjunctions, such as *before* and *after;* and more ordinary conjunctions, such as *and* and *although,* all serve to embed sentences inside other sentences. In this section, we'll discuss a symbolic language designed to clarify the structure of some embedding mechanisms. The language of sentential logic does not treat complemented verbs—indeed, their logic is extremely complex—but it can represent many of the expressions that operate on one or more sentences to produce a new sentence.

Expressions that convert single sentences into other sentences are *singulary* or *unary* sentence connectives. Those that combine two sentences into a single sentence are *binary* sentence connectives. A more general definition follows.

> **DEFINITION.** An n-*ary sentence connective* is a word or phrase that forms a single, compound sentence from *n* component sentences.

Sentence connectives are the chief study of sentential logic. Some examples of singulary connectives are *not, maybe, of course, possibly, necessarily,* and—somewhat controversially—auxiliary verbs such as *may, can, could, might, must,* and *should.* Examples of binary connectives are *and, but, however, although, if, or, unless, though, before,* and *because.*

Some of these connectives behave predictably, in a certain sense, while others are more complex. Assume that the *truth value* of a sentence is truth if the sentence is true and falsehood if it is false. The predictable connectives form compound sentences whose truth values are a function of the truth values of their components. Predictable connectives are therefore called *truth-functional.*

> **DEFINITION.** An *n*-ary sentence connective is *truth-functional* if and only if the truth values of the *n* component sentences always completely determine the truth value of the compound sentence formed by the connective.

If a connective is truth-functional, then the compounds formed with it match in truth value whenever the truth values of the components match in truth value. We simply need to know whether the component sentences are true or false to determine the truth value of the compound.

Consider, for instance, the sentence 'It's not snowing,' which contains the component 'It's snowing.' If this component is true, then the compound is false: If it's true that it's snowing, it's false that it's not snowing. Now suppose that the component is false. If it's false that it's snowing, then it's true that it's not snowing, so the compound is true. The truth value of 'It's snowing,' in other words, completely determines the truth value of 'It's not snowing.' Nothing about this depends on any special feature of the sentence 'It's snowing.' Any sentence in this role should produce much the same result. *Not* is thus a truth-functional sentence connective.

In contrast, consider *after*. This connective is not truth-functional. To see why, consider the sentence 'George resigned after the commissioner stopped the trade,' which contains the components 'George resigned' and 'the commissioner stopped the trade.' If both these sentences are true, their truth does not determine the truth of the compound. George, after all, may have resigned before the commissioner nixed the trade.

Note that it isn't enough for the truth values of the components to determine the compound's truth value only in some cases; they must suffice in every case if the connective is to be truth-functional. The truth value of the components sometimes fixes the value of the compound formed by *after:* If George didn't resign, then he didn't resign after the commissioner stopped the trade. *After,* nevertheless, is not truth-functional.

We can summarize each example readily in a small table. Consider the possible truth values of the sentences, and see whether the truth value of the compound is fully determined in every case.

It's snowing	It's not snowing
T	F
F	T

George resigned	The commissioner stopped the trade	George resigned after the commissioner stopped the trade
T	T	?
T	F	?
F	T	F
F	F	F

A connective is truth-functional if and only if it's possible to assign a determinate truth value to every row under the compound sentence. If, as with *after,* one or more rows contains neither T nor F, then the connective is not truth-functional.

Problems

Are these connectives truth-functional? Why, or why not?

1. because	**2.** or
3. before	**4.** nevertheless
▶ **5.** may	**6.** in order that
7. in spite of	**8.** regardless whether
9. implies	▶ **10.** should
11. it's improbable that	**12.** it's a necessary truth that
13. can	**14.** could
▶ **15.** maybe	**16.** it's obvious that
17. it's surprising that	**18.** when
19. provided that**	▶ **20.** if **

2.2 A SENTENTIAL LANGUAGE

In this section we'll develop a symbolic language, SL, that permits ready evaluation of arguments depending on truth-functional sentence connectives.[1] The *syntax* of a language consists of (a) its vocabulary and (b) its grammatical rules. The *semantics* of a language is its theory of meaning; its *pragmatics* is the theory of its use in context. The syntax of SL is fairly simple. The vocabulary falls into three basic categories: *sentence letters, connectives,* and *grouping indicators.*

Vocabulary

Sentence letters: p, q, r, s, with or without numerical subscripts
Connectives: ¬, &, ∨, →, ↔
Grouping indicators: (,)

The symbol ¬ has no name in the literature; we'll call it the negation sign or, more picturesquely, the hoe. The other connectives have standard names: & is the ampersand; ∨ is the wedge; → is the arrow; and ↔ is the double arrow.

The syntactical rules by which we combine vocabulary elements to form strings of the language define the notion of a *formula*. In presenting the rules, we must distinguish carefully between the language, SL, that we're defining, and the language, namely English, that we're using to define SL. The artificial language we're defining is the *object language;* the language we're using to discuss the symbolic language is the *metalanguage.*

In the metalanguage, we'll use script variables such as $\mathscr{A}$, $\mathscr{B}$, x, y, p, q, etc., to stand for items in the object language. The script variables are in the metalanguage, but they stand for items in the object language. In particular, we'll assume that $\mathscr{A}$ and $\mathscr{B}$ stand for formulas of SL. A string of symbols containing a script variable is not a formula of SL; script letters never appear as sentence letters. The string is, instead, a part of the metalanguage. We'll call script letters such as $\mathscr{A}$ and $\mathscr{B}$ *schematic letters,* and strings such as '$(\mathscr{A} \to \mathscr{B})$' *schemata.*[2]

Formation Rules

Any sentence letter is a formula.
If $\mathscr{A}$ is a formula, then $\neg \mathscr{A}$ is a formula.
If $\mathscr{A}$ and $\mathscr{B}$ are formulas, then $(\mathscr{A} \,\&\, \mathscr{B})$, $(\mathscr{A} \lor \mathscr{B})$, $(\mathscr{A} \to \mathscr{B})$, and $(\mathscr{A} \leftrightarrow \mathscr{B})$ are formulas.
Every formula can be constructed by a finite number of applications of these rules.

A formula is *atomic* if and only if it contains no connectives. Atomic formulas of SL, then, are sentence letters. We will adopt a convention so that we can use our symbolic language more easily: We'll drop the outside parentheses of a formula, since they do no further work. We can imagine them drawn in so lightly that they can't be seen.

The definition of a formula allows us to tell how a formula can be built up from sentence letters. It's easy to record this information in a *phrase structure tree,* in which the formula appears at the top and the sentence letters appear at the bottom. These trees chart the construction, for example, of $\neg p$, $(p \,\&\, q)$, $p \lor \neg (q \lor r)$ and $\neg (p \leftrightarrow \neg p)$ from the vocabulary items of SL.

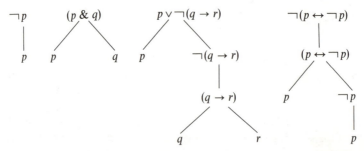

At each node of each tree is a formula. The tree always ends with atomic formulas, that is, sentence letters. Each transition from a higher node to one or more lower nodes reflects a rule used in building the formula. Any formula appearing at any node of a phrase structure tree is a *subformula* of the formula at the top of the tree.

There is an important theoretical difference between the syntax of SL and the syntax of a natural language. Some English sentences, such as 'They watched the fireworks explode on the porch,' can be constructed in more than one way. They are thus ambiguous. No such ambiguities arise in our symbolic language. Every formula is associated with one and only one phrase structure tree.

We'll close this section with two definitions that will prove extremely useful throughout the text. An occurrence of a connective creates a compound formula from one or more component formulas. The connective occurrence itself and these components constitute the *scope* of the connective occurrence in a given formula. The connective occurrence with the largest scope is the *main connective* of the formula.

> **DEFINITION.** The *scope* of a connective occurrence in a formula is the connective occurrence itself, together with the subformulas (and any grouping indicators) it links.

> **DEFINITION.** The *main connective* of a formula is the connective occurrence in the formula with the largest scope.

Connective Occurrence	Formula	Scope	Main Connective
$\neg$	$(p \to \neg q)$	$\neg q$	$\to$
$\neg$	$\neg(p \to q)$	$\neg(p \to q)$	$\neg$
$\vee$	$(p \vee (q \,\&\, r))$	$(p \vee (q \,\&\, r))$	$\vee$
$\&$	$(p \vee (q \,\&\, r))$	$(q \,\&\, r)$	$\vee$

The main connective is always the connective occurrence having the entire formula as its scope. In a phrase structure tree of the formula, a rule for the main connective is applied in descending from the top node. This rule is the first to be applied in decomposing the formula into its component parts, and the last to be applied in building it up from those parts.

Problems

Classify each of these as (a) a formula, (b) a conventional abbreviation of a formula, or (c) neither. If you answer a or b, identify the main connective.

1. $p \,\&\, \neg q$

2. $\neg p \vee \neg q$

3. $\neg(p \,\&\, \neg q)$

4. $(\neg p \vee \neg q)$

5. $(\& p \vee \& q)$

6. $(p \vee q) \to r$

7. $p \vee q \to r$

8. $(p \vee q \to r)$

9. $p \vee (q \to r)$

▶ **10.** $(p \vee q) \to r)$

11. $((p \vee q) \to d)$

12. $\neg((p \vee (q \vee r)) \,\&\, q$

13. $\neg[p \to (r \to \neg q) \leftrightarrow \neg p]$ **14.** $m \to (r \leftrightarrow (k \lor l))$

▶ **15.** $\neg((p \to (q \to r)) \lor p$

16. The rules for forming formulas of sentential logic admit no ambiguity. How could these rules be altered to admit ambiguity? Give an example of a set of rules allowing ambiguity, and specify an ambiguous formula together with at least two phrase structure trees showing its construction according to those revised rules.*

2.3 TRUTH FUNCTIONS

Truth-functional connectives, in English or in SL, yield compounds whose truth values depend solely on the truth values of their components. The key to the semantics of SL, then, is this: Each truth-functional connective represents a corresponding function from truth values into truth values. Such a function takes as inputs the truth values of the component sentences and yields the truth value of the compound sentence. These functions are called *truth functions.*

> DEFINITION. An n-*ary truth function* is a function taking n truth values as inputs and producing a truth value as output.

There are four singulary truth functions. These take a single truth value as input and produce a truth value as output. Only two inputs—truth and falsehood—are possible, and, similarly, only two outputs. One of the four singulary truth functions takes both truth and falsehood into the output truth; another takes both into the output falsehood. One takes each value into itself; another takes them into each other. Those are all the possible singulary functions. There are sixteen binary functions, and, in general, for each n, 2^{2^n} truth functions.

Since we may make n as large as we like, there are infinitely many truth functions. How can we formulate a theory to describe this infinite array of functions? Luckily, it's not very difficult. This chapter presents a few commonly used truth functions. Any truth function at all can be defined in terms of them alone. In fact, as we'll see, a single binary truth function suffices to define every truth function in this infinite collection.

The first function we'll define is singulary. It is called *negation,* and we'll use the symbol $\neg$ to represent it. In this definition, $\mathscr{A}$ is any formula.

$\mathscr{A}$	$\neg \mathscr{A}$
T	F
F	T

Negation transforms the truth value of the component sentence into its opposite.

We've already seen an English connective that has this effect: the logical particle *not*. Other English expressions having much the same impact are *it is not the case that, it's false that,* and, often, the prefixes *un-, dis-, a-, im-,* etc.

The second function, called *conjunction,* is binary. We'll represent it with the ampersand, &.

$\mathscr{A}$	$\mathscr{B}$	$(\mathscr{A}\ \&\ \mathscr{B})$
T	T	T
T	F	F
F	T	F
F	F	F

A conjunction is true just in case both its components—called *conjuncts*—are true.

English expressions functioning in this way include most grammatical conjunctions: *and, both . . . and, but, though,* and *although.* The correlates of conjunction are very faithful to it, except that *and,* in one usage, seems to express a temporal ordering. There is a difference in meaning between 'Heidi lost weight and got sick' and 'Heidi got sick and lost weight,' even though the definition of conjunction indicates that these should be true in exactly the same circumstances. Also, English treats conjunction as a *multigrade* connective, that is, one that can combine two, three, or more sentences into a single compound sentence. The logical & always links two sentences, but in English we can say 'John brought the mustard, Sally brought the pickles, and I brought the hot dogs.' We will simulate this in SL by allowing continued, ungrouped conjunctions. For example, we'll allow *p & q & r* as an abbreviation of either *((p & q) & r)* or *(p & (q & r)),* since the last two formulas are equivalent. Within conjunctions, grouping makes no difference.

The third function is also binary. Represented by $\vee$, it's called *disjunction.*

$\mathscr{A}$	$\mathscr{B}$	$(\mathscr{A} \vee \mathscr{B})$
T	T	T
T	F	T
F	T	T
F	F	F

A disjunction is true just in case either of its components—called *disjuncts*—is true.

English expressions corresponding to this function are *or* and *either . . . or.* The correlates to disjunction are also quite close to the logical definition, although *or,* like *and,* appears to be multigrade in English. As with conjunction, we will simulate this in SL by allowing continued, ungrouped disjunctions. For example, we'll allow *p ∨ q ∨ r* as an abbreviation of either *((p ∨ q) ∨ r)* or *(p ∨ (q ∨ r)).* The last two formulas are equivalent; within disjunctions, grouping makes no difference.

The fourth function, again binary, is represented by → and is called the *conditional:*

$\mathcal{A}$	$\mathcal{B}$	$(\mathcal{A} \rightarrow \mathcal{B})$
T	T	T
T	F	F
F	T	T
F	F	T

Here, for the first time, the order of the components makes a difference. The first component of a conditional is its *antecedent;* the second is its *consequent.* A conditional is true just in case it doesn't have a true antecedent and false consequent.

English expressions having, roughly, the force of the conditional truth function are '*B* if *A*,' ' if *A* then *B*,' '*A* only if *B*,' '*B* so long as *A*,' '*B* provided that *A*,' '*B* assuming that *A*,' and '*B* on the condition that *A*.' At first it seems surprising that '*A* only if *B*' and 'if *A*, then *B*' are both correlates of the conditional $(A \rightarrow B)$. But these sentences seem equivalent: 'If we don't operate, the patient will die' and 'The patient will survive only if we operate.' These, of course, are not quite the same: One is 'If not *A*, then *B*,' while the other is 'If not *B*, then *A*.' But these, according to our definitions of negation and the conditional, are true in exactly the same circumstances.

Switching the negations and the English connectives, however, yields sentences that ought to be logically equivalent to these but sound bizarre: 'If the patient survives, we operate' and 'We don't operate only if the patient dies.' The first seems to mean "We operate only when the patient is out of danger," and the second seems to mean, "We operate as long as we have a live body on which to operate." The conditional doesn't capture these differences. Whatever follows the English word *if* imposes a condition, usually causal, on the other component. The English conditional has some causal meaning, and a truth-functional rendering of the conditional can't capture it.

The English connective *unless* raises similar issues. Consider the sentence: 'The patient will die unless we operate.' This is equivalent to 'If we don't operate, the patient will die.' In general, we'll interpret '*A* unless *B*' as '*A*, if not *B*.' Both sentences seem to assert a causal connection between the operation and the patient's chances for survival. We wouldn't normally judge the sentence true just because the operation was performed or just because the patient died. In fact, a surgeon who says this knowing that the patient will die in any case seems unethical.[3]

Finally, the *biconditional* is a binary truth function, symbolized by ↔.

$\mathcal{A}$	$\mathcal{B}$	$(\mathcal{A} \leftrightarrow \mathcal{B})$
T	T	T
T	F	F
F	T	F
F	F	T

Biconditionals are true just in case their components agree in truth value.

English expressions such as *if and only if, when and only when,* and *just in case* correspond to the biconditional. The problems that plague the English correlates of the conditional, of course, occur doubly with the correlates of the biconditional.

The symbols $\neg$, &, $\vee$, $\rightarrow$, and $\leftrightarrow$ are in common use. Unfortunately, however, there is no standard logical notation. This table shows other symbols used as logical connectives:

Truth Function	Our Symbol	Other Symbols
Negation	$\neg p$	$-p$, $\sim p$, p', $\bar{p}$, Np
Conjunction	p & q	$p \wedge q$, pq, $p \cdot q$, Kpq
Disjunction	$p \vee q$	$p \veebar q$, Apq
Conditional	$p \rightarrow q$	$p \supset q$, Cpq
Biconditional	$p \leftrightarrow q$	$p \equiv q$, $p \sim q$, Epq

The expressive power of these five truth functions is sufficient to construct any truth function. Indeed, not even all five are necessary. Negation and conjunction alone suffice; so do negation and disjunction, or negation and the conditional. Any set of truth functions that, like $\{\neg, \&\}$, $\{\neg, \vee\}$, and $\{\neg, \rightarrow\}$, allows us to define every other truth function is *functionally complete.*

Two binary truth functions are functionally complete all by themselves. They are called *Sheffer functions,* after Henry Sheffer (1883–1964), a Harvard logician.[4] These functions are *Sheffer's stroke* (/) and *nondisjunction* ($\downarrow$):

$\mathscr{A}$	$\mathscr{B}$	$(\mathscr{A}\,/\,\mathscr{B})$	$(\mathscr{A} \downarrow \mathscr{B})$
T	T	F	F
T	F	T	F
F	T	T	F
F	F	T	T

An English correlate of Sheffer's stroke is *not both . . . and;* the chief correlate of nondisjunction is *neither . . . nor.* The existence of Sheffer functions is intriguing; computer logic gates equivalent to them (called NAND and NOR gates) have been extremely important in computer logic design. Nevertheless, we'll generally restrict our sentential language to the five connectives first presented.

Problems

Explain how a connective in each of these sentences deviates somewhat from the truth function with which it is generally correlated.

1. Mary fell down and got up.

2. My heart rate approaches 200 if and only if I do heavy exercise.

3. Only if you come to me on your knees will I let you out of the contract.

4. Give me a place to stand, and I will move the earth (Archimedes).

▶ 5. If the moon smiled, she would resemble you (Sylvia Plath).

6. Why, if 'tis dancing you would be, there's brisker pipes than poetry (A. E. Housman).

7. It is not the case that if God is dead life is meaningless.

8. I'll leave only if you have somebody to take my place.

9. If Mike clears his throat once more, I'll strangle him.

▶ 10. Fame and rest are utter opposites (Richard Steele).

11. Show me a good loser, and I'll show you an idiot (Leo Durocher).

12. Lajoie chews Red Devil tobacco/Ask him if he don't (Queen City Tobacco Co. advertisement, circa 1900).

2.4 SYMBOLIZATION

We can use the symbolic language we've constructed to help evaluate arguments in natural language only by symbolizing English arguments. To symbolize an English discourse in SL,

1. Identify English sentence connectives and replace them with symbolic connectives.

2. Identify the smallest sentential components of the English sentences and replace each distinct component with a distinct sentence letter. (A record of which sentence letter symbolizes which English sentence component is called a *dictionary*.)

3. Use the structure of the English sentence to determine grouping.

Our objective in symbolization is to devise a formula that would be true exactly when the corresponding sentence is true, and false exactly when that sentence is false. In the process, we want our symbolism to represent as much of the logical structure of the corresponding sentence as possible.

Several factors complicate these steps. The first step—identify and replace sentence connectives—relies on the correlation we have outlined between truth functions and certain English sentence connectives.

Step two—identify and replace atomic sentences—is complicated chiefly by the fact that English arguments try not to be repetitive or dull. Authors rarely use even a component sentence twice in the same passage. Even if the meaning is the same, the wording may vary. To allow for this, we'll judge atomic sentences to be the same if they clearly have the same meaning.

The third step—determining the grouping—is, in a sense, the trickiest of all. To avoid ambiguity in our symbolic language, we use parentheses as grouping indicators. Parentheses in English don't have this meaning, and so natural language sentences are sometimes ambiguous.

English, however, offers a number of devices for making grouping clear. The first is the use of commas. The English sentence 'John will come and Fred will leave only if you sign' has no clear grouping. Suppose we use this dictionary:

p: John will come
q: Fred will leave
r: You sign

We can symbolize this sentence as either $((p \ \& \ q) \to r)$ or $(p \ \& \ (q \to r))$. But a comma can make it clear that *and* is the main connective: 'John will come, and Fred will leave only if you sign' should be symbolized as $(p \ \& \ (q \to r))$. Basically, a comma emphasizes a break in the sentence, stressing the combination of two different phrases. Commas, therefore, tend to suggest that the nearest connective has some priority.

Second, English offers coordinate phrases such as *either . . . or, both . . . and,* and *if . . . then.* Coordinate phrases make the grouping clear because they identify the components clearly. 'Either Bill brought Mary and Susan brought Sam, or Susan brought Bob' makes the intended grouping clear. Using the dictionary

p: Bill brought Mary
q: Susan brought Sam
r: Susan brought Bob

we can symbolize the English as $((p \ \& \ q) \lor r)$. The coordinate connective takes precedence over any connective in the coordinated sections. Notice that the sentence without the coordinate phrase—'Bill brought Mary and Susan brought Sam or Susan brought Bob'—is far less clear.

Third, English allows a device that logicians have called "telescoping" and that linguists call "conjunction reduction." 'Susan brought Sam or Susan brought Bob,' for example, can be reduced to the shorter sentence 'Susan brought Sam or Bob.' Similarly, 'Fred likes Wanda and Kim likes Wanda' can be reduced to 'Fred and Kim like Wanda.' Telescoping, too, can clarify grouping. We could group the last example in this way: 'Bill brought Mary and Susan brought Sam or Bob.'

Consider this complex passage from the Magna Carta, which limits inheritance taxes:

> If any of our earls, or barons, or others who hold of us in chief by military service, shall die, and at the time of his death his heir shall be of full age, and owe a relief, he shall have his inheritance by the ancient relief. . . .

First, we can identify English connectives and replace them with their symbolic correlates:

> any of our earls ∨ barons ∨ others who hold of us in chief by military service, shall die, & at the time of his death his heir shall be of full age, & owe a relief, → he shall have his inheritance by the ancient relief. ...

The only difficulty here is determining where the symbol → should go. We've placed it in the only position where the English word *then* would make sense.

Second, we can identify atomic sentence components and replace distinct components with distinct sentence letters, as in this dictionary:

p: An earl dies
q: A baron dies
r: Another who holds of us in chief by military service dies
s: At the time of death the heir is of full age
p_1: At the time of death the heir owes a relief
q_1: The heir has his inheritance by the ancient relief

$$p \lor q \lor r, \& \ s, \& \ p_1, \to q_1$$

Third, we must determine grouping. The telescoping of the disjuncts in 'If any of our earls, or barons, or others who hold of us in chief by military service, shall die' tells us that they should be grouped together.

$$(p \lor q \lor r) \& \ s, \& \ p_1, \to q_1$$

Our placement of the arrow tells us that it must be the main connective:

$$((p \lor q \lor r) \& \ s, \& \ p_1) \to q_1$$

Finally, the comma after *of full age* suggests that we should group s with the disjunction:

$$(((p \lor q \lor r) \& \ s) \& \ p_1) \to q_1$$

However, the symbolization

$$((p \lor q \lor r) \& \ (s \& \ p_1)) \to q_1$$

is equivalent.

Problems

Symbolize each of these sentences in SL.

1. Kindness is in our power, but fondness is not (Samuel Johnson).

2. My father taught me to work, but not to love it (Abraham Lincoln).

3. And April's in the west wind, and daffodils (John Masefield).

4. But if you wisely invest in beauty, it will remain with you all the days of your life (Frank Lloyd Wright).

▶ 5. If you get simple beauty and naught else, you get about the best thing God invents (Robert Browning).

6. A man can hardly be said to have made a fortune if he does not know how to enjoy it (Vauvenargues).

7. It is not poetry, if it make no appeal to our passions or our imagination (Samuel Taylor Coleridge).

8. The forces of a capitalist society, if left unchecked, tend to make the rich richer and the poor poorer (J. Nehru).

9. Honesty pays, but it don't seem to pay enough to suit a lot of people (K. Hubbard).

▶ 10. If you don't get what you want, it is a sign either that you did not seriously want it, or that you tried to bargain over the price (R. Kipling).

11. Honesty is praised, but it starves (Latin proverb).

12. It does not do to leave a live dragon out of your calculations, if you live near him (J. R. R. Tolkien).

13. I am an idealist. I don't know where I'm going but I'm on my way (Carl Sandburg).

14. Size is not grandeur, and territory does not make a nation (Thomas H. Huxley).

▶ 15. And the light shineth in the darkness; and the darkness comprehendeth it not (John 1:5).

16. If you are patient in one moment of anger, you will escape 100 days of sorrow (Chinese proverb).

17. It is only by painful effort, by grim energy and resolute courage, that we move on to better things (Theodore Roosevelt).

18. If we are to preserve civilization, we must first remain civilized (Louis St. Laurent).

19. Miracles sometimes occur, but one has to work terribly hard for them (Chaim Weizmann).

▶ 20. Happiness is not the end of life, character is (Henry Ward Beecher).

21. A sense of duty is useful in work but offensive in personal relations (Bertrand Russell).

22. Loafing needs no explanation and is its own excuse (Christopher Morley).

23. So then neither is he that planteth any thing, neither he that watereth; but God that giveth the increase (I Corinthians 3:7).

24. In cases of difficulty and when hopes are small, the boldest counsels are the safest (Livy).

▶ 25. The mind itself, like other things, must sometimes be unbent; or else it will be either weakened or broken (Philip Sidney).

26. I could not love thee, dear so much/Loved I not honor more (Richard Lovelace).

27. Nothing will ever be attempted if all possible objections must first be overcome (Jules W. Lederer).

28. A man doesn't need brilliance or genius; all he needs is energy (Albert M. Greenfield).

29. It's not good enough that we do our best; sometimes we have to do what's required (Winston Churchill).

▶ 30. Action does not always bring happiness, but there is no happiness without it (Benjamin Disraeli).

31. If you build a castle in the air, you won't need a mortgage (Philip Lazarus).

32. So long as it [baseball] remains our national game, America will abide no monarchy, and anarchy will be too slow (Allen Sangree).

33. Without a plan for completion, it just won't happen (Robert Half).

34. It hinders the creative work of the mind if the intellect examines too closely the ideas as they pour in (Johann Schiller).

▶ 35. If the spirit of business adventure is dulled, this country will cease to hold the foremost position in the world (Andrew Mellon).

36. It's a great time to be in the [oil] business if you're not broke (J. P. Cullen).

37. If the power to do hard work is not talent, it is the best possible substitute for it (James A. Garfield).

38. People have a perception that money will make everything wonderful. It doesn't. If your Mercedes Benz has a dead battery, it's even more frustrating than a dead battery in your Chevy (George Lucas).

39. In actuality the very wealthy in some ways have much in common with the very poor. Both classes migrate with the seasons, often wear tattered shoes with no socks, drink too much and have plenty of time to hang around (Earl Brechlin).

40. Arguments are to be avoided; they are always vulgar and often convincing (Oscar Wilde).

2.5 VALIDITY

Our tactic for analyzing English arguments and sentences is twofold. First, we will symbolize the sentences or arguments in SL. Second, we will evaluate the results for validity, satisfiability, and so on. Sequences of sentences, when symbolized in SL, become sequences of formulas, or *argument forms*.

> **DEFINITION.** An *argument form* consists of a finite sequence of formulas, called its *premise formulas,* together with another formula, its *conclusion formula.*

An argument is valid if its conclusion is true in every circumstance in which its premises are true. An argument form is simply a symbolized argument. An argument form, then, is valid if its conclusion formula is true in every circumstance in which its premise formulas are true.

In this section we'll develop a precise counterpart to our talk about circumstances. To judge validity, we must somehow survey every possible circumstance. How can we tell whether we've done this? In any given circumstance, a sentence or formula is either true or false. In the analysis we are currently prepared to do, we consider nothing about a sentence or sentence letter but its truth value. So, in truth-functional sentential logic, we need to survey all possible combinations of truth and falsehood.

In analyzing formulas and argument forms for validity, therefore, we'll speak about possible combinations of truth values, or *interpretations*. We can interpret a formula, in general, by saying whether the sentence letters it contains are true or false.

> **DEFINITION.** An *interpretation of a sentence letter* is an assignment of a truth value to the sentence letter. An *interpretation of a formula* of sentential logic is an assignment of truth values to its sentence letters. An *interpretation of an argument form* or set of formulas in sentential logic is an assignment of truth values to all the sentence letters in the argument form or set.

Using the concept of an interpretation, we can easily formulate definitions of validity and other semantic notions for formulas and argument forms. A formula is valid if and only if it's true on every interpretation of it. A formula is (a) contradictory if and only if it's false on every interpretation of it, (b) satisfiable if and only if it's true on at least one interpretation of it, and (c) contingent if and only if it's neither valid nor contradictory. An argument form is valid if and only if no interpretation of it makes its premise formulas all true and its conclusion formula false. A set of formulas, S, implies a formula, $\mathscr{A}$, if and only if no interpretation of S together with $\mathscr{A}$ makes every member of S true but $\mathscr{A}$ false. Two formulas are equivalent if and only if they agree in truth value on every interpretation of them. And a set of formulas is satisfiable if and only if, on some interpretation of it, every member of the set is true; the set is contradictory otherwise.

The definitions of the logical connectives specify the values of complex formulas on the basis of the values of their components. Thus, we can compute the truth value of a formula of any length, given the values of the atomic formulas—that is, sentence letters—appearing in the formula.[5]

Consider this simple example: Suppose that p and q are true, but r is false. What is the truth value of the formula $((p \to q) \to r)$? The main connective of this formula is the second arrow. The antecedent of this conditional is another conditional, both of whose components are true; the definition of the conditional indicates that this smaller conditional is true. The larger conditional, then, has a true antecedent, but its consequent, r, is false. From the definition of the conditional we can deduce that the larger conditional is false.

Approaching this problem more systematically, we might list the values of the sentence letters first and then proceed to generate the values of subformulas until we reach the formula as a whole. We might, in other words, work our way up the phrase structure tree for the formula. The phrase structure tree on the left leads us to the table on the right, and, perhaps, to the more compressed table below it:

$$((p \to q) \to r)$$

p	q	r	$(p \to q)$	$((p \to q) \to r)$
T	T	F	T T T	T T T F F

p	q	r	$((p \to q) \to r)$
T	T	F	T T T F F

A boldface letter in the tables represents the truth value of the entire formula above it.

Notice that, in constructing these tables, we begin with the sentence letters, which appear at the bottom of any phrase structure tree. We then assign values to items one level up from the lowest sentence letters in the tree and work our way gradually to the top level, where the formula itself appears. In effect, we work from inside parentheses out. When two subformulas are equally "deep" inside parentheses—whenever, alternatively, they occupy the same level of the tree—it makes no difference which we attack first. Our definitions thus allow us to compute the value of a formula, given some interpretation assigning values to its sentence letters. This is the central idea behind the method of *truth tables*, which we'll develop in the next section.

Problems

Classify these sentences as correct or incorrect, and explain.

1. Valid argument forms having valid conclusion formulas are valid.

2. Some argument forms with contradictory premise formulas aren't valid.

3. There is a formula that implies every other formula.

4. There is a formula that is equivalent to every other formula.

▶ 5. No satisfiable sets of formulas imply every formula.

6. Any formula that follows from a satisfiable formula is satisfiable.

7. Any formula that implies a contingent formula is not valid.

8. Any formula that follows from a contingent formula is contingent.

9. Any formula that follows from a valid formula is valid.

▶ 10. Any formula that implies a valid formula is valid.

11. All contradictory formulas imply one another.

12. All contingent formulas imply one another.

13. All valid formulas imply one another.

14. If a set of formulas is satisfiable, no member of that set implies a contradictory formula.

▶ 15. If no member of a set of formulas implies a contradictory formula, that set is satisfiable.

16. If a set of formulas implies no contradictory formula, that set is satisfiable.

17. Some valid argument forms have contradictory formulas as conclusion formulas.

18. No formula implies its own negation.

19. Any formula that implies its own negation is contradictory.

▶ 20. Any formula implied by its own negation is valid.

Using the interpretations listed, calculate the truth values of these formulas.

21. $(p \lor q) \lor (p \lor \neg r)$ (p and q false; r true)

22. $(p \rightarrow (q \leftrightarrow r)) \rightarrow (q \lor \neg r)$ (p and r true; q false)

23. $((r \mathbin{\&} q) \rightarrow \neg p) \mathbin{\&} (\neg q \leftrightarrow p)$ (p true; q and r false)

24. $\neg((p \rightarrow \neg q) \lor (q \lor \neg r))$ (p false; q and r true)

▶ 25. $(p \leftrightarrow \neg r) \rightarrow (r \rightarrow q)$ (p, q, and r true)

26. $\neg(p \lor q) \mathbin{\&} \neg(p \rightarrow r)$ (p, q, and r false)

27. $(p \lor (q \mathbin{\&} r)) \rightarrow (p \mathbin{\&} q)$ (p and q true; r false)

28. $(p \mathbin{\&} q) \leftrightarrow \neg r$ (p and r true; q false)

29. $\neg(\neg(\neg p \mathbin{\&} \neg q) \mathbin{\&} \neg r)$ (p true; q and r false)

▶ 30. $\neg((q \lor \neg p) \rightarrow \neg r)$ (p false; q and r true)

31. $(p \leftrightarrow \neg(q \leftrightarrow r)) \leftrightarrow \neg p$ (p and q false; r true)

32. $(p \lor \neg q) \leftrightarrow (p \lor \neg r)$ (p and r false; q true)

33. $(\neg(p \mathbin{\&} r) \to \neg q) \to \neg q$ (p, q, and r false)

34. $\neg((p \to \neg q) \leftrightarrow \neg r)$ (p, q, and r true)

▶ **35.** $(p \to \neg r) \to \neg(p \mathbin{\&} q)$ (p true; q and r false)

Bernard Bolzano (1781–1848) called an argument form *exact* or *adequate* if and only if it is valid and each of the premise formulas is necessary for the validity of the argument form (that is, omitting any premise formula or formulas makes the argument form invalid). In such an argument form, the premise formulas *exactly imply* the conclusion formula. Are these assertions about Bolzano's concept correct or incorrect? Explain.

36. No exact argument form has a valid formula as a conclusion.*

37. Some exact argument forms have valid formulas as premises.*

38. Some exact argument forms have contradictory conclusion formulas.*

39. Some exact argument forms have contradictory premise formulas.*

▶ **40.** Some exact argument forms have contingent conclusions.*

41. Every formula exactly implies itself.*

42. Some formula exactly implies every formula.*

43. Some formula is exactly implied by every formula.*

44. Sets of contingent formulas exactly imply only contingent formulas.*

▶ **45.** No formula exactly implies its own negation.*

2.6 TRUTH TABLES FOR FORMULAS

Given an interpretation, we can now compute the truth value of a formula on that interpretation. This in itself is interesting only if we have some reason for singling out a particular interpretation—for instance, because we think it corresponds to circumstances as they actually are. But the validity of a formula amounts to its truth on every interpretation of it. We can thus determine validity by surveying the set of all possible interpretations.

A truth table is, in essence, a computation of the truth value of a formula under each of its possible interpretations. If the formula is valid, then the main column of the truth table technique will be a string of T's. If it's contradictory, the main column will be a string of F's. If the formula is contingent, finally, the main column will be a string containing both T's and F's. Truth tables amount to nothing more than several simple tables, of the kind we saw in the last section, done at once. But they allow us to evaluate any formula of sentential logic as valid, contradictory, or contingent.

A *truth table* for a formula consists of four elements: (a) a listing of the sentence letters of the formula, (b) the formula itself, (c) a list of all possible interpretations of the formula, and (d) a computation of the truth value of the formula on each interpretation. We will write truth tables in the following configuration, which is very similar to the one we developed to evaluate a formula on a single interpretation:

Sentence letters	Formula
List of interpretations	Computation

The column heads of a truth table contain the formula to be evaluated, preceded by the sentence letters it contains. To see how this works in practice, consider an example. Is the formula $(p \vee (q \rightarrow p))$ valid, contradictory, or contingent? We can set up column heads of the table by listing the sentence letters in the formula, followed by the formula.

$$p \quad q \qquad (p \vee (q \rightarrow p))$$

The table lists all possible interpretations of the sentence letters under their respective column heads. If there is one sentence letter, then there are obviously only two interpretations: truth and falsehood. If there are two letters, then there are four interpretations; three letters have eight possible interpretations; and so on. In general, n sentence letters can be given 2^n interpretations. One way to list all these possibilities in a standard order is to count backward in base 2 from $2^n - 1$ to 0, replacing 0 and 1 with F and T, respectively. Alternatively, imagine an array of the following kind, which is based on the notion that the truth value of any sentence letter is logically independent of the truth value of any other. The second letter, for instance, could be either true or false whether the first letter happened to be true or happened to be false.

```
                    T T
              T     F
                  F T
        T           F
                  T T
              F   F
                  F T
                    F
                  T T
              T   F
                  F T
        F           F
                  T T
              F   F
                  F T
                    F
```

This table suggests that the listings of possible interpretations of cases involving one, two, three, or four sentence letters should look like this:

```
T     T T   T T T     T T T T
F     T F   T T F     T T T F
      F T   T F T     T T F T
      F F   T F F     T T F F
            F T T     T F T T
            F T F     T F T F
            F F T     T F F T
            F F F     T F F F
                      F T T T
                      F T T F
                      F T F T
                      F T F F
                      F F T T
                      F F T F
                      F F F T
                      F F F F
```

If the formula contains more than four sentence letters, the table will have thirty-two or more rows: one for each possible interpretation of the formula. In such cases, the method works but is so cumbersome that other techniques are probably better suited to the task.

Our example, however, has only two sentence letters. We can list all possible interpretations as follows:

p	q	$(p \vee (q \to p))$
T	T	
T	F	
F	T	
F	F	

Here is a step-by-step breakdown of the computation portion: First, copy the interpretations of each letter under its occurrences in the formula. This gives the values for the bottom nodes of the formula's phrase structure tree.

p	q	$(p \vee (q \to p))$
T	T	T T T
T	F	T F T
F	T	F T F
F	F	F F F

Then, begin searching for subformulas one level up from the bottom of the tree; look, that is, for subformulas as far inside parentheses as possible.

Negations of single sentence letters, although they might not be at this level of the tree, are always safe as well, since the value of the negated letter depends on nothing but the value of the letter itself. To systematize this process, we can decide to compute values of negations of single sentence letters first. Then, compute values of subformulas, working from inside parentheses out.

In our example, we begin with the conditional:

p	q		(p	∨	(q	→	p))
T	T		T		T	T	T
T	F		T		F	T	T
F	T		F		T	F	F
F	F		F		F	T	F

This brings us up the phrase structure tree until, finally, we reach the formula as a whole. The last computation should be for the main connective of the entire formula.

In summary: To compute the truth value of a formula under each of its interpretations:

1. Copy the interpretations of each sentence letter under its occurrences in the formula

2. Compute values of negations of single sentence letters

3. Compute values of subformulas, working from inside parentheses out

So, we conclude the example by computing the value of the larger disjunction. The ∨ is the main connective of the formula, so the truth values we are computing are those of the formula as a whole.

p	q		(p	∨	(q	→	p))
T	T		T	**T**	T	T	T
T	F		T	**T**	F	T	T
F	T		F	**F**	T	F	F
F	F		F	**T**	F	T	F

Truth tables thus represent a "bottom-up" strategy; we begin with the letters at the bottom of the phrase structure tree and work up the tree to compute values for more complex units.

Under each column head giving the sentence letters and connectives of the formula, a completed table has a column of T's and F's. These represent the truth values of the formula or subformula under each interpretation of it. The column for the entire formula itself—the last to be filled in—is the *main column* of the table. The main column specifies the truth value of the formula on each interpretation. In the table above, and in those we'll construct throughout this chapter, the entries in the main column are in boldface.

> DEFINITION. The *main column* of a truth table is under the column head showing the main connective of the formula.

By now, it may be obvious how to use truth tables to evaluate formulas for validity, contradiction, and so on. A valid formula is true on every interpretation of it; the main column of the formula's table, therefore, should contain all T's. A contradictory formula is false on every interpretation of it; the main column of its table, therefore, should contain all F's. Satisfiable formulas, which are true on at least one interpretation of them, give rise to tables whose main columns contain at least one T. The main columns of tables for contingent formulas, finally, contain both T's and F's. The main column in the table for $(p \vee (q \rightarrow p))$ contains both T's and F's, thus $(p \vee (q \rightarrow p))$ is contingent. To summarize:

Main Column	Formula
All T's	Valid
All F's	Contradictory
T's and F's	Contingent
At least one T	Satisfiable

Let's look at another, more complex example. To evaluate $\neg((p \rightarrow q) \vee (q \rightarrow r))$, we begin a truth table by writing as column heads the three sentence letters in the formula, followed by the formula itself. Then we list all possible interpretations, copying these down under the occurrences of the sentence letters in the formula:

p q r	$\neg((p \rightarrow q) \vee (q \rightarrow r))$
T T T	T T T T
T T F	T T T F
T F T	T F F T
T F F	T F F F
F T T	F T T T
F T F	F T T F
F F T	F F F T
F F F	F F F F

Now, we begin to compute the values of subformulas, working our way up the phrase structure tree. There are no negations of single sentence letters—the negation here is, in fact, the main connective—so we begin as far inside parentheses as possible, with the two conditionals. We can do these in either order, ending up with this table:

p q r	¬((p → q) ∨ (q → r))
T T T	T T T T T T
T T F	T T T T F F
T F T	T F F F T T
T F F	T F F F T F
F T T	F T T T T T
F T F	F T T T F F
F F T	F T F F T T
F F F	F T F F T F

After we compute the truth values of the two conditionals, we are in a position to calculate the values of the disjunction, which would be immediately above them on a phrase structure tree:

p q r	¬((p → q) ∨ (q → r))
T T T	T T T T T T T
T T F	T T T T T F F
T F T	T F F T F T T
T F F	T F F T F T F
F T T	F T T T T T T
F T F	F T T T T F F
F F T	F T F T F T T
F F F	F T F T F T F

The disjunction is valid; it is true on every interpretation. Finally, then, we use the definition of negation to calculate the values of the entire formula:

p q r	¬((p → q) ∨ (q → r))
T T T	**F** T T T T T T T
T T F	**F** T T T T T F F
T F T	**F** T F F T F T T
T F F	**F** T F F T F T F
F T T	**F** F T T T T T T
F T F	**F** F T T T T F F
F F T	**F** F T F T F T T
F F F	**F** F T F T F T F

Because the formula is false on every row—that is, on every interpretation of it—it is contradictory.

Problems

Construct truth tables for these formulas and determine whether they are valid, contradictory, or contingent.

1. p **2.** $p \mathbin{\&} \neg p$ **3.** $p \vee \neg p$

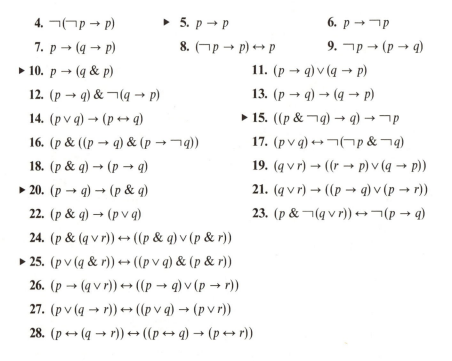

4. $\neg(\neg p \rightarrow p)$ ▸ 5. $p \rightarrow p$ 6. $p \rightarrow \neg p$

7. $p \rightarrow (q \rightarrow p)$ 8. $(\neg p \rightarrow p) \leftrightarrow p$ 9. $\neg p \rightarrow (p \rightarrow q)$

▸ 10. $p \rightarrow (q \,\&\, p)$ 11. $(p \rightarrow q) \vee (q \rightarrow p)$

12. $(p \rightarrow q) \,\&\, \neg(q \rightarrow p)$ 13. $(p \rightarrow q) \rightarrow (q \rightarrow p)$

14. $(p \vee q) \rightarrow (p \leftrightarrow q)$ ▸ 15. $((p \,\&\, \neg q) \rightarrow q) \rightarrow \neg p$

16. $(p \,\&\, ((p \rightarrow q) \,\&\, (p \rightarrow \neg q)))$ 17. $(p \vee q) \leftrightarrow \neg(\neg p \,\&\, \neg q)$

18. $(p \,\&\, q) \rightarrow (p \rightarrow q)$ 19. $(q \vee r) \rightarrow ((r \rightarrow p) \vee (q \rightarrow p))$

▸ 20. $(p \rightarrow q) \rightarrow (p \,\&\, q)$ 21. $(q \vee r) \rightarrow ((p \rightarrow q) \vee (p \rightarrow r))$

22. $(p \,\&\, q) \rightarrow (p \vee q)$ 23. $(p \,\&\, \neg(q \vee r)) \leftrightarrow \neg(p \rightarrow q)$

24. $(p \,\&\, (q \vee r)) \leftrightarrow ((p \,\&\, q) \vee (p \,\&\, r))$

▸ 25. $(p \vee (q \,\&\, r)) \leftrightarrow ((p \vee q) \,\&\, (p \,\&\, r))$

26. $(p \rightarrow (q \vee r)) \leftrightarrow ((p \rightarrow q) \vee (p \rightarrow r))$

27. $(p \vee (q \rightarrow r)) \leftrightarrow ((p \vee q) \rightarrow (p \vee r))$

28. $(p \leftrightarrow (q \rightarrow r)) \leftrightarrow ((p \leftrightarrow q) \rightarrow (p \leftrightarrow r))$

2.7 OTHER USES OF TRUTH TABLES

An argument form is valid just in case every interpretation making its premise formulas true makes its conclusion formula true as well. Truth tables let us calculate the truth values of formulas under each of their interpretations, and so it's easy to set up a truth table to give us the information needed to evaluate an argument form.

We can set up a table computing the values of each of the premise formulas, and the conclusion formula, separately, and then interpret these tables as evaluating argument forms. To do so, we

1. List the sentence letters appearing in the argument form.

2. Beneath them, list all possible interpretations of them.

3. List each premise formula, and then the conclusion formula.

4. Compute the value of each formula separately.

This gives us a main column for each formula. An argument form is valid, of course, just in case no interpretation makes the premise formulas all true but the conclusion formula false.

Here are some sample argument forms:

(1)	(2)	(3)
p	$p \to q$	$(p \vee q) \to \neg(q \,\&\, r)$
$p \to q$	$q \leftrightarrow r$	q
$\therefore q$	$\therefore (r \,\&\, q) \to p$	$\therefore \neg r$

It's easy to evaluate them for validity.

p q	p	$(p \to q)$	q	
T T	T	T T T	T	
T F	T	T F F	F	
F T	F	F T T	T	
F F	F	F T F	F	(Valid)

Argument form (1) is valid; on no row are the premises true and the conclusion false.

p q r	$(p \to q)$	$(q \leftrightarrow r)$	$((r \,\&\, q) \to p)$	
T T T	T T T	T T T	T T T T T	
T T F	T T T	T F F	F F T T T	
T F T	T F F	F F T	T F F T T	
T F F	T F F	F T F	F F F T T	
F T T	F T T	T T T	T T T F F	(Invalid)
F T F	F T T	T F F	F F T T F	
F F T	F T F	F F T	T F F T F	
F F F	F T F	F T F	F F F T F	

Argument form (2) is invalid; in the interpretation

p	q	r
F	T	T

the conclusion formula is false while the premise formulas are true.

p q r	$((p \vee q) \to \neg(q \,\&\, r))$	q	$\neg r$	
T T T	T T T F F F T T T	T	F T	
T T F	T T T T T T T F F	T	T F	
T F T	T T F T T T F F T	F	F T	
T F F	T T F T T T F F F	F	T F	
F T T	F T T F F F T T T	T	F T	
F T F	F T T T T T T F F	T	T F	
F F T	F F F T T T F F T	F	F T	
F F F	F F F T T T F F F	F	T F	(Valid)

When a truth table shows an argument form to be invalid, it also specifies an interpretation making the premise formulas true and the conclusion formula false. The table, in other words, not only indicates that there is such an interpretation but also tells what it is. If there are several interpretations, the table specifies them all. To see under what circumstances the premise formulas would be true while the conclusion formula would be false, look at those rows of the table that yield that combination of truth values. The rows detail interpretations.

Argument form validity is tantamount to implication. A set of formulas implies a given formula just in case every interpretation making every member of the set true makes the given formula true as well. This is exactly the relation that holds between the premise formulas of an argument form and the conclusion formula. Consequently, our method also works for implication problems. To find out whether a set of formulas, $\{\mathscr{A}_1, \ldots, \mathscr{A}_n\}$, implies a formula, $\mathscr{B}$, we can construct the table we'd use to evaluate the argument form

$$\mathscr{A}_1$$
$$\vdots$$
$$\mathscr{A}_n$$
$$\therefore \mathscr{B}$$

Equivalence, of course, is just implication in both directions. $\mathscr{A}$ and $\mathscr{B}$ are equivalent just in case they have the same truth value on every interpretation of them.

To show that p is equivalent to $\neg\neg p$, for example, we can construct this table:

p	p	$\neg\neg p$
T	T	T F T
F	F	F T F

The table shows that p and $\neg\neg p$ agree in truth value on every interpretation of them, so they are equivalent.

Finally, we can use truth tables to evaluate satisfiability. We have already done so with individual formulas. Even when we are concerned with the satisfiability of sets of formulas, however, truth tables offer a simple test. A set of formulas is satisfiable if and only if its members are all true on some interpretation of them.

Consider, for instance, the set of formulas $\{\neg q, p \vee q, p \rightarrow q\}$. Is this set satisfiable? To find out, we can construct this table:

p q	$\neg q$	$(p \vee q)$	$(p \rightarrow q)$
T T	F T	F T T	T T T
T F	T F	T T T	T F F
F T	F T	F T T	F T T
F F	T F	F F F	F T F

No interpretation makes every member of the set true. So, the set is contradictory.

The truth table technique is a simple but remarkably powerful tool for solving problems in sentential logic. We can use this technique to evaluate argument form validity, formula validity, implication, equivalence, formula satisfiability, and set satisfiability. We can do all this, furthermore, in a clear and easily understandable way. With a truth table, we can survey all possible interpretations of a formula, argument form, and the like and compute truth values on each interpretation.

Truth tables constitute a *decision procedure* for validity in SL.

> **DEFINITION.** A *decision procedure* for a property is a mechanical method for determining, in a finite time, whether any given thing has that property.

Decision procedures must be completely mechanical; using them must involve nothing more than following rules. No ingenuity or creativity is required. Also, a decision procedure applied to an object always gives a yes or no answer after a finite time. Truth tables, clearly, are decision procedures for validity, equivalence, and so on. We can construct them just by following rules, and they always have a finite size. Whenever there is a decision procedure for a property, that property is *decidable*. Formula and argument form validity, formula and set satisfiability, implication, and equivalence are all decidable in SL.

The clarity of the truth table technique comes at a price. When the formula or argument form becomes very complex, the number of possible interpretations may be very large. In the next chapter, therefore, we'll turn toward a method of answering a question about validity, satisfiability, etc., without running through every possible way of assigning truth values to sentence letters.

Truth tables are decision procedures for formula and argument form validity. What is the link between the validity of an argument, however, and the validity of its symbolization in SL? If an argument form that results from our process of symbolization is valid, then the original argument it symbolizes is valid. But an invalid argument form doesn't show that the original argument is invalid. Many English arguments depend on logical relationships that sentential logic doesn't capture. Consider, for example, the argument

All captains are officers.
∴ No platoon that contains no officers contains a captain.

This argument contains no sentence connectives: We can symbolize it in SL, at best, as

p
∴ *q*

We can conclude, not that the argument is invalid, but that it's invalid in SL. A form for the argument showing as much logical structure as possible within SL is invalid, even though the argument is valid. Because of the nature of our method, therefore, "valid" verdicts merit more trust than "invalid" verdicts. The same holds of other concepts. "Contradictory" verdicts merit more trust than "satisfiable" verdicts, and "implies" and "are equivalent" verdicts merit more trust than their opposites. If a set of formulas is satisfiable, then we can conclude only that the set of sentences they symbolize is satisfiable in SL. It's always possible that, to establish the validity of a formula or argument, the equivalence of two formulas, and so on, we would have to bring out bigger guns than are available in our logical theory.

Problems

Symbolize these arguments in SL and evaluate the corresponding argument forms as valid or invalid.

1. Sandra mailed the grant proposal I placed on her desk. So, she either mailed the grant application or threw it away.

2. You are either a knave or a fool. You are a knave; so, you're no fool.

3. If I'm right, then I'm a fool. But if I'm a fool, I'm not right. Therefore, I'm no fool.

4. If I'm right, then I'm a fool. But if I'm a fool, I'm not right. Therefore, I'm not right.

▶ 5. Unless I'm mistaken, I'm a fool. But if I am a fool, I must be mistaken. So I'm mistaken.

6. If Einstein's theory of relativity is correct, light bends in the vicinity of the sun. Light does indeed bend in the vicinity of the sun. It follows that Einstein's theory is correct.

7. The thief is either in Tewkesbury or in Bristol. He is not in Tewkesbury; therefore, he is in Bristol.

8. Aristotle was a brilliant thinker. If his theory of remembering is right, then modern psychological accounts are wrong. So, either modern accounts of memory are wrong, or Aristotle wasn't so brilliant after all.

9. Jessica meows just in case she is hungry. She is meowing, but she isn't hungry. Therefore, the end of the earth is at hand.

▶ 10. If Socrates died, he died either while he was living or while he was dead. But he did not die while living; moreover, he surely did not die while he was already dead. Hence, Socrates did not die.

11. Nothing can be conceived as greater than God. If God existed in our imaginations, but not in reality, then something would be conceivable as

greater than God (namely, the same thing, except conceived as existing in reality). Therefore, if God exists in our imaginations, He exists in reality.

12. You believe this only if you're a turkey. You don't believe this. So, you are no turkey.

13. If you have a cake, just looking at it will make you hungry; if looking at it makes you hungry, you will eat it. So, you can't both have your cake and fail to eat it.

14. A man cannot serve both God and Mammon. But if a man does not serve Mammon, he starves; if he starves, he can't serve God. Therefore, a man cannot serve God.

▶ 15. Either we ought to philosophize or we ought not. If we ought, then we ought. If we ought not, then also we ought (to justify this view). Hence in any case we ought to philosophize (Aristotle).

Chrysippus regarded the following argument forms as basic to sentential logic. In some cases, however, he thought of connectives as having meanings different from those we've associated with them. Which of these are valid in SL?

16. $p \rightarrow q; p; \therefore q$

17. $p \rightarrow q; \neg q; \therefore \neg p$

18. $\neg(p \& q); p; \therefore \neg q$

19. $p \vee q; p; \therefore \neg q$

▶ 20. $p \vee q; \neg q; \therefore p$

21. Cicero (106–43 B.C.), the famous Roman orator, summarized Stoic logic by citing seven principles: the five of Chrysippus, above, a repeat of #18, and $\neg(p \& q); \neg p; \therefore q$. Is this valid in SL?

Sextus Empiricus, a Greek sceptic who wrote in the third century A.D., preserved some argument forms whose validity the Stoics thought followed from the basic principles of Stoicism. Which of these are valid in SL?

22. $p \rightarrow (p \rightarrow q); p; \therefore q$

23. $(p \& q) \rightarrow r; \neg r; p; \therefore \neg q$

24. $p \rightarrow q; p \rightarrow \neg q; \therefore \neg p$

25. $p \rightarrow p; \neg p \rightarrow p; \therefore p$

Determine whether the formulas in each pair are equivalent. If they are not, say whether either formula implies the other.

26. $p \rightarrow q$ and $p \leftrightarrow (p \& q)$

27. $\neg(p \vee q)$ and $\neg p \vee \neg q$

28. $\neg(p \& q)$ and $\neg p \& \neg q$

29. $\neg(p \rightarrow q)$ and $\neg p \rightarrow \neg q$

▶ 30. $\neg(p \leftrightarrow q)$ and $\neg p \leftrightarrow \neg q$

31. $\neg(p \leftrightarrow q)$ and $\neg p \leftrightarrow q$

32. $\neg(p \rightarrow q)$ and $p \& \neg q$

33. $\neg(p \& q)$ and $\neg p \vee \neg q$

34. $\neg(p \vee q)$ and $\neg p \& \neg q$

▶ 35. $p \& q$ and $(p \vee q) \& (p \leftrightarrow q)$

36. $p \& q$ and $(p \vee q) \& (p \rightarrow q)$

37. $p \& q$ and $(p \vee q) \& (q \rightarrow p)$

38. $p \vee q$ and $\neg p \to q$ **39.** $p \vee q$ and $\neg q \to p$

40. $p \to q$ and $p \leftrightarrow (p \vee q)$ **41.** $p \to q$ and $q \leftrightarrow (p \vee q)$

42. $p \leftrightarrow q$ and $(p \to q) \& (q \to p)$

43. $p \leftrightarrow q$ and $(p \& q) \& (\neg p \& \neg q)$

44. $p \leftrightarrow q$ and $\neg p \leftrightarrow \neg q$

▶ **45.** $p \leftrightarrow q$ and $(p \leftrightarrow q) \leftrightarrow p$

Evaluate these sets of formulas as satisfiable or unsatisfiable. For each satisfiable set, specify an interpretation making every member of the set true.

46. $p, (p \vee q) \to q, \neg(p \& q)$

47. $p \to (q \to r), \neg((p \& \neg r) \to \neg q)$

48. $p \to (q \vee r), p \to \neg q, p \to \neg r$

49. $p \to q, p \to \neg q$

▶ **50.** $q \leftrightarrow (p \to (q \vee \neg r)), \neg(p \to (q \vee r)), r$

51. $p, q \to \neg r, r$

52. $p \to \neg q, (r \& p) \to q, (r \& \neg p) \to p$

53. $\neg(p \to (q \& \neg r)), p \vee (q \& \neg r), r \to (p \vee q)$

54. $(p \vee \neg q) \leftrightarrow (p \to r), p \leftrightarrow (q \& (r \to \neg p)), q$

▶ **55.** $\neg(p \leftrightarrow \neg q), p \leftrightarrow (r \leftrightarrow q), q \to (p \leftrightarrow r)$

56. $p, (q \vee r) \to (p \vee \neg q), p \leftrightarrow (q \leftrightarrow \neg r)$

57. $q \to (p \vee r), p \to (r \& \neg q), r \to (q \& p)$

58. $(r \& q) \to \neg(q \& p), (p \vee q) \leftrightarrow (q \vee r), (p \leftrightarrow r) \to (q \leftrightarrow \neg p)$

59. $p \vee (q \vee (\neg p \& r)), \neg((p \& \neg q) \to (r \vee p)), \neg(p \leftrightarrow \neg r)$

▶ **60.** $((p \to q) \to p) \to r, ((r \to p) \to r) \to q, ((q \to r) \to q) \to p$

Consider the following six formulas: $p \vee q, p \& q, p \to q, p \leftrightarrow q, \neg p$, and $\neg q$. Where $\mathcal{A}$ and $\mathcal{B}$ are different formulas from among this collection, there are 30 possible statements of the form $\mathcal{A}$ implies $\mathcal{B}$. Of these, only five are true. Show that these are the five.

61. $p \& q$ implies $p \vee q$ **62.** $p \& q$ implies $p \to q$

63. $p \& q$ implies $p \leftrightarrow q$ **64.** $p \leftrightarrow q$ implies $p \to q$

65. $\neg p$ implies $p \to q$

Chrysippus held as a basic valid argument form $p \vee q$; p; $\therefore \neg q$, which indicates that he did not think of $\vee$ as having the truth table we've associated with it. Subsequent logicians advanced a number of principles, some of which are invalid in SL. Do any of these follow from the formula associated with Chrysippus's argument form, $((p \vee q) \mathbin{\&} p) \rightarrow \neg q$?

66. $((p \rightarrow \neg q) \mathbin{\&} p) \rightarrow \neg (p \rightarrow q)$ (Stoics)

67. $((\neg p \rightarrow q) \mathbin{\&} \neg q) \rightarrow \neg (p \rightarrow q)$ (Stoics)

68. $(\neg (p \mathbin{\&} q) \mathbin{\&} \neg p) \rightarrow q$ (Cicero)

69. $(p \rightarrow \neg q) \leftrightarrow \neg (p \rightarrow q)$ (Boethius)

▶ **70.** $((p \rightarrow (q \rightarrow r)) \mathbin{\&} (q \rightarrow \neg r)) \rightarrow \neg p$ (Boethius)

Notes

[1] Aristotle, the first logician, was also the first logician to use symbols. The Stoics made the practice common, using numbers to stand for sentences. Nevertheless, no one developed a fully symbolic logical language until the nineteenth century, when Englishman George Boole (1815–1864) saw that logic could profit from mathematical analysis. (His greatest work, in fact, is entitled *The Mathematical Analysis of Logic*.)

[2] We'll assume that sentence connectives and other vocabulary items of our symbolic language are names of themselves. $(\mathscr{A} \rightarrow \mathscr{B})$ thus refers to the formula resulting from concatenating a left parenthesis, the formula $\mathscr{A}$, an occurrence of the conditional connective, the formula $\mathscr{B}$, and a right parenthesis. Notice that we can't refer to this formula by using ordinary quotation. '$(\mathscr{A} \rightarrow \mathscr{B})$' signifies the string of symbols quoted, that is, a left parenthesis, followed by the letter $\mathscr{A}$, followed by an occurrence of the conditional, followed by the letter $\mathscr{B}$, followed by a right parenthesis.

[3] These problems with the truth-functional rendering of the conditional have been recognized since the third century B.C., when the Greek logician Philo of Megara—a classmate of Zeno of Citium (336–264 B.C.), the founder of Stoicism—first proposed the analysis reflected in the table above. Diodorus, Philo's teacher, held that conditionals involve necessity, and Chrysippus (279–206 B.C.), the third head of the Stoic school and widely acclaimed as the greatest logician of his time, held that the sort of necessity involved is specifically logical, as opposed, for instance, to causal necessity. The controversy among their followers became so intense that Callimachus wrote that "even the crows on the roofs caw about the nature of conditionals."

[4] They were actually discovered by the American philosopher Charles Sanders Peirce (1839–1914) around 1880, but went unnoticed until Sheffer reproduced the discovery in 1913.

[5] In the history of logic, this was not so easy to see: Boole and Gottlob Frege (1848–1925), a German mathematician and logician who first invented quantification theory, both defined connectives using tables like those above in the nineteenth century. But not until around 1920 did the Austrian philosopher Ludwig Wittgenstein (1889–1951), the Polish logician Jan Lukasiewicz (1878–1956), and the American logician Emil L. Post recognize, independently, that the tabular definitions suffice for a computation of the truth values of complex formulas.

3

SEMANTIC
TABLEAUX

The method of truth tables constitutes a decision procedure for the validity of both formulas and argument forms, for implication and equivalence, and for the satisfiability of formulas and sets of formulas. Nevertheless, the technique has failings. Truth tables quickly become very large and tedious: A truth table for evaluating an argument form containing six distinct sentence letters has 64 rows, and one for evaluating an argument form with ten distinct sentence letters has 1,024 rows. Englishman Charles Dodgson (1832–1898), who was a mathematician and logician as well as—under the name Lewis Carroll—the author of *Alice in Wonderland,* devised "Froggy's Problem." This problem has eighteen distinct letters, requiring a table of 262,144 rows! And each row would have to be quite long: the table would have to contain over 31 million T's and F's. A person filling in one symbol per second and working nonstop would take almost a year to complete it. Furthermore, truth tables serve as a decision procedure for sentential logic, but don't extend easily to other, more comprehensive logical systems. Once our symbolic language becomes powerful enough to contain resources corresponding to the English words *some* and *all,* the truth table method breaks down.

In this chapter, therefore, we'll develop a decision procedure for sentential logic that offers many practical advantages. The procedure reflects intuitive ways of thinking about arguments more nearly than truth tables do. It evaluates arguments much more efficiently. And it extends readily to more comprehensive logical systems.

This method is that of semantic tableaux. *Semantic tableaux* are treelike diagrams that serve as tests for validity, implication, satisfiability, and so on.[1] We can test for these properties and relations by using semantic tableaux to conduct searches for interpretations.

Tableaux, mathematically speaking, are *trees*. At the top of each tree is the tree's *root;* at the bottom are the tree's *tips,* or *leaves.* A path going directly from the root to a leaf is a *branch.* Trees with more than one branch *split* where these paths diverge. A tableau has exactly as many branches as leaves. The portion of a tree above any splitting is the tree's *trunk.*

A semantic tableau is a tree with formulas appearing on it. A formula may appear on either the left side or the right side of a branch. Any formula on the trunk of a tableau appears on every branch. Some formulas, furthermore, may be marked with a *dispatch mark,* ✓, to indicate that a rule has been applied to them. The dispatch mark signals that we've used the information in the formula to extend the tableau; we can safely ignore dispatched formulas, because we've already taken account of the information they provide. Dispatched formulas are *dead;* undispatched formulas are *live.*

Branches that have the same formula appearing live on both sides are *closed.* Tableaux with all their branches closed are also *closed.*

> **DEFINITION.** A *tableau branch is closed* if and only if a live formula appears on both sides of the branch. Otherwise, the branch is *open.*

> **DEFINITION.** A *tableau is closed* if and only if every branch of the tableau is closed. Otherwise, it is *open.*

All the tableaux we'll consider in this text have special features. We will create them by using certain explicit rules. All the tableaux will be finite. And they will be *binary* in the sense that, whenever they split, one path will divide into two. These, then, are examples of semantic tableaux:

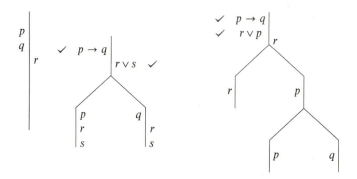

The left side of a tableau branch is the *truth* side; we will assume that formulas on the left are true. The right side is the *falsehood* side. The leftmost tableau above, then, corresponds to the assumption that p and q are true, while r is false. It has only one branch, which is open. We can record the information on the tableau by writing:

True: p, q False: r

The center tableau has two branches: The left branch has $p \to q$ on its left and $r \lor s$, p, r, and s on its right. The right branch has $p \to q$ and q on its left and $r \lor s$, r, and s on its right. Only the atomic formulas are live. Both branches, clearly, are open. We can record the information on this tableau by writing:

    (1) True: False: p, r, s
    (2) True: q False: r, s

The rightmost tableau has three branches. The left branch has $p \to q$, $r \lor p$, and r on its left and r on its right. The center branch has $p \to q$, $r \lor p$, and p on its left and r and p on its right. The right branch has $p \to q$, $r \lor p$, p, and q on its left and r on its right. Again, only the atomic formulas are live. We can record this by writing:

    (1) True: r False: r
    (2) True: p False: r, p
    (3) True: p, q False: r

Branches (1) and (2) are closed. Branch (1) has r on both sides, while branch (2) has p on both sides. Branch (3), however, is open, so the tableau as a whole is open.

Problems

Record the information on each branch of each of the following tableaux. Say whether each branch, and each tableau, is open or closed.

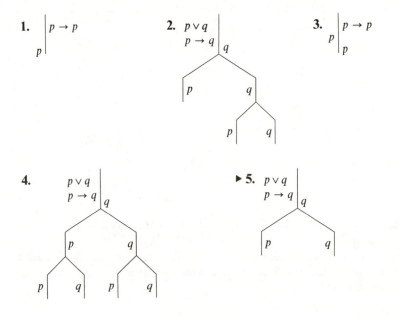

3.1 Rules for Negation, Conjunction, and Disjunction

Before we develop rules for each connective, two aspects of the general strategy underlying the use of semantic tableaux are worth noting. First, a tableau branch represents an interpretation or a possible circumstance. It corresponds to a row of a truth table. A tableau begins with a set of assumptions about the truth or falsehood of certain formulas; these assumptions are reflected in the placement of the formulas on the left or right side of the tableau. By way of the rules, other formulas are generated, extending the branches. In this process, the tableau gives us a picture of how the world could be, given the initial assumptions. In other words, the tableau is an attempt to depict some possible circumstances—that is, some interpretations— under which the initial assumptions would hold. When a branch closes, such an attempt ends in a contradiction; some formula is assigned both truth and falsehood. When all branches close, then every attempt ends in contradiction, allowing us to conclude that under no possible circumstances would the initial assumptions hold.

Second, given some formulas placed on the tableau, we try to generate some pictures of possible circumstances by determining what those formulas imply about the truth values of the sentence letters they contain. The strategy is the opposite of that used in constructing truth tables. We begin a truth table by assigning truth values to sentence letters; the table generates corresponding values for entire formulas. We begin a tableau, in contrast, by assigning truth values to formulas and use the tableau in an effort to find corresponding assignments to sentence letters.

Truth tables reflect a "bottom-up" strategy. In the table, we assign values to sentence letters, which appear at the bottom of phrase structure trees, and gradually, on the basis of these values, we assign values to more complex parts of the formula. Semantic tableaux reflect a "top-down" strategy. We begin by assigning a value for an entire formula (or set of formulas) and proceed to assign values to parts of formulas until the tableau closes, or reaches the bottom level of sentence letters.

Tableau rules come in pairs. Since we follow a "top-down" strategy, we are interested in decomposing formulas into their parts. We always begin, therefore, by applying a rule to the main connective of a formula. The formula could be on the left or right side of a branch. We need, then, two rules for each connective: one to handle formulas appearing on the left having that connective as a main connective, the other to handle similar formulas on the right. In this section we'll develop six rules: ¬L (negation left), ¬R (negation right), &L (conjunction left), &R (conjunction right), ∨L (disjunction left) and ∨R (disjunction right).

All these rules reflect the definitions of the truth functions in Chapter 2. We can see this more clearly, in the case of negation, by expressing the tabular definition differently. (Here, *iff* abbreviates *if and only if*.)

$\neg \mathscr{A}$ is true on an interpretation iff $\mathscr{A}$ is false on that interpretation.

Negation Left ($\neg$L)

This rule applies to formulas with a negation sign as main connective, appearing on the left side of a tableau branch. The left side represents truth. So, the question becomes: If a formula, $\neg \mathscr{A}$, is true, what is the truth value of $\mathscr{A}$? The answer, clearly, is falsehood. So, this rule takes the form:

Since $\mathscr{A}$ must be false, we check $\neg \mathscr{A}$ and write $\mathscr{A}$ on the right side of the tableau branch. In our statement of the rule, the result of applying the rule is in boldface; that to which the rule is applied is not. The check mark beside the decomposed formula on the left signals that a rule has been applied to this formula. Formulas dispatched with a check may be ignored for the rest of the tableau.

Negation Right ($\neg$R)

This rule applies to negated formulas appearing on the right side of a branch, which represents falsehood. If a formula, $\neg \mathscr{A}$, is false, then $\mathscr{A}$ must be true, so we write $\mathscr{A}$ on the left side of the tableau branch:

$$\mathscr{A} \;\Big|\; \neg \mathscr{A} \;\checkmark$$

Rules for conjunction also derive from that truth function's definition. Putting the tabular definition of Chapter 2 into a different form:

$\mathscr{A}$ & $\mathscr{B}$ is true on an interpretation iff $\mathscr{A}$ is true on that interpretation, and so is $\mathscr{B}$.

The rules mirror this definition directly.

Conjunction Left (&L)

This rule applies to formulas appearing on the left of a branch, with conjunctions as their main connectives. Under what circumstances is a formula, ($\mathscr{A}$ & $\mathscr{B}$), true? Obviously, circumstances in which $\mathscr{A}$ and $\mathscr{B}$ are both true.

So we can write both $\mathscr{A}$ and $\mathscr{B}$ on the left side of the branch:

$$
\begin{array}{cc}
\checkmark & \mathscr{A} \,\&\, \mathscr{B} \ \Big| \\
& \mathscr{A} \\
& \mathscr{B} \ \ \Big|
\end{array}
$$

Conjunction Right (&R)

This rule applies to formulas appearing on the right of a branch, with conjunctions as their main connectives. Under what circumstances is a formula, ($\mathscr{A}$ & $\mathscr{B}$), false? In essence, there are two possibilities: The formula is false when either $\mathscr{A}$ or $\mathscr{B}$ is false. To capture the idea that there are two options, we split the branch. On one of the newly created branches, we want to reflect the possibility that $\mathscr{A}$ is false, so we write $\mathscr{A}$ on the right. On the other, we write $\mathscr{B}$ on the right to reflect the possibility that $\mathscr{B}$ is false. The rule thus takes the form:

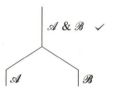

It might seem that we would need a third branch to reflect the possibility that both $\mathscr{A}$ and $\mathscr{B}$ are false. But these branches already take care of that possibility. Consider the left branch, which represents $\mathscr{A}$ as false. It represents nothing about $\mathscr{B}$; $\mathscr{B}$ could be either true or false. This branch alone, then, really captures two possibilities, when we consider the truth value of $\mathscr{B}$. It reflects the possibility that $\mathscr{A}$ is false and $\mathscr{B}$ is true, and the possibility that $\mathscr{A}$ and $\mathscr{B}$ are both false.

The rules for disjunction reflect its definition. We can express the content of our tabular definition in the last chapter in somewhat different form:

> $\mathscr{A} \lor \mathscr{B}$ is true on an interpretation iff either $\mathscr{A}$ is true on that interpretation, or $\mathscr{B}$ is true on it.

Given this formulation, we easily see what the rules should be.

Disjunction Left (∨L)

This rule, relevant to formulas appearing on the left with disjunctions as main connectives, asks us to consider when formulas of the form ($\mathscr{A} \lor \mathscr{B}$) are true.

The answer, clearly, is that they are true whenever either $\mathscr{A}$ or $\mathscr{B}$ is true. Again we must split the branch to reflect these two possibilities:

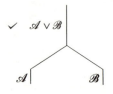

Disjunction Right (∨R)

This rule applies to formulas on the right with disjunctions as main connectives. When is a formula of the form $(\mathscr{A} \vee \mathscr{B})$ false? When both $\mathscr{A}$ and $\mathscr{B}$ are false. Dispatching the disjunction, therefore, we write both disjuncts on the right side of the branch:

$$
\begin{array}{l}
\mathscr{A} \vee \mathscr{B} \quad \checkmark \\
\mathscr{A} \\
\mathscr{B}
\end{array}
$$

Several questions arise about applying these rules. First, when a rule directs us to enter a formula on a given side of a branch, where on that branch do we put it? After all, there may already be formulas under the one to which we are applying the rule. To be consistent, we'll follow this rule: Always write the new formula at the bottom of the branch, *underneath* all the formulas already appearing there. For example, we may want to apply &L to the second formula in the tableau on the left below. The rule tells us to enter two new formulas (the conjuncts of the conjunction) on the left. We write these below the other formulas, resulting in the tableau on the right.

$$
\begin{array}{ll}
p \vee q & \\
q \,\&\, r & \\
r \to s &
\end{array}
\qquad\qquad
\begin{array}{ll}
& p \vee q \\
\checkmark & q \,\&\, r \\
& r \to s \\
& q \\
& r
\end{array}
$$

Second, what happens when the tableau splits? Suppose that we want to apply a rule to a formula above a split in the tableau. The rule directs us to enter a new formula on the branch. But which branch? What was one branch before is now two (or four, or eight, and so on). To capture the meaning of the formula on every branch on which it appears, we must make the required entries on every such branch. We'll follow this rule:

When applying a rule to a formula, make the entries it calls for on *every* branch on which that formula appears.

Normally, this means that we must make the entries on each branch that, below the formula in question, splits from what was once the single branch on which the formula appeared. As an example, the progress on a tableau beginning with $(p \& q)$ on the right and $(q \lor r)$ on the left might take this form:

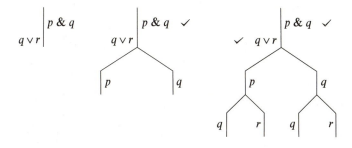

Applying the rule &R to $(p \& q)$ results in the center tableau. At that point, we apply $\lor$L to $(q \lor r)$, which results in the rightmost tableau. Notice that we had to make the entries—and perform the required split—on both branches that resulted from applying &R.

We will follow two other policies in applying rules. The order in which we apply rules, within the bounds of sentential logic, makes no difference in results. But it does make a difference in efficiency. To create as few branches, and as little work for ourselves, as possible, we'll follow this rule:

Apply rules that don't require any splitting *before* those that do.

So we'll apply $\neg$L, $\neg$R, &L, and $\lor$R before &R and $\lor$L. In addition, closed branches merit no further interest, since they contain a contradiction. No matter what rules are applied to them later, they will remain closed; they cannot depict any possible circumstance. We'll therefore follow this rule:

Abandon closed branches as soon as they are closed, marking them with the notation 'Cl' to indicate why we've done so.

This, too, adds to efficiency. To see why these rules are worth following, compare these two tableaux. They begin with the same formulas and contain the same logical information, but one is much simpler than the other.

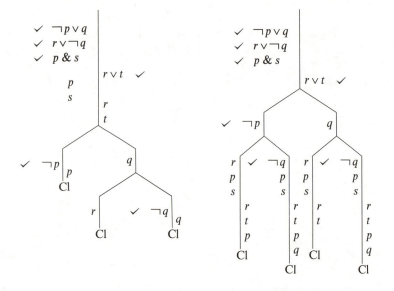

Though the more complicated tableau is just as good, theoretically, as its more elegant cousin, we are interested in tableaux largely for their ease and efficiency.

Problems

Write and apply the relevant rules to the tableau with these initial formula placements. Identify each branch as either open or closed.

   1. Left: $\neg p$ Right: $p \vee (q \mathbin{\&} p)$

   2. Right: $p \vee \neg p$

   3. Left: $p \mathbin{\&} q$ Right: $r \vee p$

   4. Left: $(p \vee r) \mathbin{\&} q$ Right: $p \vee q$

▸ **5.** Left: $p \vee q$ Right: $\neg(\neg p \mathbin{\&} \neg q)$

   6. Left: $p \mathbin{\&} \neg r, s \vee \neg p$ Right: $s \mathbin{\&} \neg r$

   7. Left: $\neg(p \mathbin{\&} r), r \vee q$ Right: $p \mathbin{\&} q$

   8. Left: $\neg(\neg p \vee \neg r)$ Right: $(p \mathbin{\&} r)$

   9. Left: $p, \neg p \vee q, (q \mathbin{\&} s) \vee \neg q$ Right: $\neg p$

▸ **10.** Left: $(p \mathbin{\&} \neg q) \vee (q \mathbin{\&} \neg p), \neg p \vee s$ Right: $\neg(\neg q \mathbin{\&} \neg s)$

3.2 Rules for the Conditional and Biconditional

So far we've developed rules for three of our five connectives. Rules for the conditional and biconditional are similar, though perhaps slightly less intuitive. Like the rules for negation, conjunction, and disjunction, they come in pairs. Also, like those rules, they mirror the definitions of the conditional and biconditional truth functions. We can express the definition of the conditional in this form:

> $\mathscr{A} \to \mathscr{B}$ is true on an interpretation iff either $\mathscr{A}$ is false on that interpretation or $\mathscr{B}$ is true on it.

Conditional Left (→L)

This rule applies to conditional formulas appearing on the left of a branch. Under what circumstances is a conditional formula true? A look at the truth table definition of the conditional indicates that a formula of the form $(\mathscr{A} \to \mathscr{B})$ is false only when $\mathscr{A}$ is true and $\mathscr{B}$ is false. So such a formula is true whenever this does not occur; whenever, that is, $\mathscr{A}$ is false or $\mathscr{B}$ is true. To account for these two possibilities, we must split the branch. So, the rule takes the form:

Conditional Right (→R)

This rule applies to conditionals on the right side of a branch. These, according to the tableau branch, are false. Something of the form $(\mathscr{A} \to \mathscr{B})$ is false just in case $\mathscr{A}$ is true and $\mathscr{B}$ is false, as we've seen. So, to apply the rule we enter $\mathscr{A}$ on the left and $\mathscr{B}$ on the right.

$$\mathscr{A} \quad \Big| \quad \begin{array}{c} (\mathscr{A} \to \mathscr{B}) \;\checkmark \\ \mathscr{B} \end{array}$$

The rules for the biconditional also reflect its definition:

> $\mathscr{A} \leftrightarrow \mathscr{B}$ is true on an interpretation iff $\mathscr{A}$ and $\mathscr{B}$ agree in truth value on that interpretation.

Biconditional Left (↔L)

This rule applies to biconditionals on the left side of a branch. If a formula of the form $(\mathscr{A} \leftrightarrow \mathscr{B})$ is true, then $\mathscr{A}$ and $\mathscr{B}$ must have the same truth value. Both must be true, or both must be false. Since there are two possibilities, the rule requires splitting the branch:

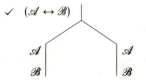

Biconditional Right (↔R)

This rule applies to biconditionals appearing on the right side of a branch. If $(\mathscr{A} \leftrightarrow \mathscr{B})$ is false, then $\mathscr{A}$ and $\mathscr{B}$ must have opposite truth values. That is, either $\mathscr{A}$ is true and $\mathscr{B}$ is false, or $\mathscr{B}$ is true and $\mathscr{A}$ is false. Again, the rule forces splitting to reflect these two possibilities:

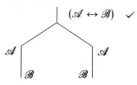

To see how these rules function, let's look at two examples. In the first tableau, we begin with the assumption that $(p \rightarrow q)$ is true, but $(p \leftrightarrow q)$ is false. We decompose these formulas to find out what these assumptions mean about the truth values of p and q alone. The order in which the rules are applied here makes no difference, even to efficiency, since both conditional left and biconditional right split the tableau. We end up, therefore, with four branches; two are closed, and two remain open:

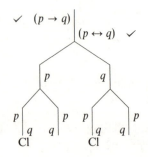

We start with the initial conditions that $(p \rightarrow q)$ is true, but $(p \leftrightarrow q)$ is false. This tableau lets us consider a variety of possibilities for assignments to the sentence letters p and q. The tableau, in effect, is a representation of a possible circumstance in which these formulas would have this particular combination of truth values. The two branches that close do not describe possible circumstances that would make $(p \rightarrow q)$ true and $(p \leftrightarrow q)$ false, but the open branches do. As it happens, the open branches describe the same interpretation: one in which q is true and p is false. To see this, look at each open branch, and record the atomic formulas appearing on each side:

Left open branch: True: q False: p, p
Right open branch: True: q, q False: p

The duplication of appearances makes no difference. These branches, therefore, describe the same interpretation. Both branches describe a circumstance in which q is true and p is false. The tableau, then, tells us the following: There is an interpretation of the letters p and q making $(p \rightarrow q)$ true and $(p \leftrightarrow q)$ false. That interpretation assigns truth to q and falsehood to p.

To turn to the second example, let's use a tableau to see whether there is an interpretation making $(p \leftrightarrow q) \leftrightarrow p$ true but $(p \rightarrow q)$ false. Here, applying the conditional right rule first is efficient since doing so saves us from splitting the tableau:

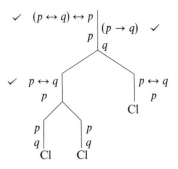

Every branch on this tableau is closed, so the tableau itself is closed. In other words, there is no interpretation meeting our initial conditions; no interpretation makes $(p \leftrightarrow q) \leftrightarrow p$ true but $(p \rightarrow q)$ false. No matter how we turn the tableau to describe a possible circumstance meeting these conditions, we run into a contradiction.

A special problem arises when a branch doesn't contain all sentence letters involved in the tableau. Suppose, for example, that we are interested in discovering whether there is an interpretation making $p \vee (q \,\&\, s)$ and $p \rightarrow q$ true but $q \,\&\, r$ false. We begin by placing the first two formulas on the left and the last on the right. Then we proceed to apply rules:

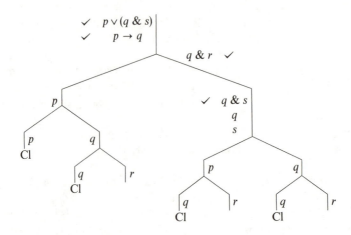

There are three open branches, indicating that there is such an interpretation. On the leftmost open branch, p and q both appear on the left, indicating that both p and q must be true. Moreover, r appears on the right, indicating that r must be false. But the branch says nothing at all about s. This means that an interpretation making p and q true and r false meets the initial conditions, no matter what it assigns s. So the branch reveals two interpretations of the letters p, q, r, and s that make $p \vee (q \& s)$ and $p \rightarrow q$ true but $q \& r$ false. One makes p true, q true, r false, and s true; the other makes p true, q true, r false, and s false.

Problems

Use tableaux to determine whether any interpretation meets these sets of conditions. If such an interpretation exists, specify it.

1. True: $\neg(p \rightarrow q)$ False: $\neg p \rightarrow \neg q$

2. True: $\neg(p \leftrightarrow q)$ False: $\neg p \leftrightarrow q$

3. True: $\neg p \leftrightarrow \neg q$ False: $\neg(p \leftrightarrow q)$

4. True: $p \rightarrow q$ False: $p \leftrightarrow (p \& q)$

▶ **5.** True: $p \rightarrow q$ False: $\neg p \vee q$

6. True: $q \vee r, p \rightarrow \neg q$ False: $p \rightarrow r$

7. True: $p \vee (q \rightarrow r)$ False: $q \rightarrow (p \vee r)$

8. True: $(p \vee q) \rightarrow q, \neg(p \& q)$ False: $\neg p$

9. True: $p \rightarrow (q \rightarrow r)$, $\neg((p \ \& \ \neg r) \rightarrow \neg q)$

▶ **10.** False: $((p \rightarrow q) \rightarrow p) \rightarrow p$

11. True: $(p \rightarrow q) \rightarrow \neg r$ False: $(p \vee r) \rightarrow q$

12. True: $\neg((p \leftrightarrow \neg q) \rightarrow r)$ False: $(p \leftrightarrow r) \rightarrow r$

13. True: $p \vee (q \vee (r \rightarrow p))$ False: $\neg p \vee (q \rightarrow r)$

14. True: $\neg(p \rightarrow (q \vee r))$ False: $p \leftrightarrow (q \ \& \ \neg r)$

▶ **15.** True: $(p \leftrightarrow q) \leftrightarrow (p \leftrightarrow r)$ False: $q \leftrightarrow r$

16. True: $q \leftrightarrow r$ False: $(p \leftrightarrow r) \leftrightarrow (p \leftrightarrow q)$

17. True: $r \rightarrow (p \rightarrow q)$ False: $p \rightarrow (q \rightarrow r)$

18. True: $p \rightarrow r$ False: $(q \rightarrow p) \rightarrow (q \rightarrow r)$

19. True: $(q \rightarrow p) \rightarrow (q \rightarrow r)$ False: $p \rightarrow r$

20. True: $(p \rightarrow q) \vee (p \rightarrow r)$ False: $p \rightarrow (q \vee (q \rightarrow r))$

3.3 DECISION PROCEDURES

A semantic tableau, we have seen, is a treelike depiction of the search for an interpretation meeting certain initial conditions. Closed branches represent contradictions; they correspond to blind alleys in this search. A branch with the same formula on both sides means that the formula is both true and false, an absurdity. If all branches of a tableau are closed, then, no matter what choices you might make in your search, you face contradictions. In a closed tableau, all paths leads to absurdity. A closed tableau hence demonstrates that the initial conditions can't be met: that no interpretation will produce the truth value assignment given to formulas at the beginning of the tableau.

The usefulness of tableaux in searching for interpretations makes them ideal for testing argument forms for validity. An argument form is valid, after all, just in case no interpretation makes its premise formulas true but its conclusion formula false. This suggests an easy method: Search for such an interpretation. If we find one, then the argument form is invalid; some truth value assignment does indeed make the premise formulas all true but the conclusion formula false. If the tableau closes, however, there's no such interpretation, and so the argument form is valid.

In a sense, then, tableaux force us to think backward about argument form validity. We begin by assuming that an argument form is invalid and see what develops from this assumption. If the tableau closes, we have reached contradictions; thus, the argument form must be valid. If the tableau remains open, we have found an interpretation showing that the argument form is invalid.

Since the validity of argument forms is a kind of implication, furthermore, the same test is a decision procedure for implication. To see whether a set of formulas implies a given formula, assume that every formula in the set is true, but the given formula is false. Then use a tableau to search for an interpretation with this effect. If there is one, a branch remains open, specifying the interpretation. If not, the tableau will close.

The tableau test for validity (and implication), then, is this. Suppose that our argument form has premise formulas $\mathscr{A}_1, \ldots, \mathscr{A}_n$ and conclusion formula $\mathscr{B}$. We list the premise formulas on the left and place the conclusion formula on the right. If the tableau closes, the argument form is valid:

Test for Argument Form Validity (and Implication)

Closes: Valid (Argument Form); Implied (Implication)
Open: Invalid (Argument Form); Not Implied (Implication)

As an example, consider the argument form

$p \to q$
$\neg p \to r$
$\neg r$
$\therefore q$

We begin by writing the premise formulas on the left and the conclusion formula on the right. At this point, we have three choices, but only one—using $\neg$L—doesn't force us to split the tableau. So, we begin by applying $\neg$L, and then move to $\to$L. It makes no difference which conditional we take first; suppose we apply the rule to the first premise. We obtain this tableau:

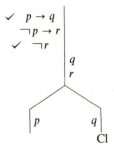

Notice that the right branch has already closed; q appears on both sides. We can finish the tableau by applying →L again, to the second premise, and then applying ¬R to the leftmost branch. This gives us the tableau:

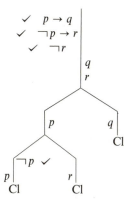

All three branches close, and so the tableau closes. Our assumption that the premise formulas were true but the conclusion formula false thus leads to absurdity at every turn; the argument form is valid.

Since semantic tableaux constitute a decision procedure for implication, they also amount to a test for equivalence. Equivalence, after all, is just implication in two directions. Equivalent formulas have the same truth value on every interpretation of them. This means that there should be no interpretation making one true but the other false. To test for this, we can construct two tableaux. Suppose that the two formulas are $\mathcal{A}$ and $\mathcal{B}$. We begin one tableau by assuming that $\mathcal{A}$ is true and $\mathcal{B}$ is false; the other by assuming that $\mathcal{B}$ is true and $\mathcal{A}$ is false. If both close, there are no interpretations on which the formulas disagree in truth value. If at least one tableau remains open, however, that tableau will specify an interpretation making one formula true and the other false.

The test, then, is this:

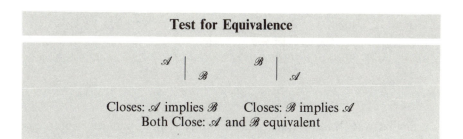

For example, we can use tableaux to show that $p \mathbin{\&} q$ is equivalent to $\neg(\neg p \vee \neg q)$.

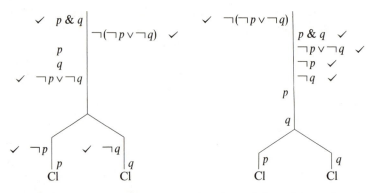

Tableaux also offer an elegant test for the validity of formulas. A valid formula comes out true on every interpretation of it. There is no assignment of truth values that makes a valid formula false. To determine whether a formula, $\mathscr{A}$, is valid, therefore, we can search for an interpretation making it false. If the search succeeds—if, that is, the tableau we construct remains open—then we have an interpretation making the formula false. If the search fails, the tableau closes, and the formula is valid.

Test for Formula Validity

$$| \quad \mathscr{A}$$

Closes: Valid
Open: Contingent or Contradictory

For example, this tableau shows that $((p \to q) \to p) \to p$ ("Peirce's law") is valid:

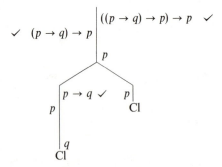

Similarly, tableaux present a method for testing formulas for contradictoriness or satisfiability. A satisfiable formula has an interpretation that makes it true; a contradictory formula doesn't. To test for contradictoriness, then, simply search for an interpretation making the formula in question true. An open branch specifies such an interpretation, establishing satisfiability; a closed tableau establishes contradictoriness.

Test for Formula Contradictoriness or Satisfiability

$$\mathscr{A} \quad |$$

Closes: Contradictory
Open: Satisfiable

These last two tests allow us to classify any formula as valid, contingent, or contradictory. One test, however, does not always determine the class of a given formula. If the formula is contingent, then testing it for validity will result in an open tableau. This means that the formula has an interpretation making it false, and so it must be either contradictory or contingent. To determine which, we must apply the other test.

For example, consider the formula $\neg(p \to \neg p)$, which tends to strike many people as valid. One can easily be misled into reading it as "no sentence implies its own negation," which might appear, at first glance, to be true. (To see why it isn't, think of contradictory sentences.) To test it for validity, we would begin a tableau with this formula on the right. We would then apply $\neg R$, $\to L$, and $\neg R$ (in that order) to produce:

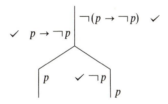

Both branches of this tableau are open. On both, p appears on the right, indicating that the formula has an interpretation making it false—namely, that of assigning falsehood to p. But this tableau does not determine whether the formula is contradictory or contingent. To find out, we need to apply the other test by beginning a tableau with $\neg(p \to \neg p)$ on the left, as follows:

$$
\begin{array}{c|c}
\checkmark \quad \neg(p \to \neg p) & \\
 & p \to \neg p \quad \checkmark \\
p & \\
 & \neg p \quad \checkmark \\
p &
\end{array}
$$

This tableau, too, is open; the formula is therefore satisfiable. Since it is neither contradictory nor valid, it must be contingent. Indeed, the tableau provides an interpretation making the formula true: that of assigning truth to p. Surprisingly, then, the formula $\neg(p \to \neg p)$ has the same truth value as p, whether p is true or false; $\neg(p \to \neg p)$ and p are equivalent.

Finally, tableaux offer a simple test for the satisfiability of sets of formulas. A set is satisfiable just in case there is an interpretation of it making all its members true. We can again search for such an interpretation. An open tableau specifies one, indicating that the set is indeed satisfiable. A closed tableau indicates contradictoriness, as with formulas alone. Given a set of formulas $\{\mathscr{A}_1, \ldots, \mathscr{A}_n\}$, then, the test is:

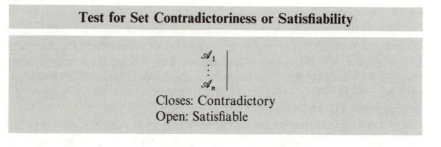

Test for Set Contradictoriness or Satisfiability

$$\mathscr{A}_1$$
$$\vdots$$
$$\mathscr{A}_n$$

Closes: Contradictory
Open: Satisfiable

For example, the set $p \to q$, $q \vee (p \to r)$, $\neg(q \,\&\, r)$ is satisfiable, as this tableau shows.

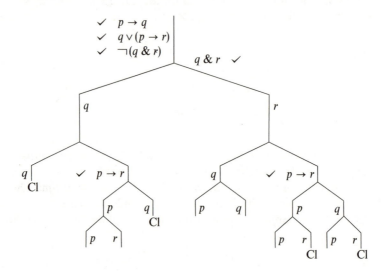

This tableau has six open branches, so the set is satisfiable. The branches give us the information

Branch	True	False
1		q, p, p
2	r	q, p
3	q	r, p
4	q	r
5		r, p, p
6	q	r, p

These branches indicate that four interpretations make all three formulas in the set true:

p	q	r
T	T	F
F	T	F
F	F	T
F	F	F

Semantic tableaux, then, provide a simple and efficient way of testing argument forms for validity; formulas for validity or contradictoriness; pairs of formulas for equivalence; sets of formulas for satisfiability; and sets of formulas, paired with formulas, for implication. Tableaux always terminate after a finite number of steps. Their construction, furthermore, requires no ingenuity, but the mechanical following of rules. Tableaux are thus decision procedures for all the significant logical properties and relations we've discussed. As later chapters will demonstrate, the tableau technique extends readily to more complex and comprehensive logical languages.

Problems

Evaluate these sets of formulas as satisfiable or contradictory.

1. $\neg p \lor q, \neg q, \neg p \leftrightarrow r, r \leftrightarrow q$

2. $(p \lor \neg q) \rightarrow r, p \leftrightarrow \neg r, (q \& r) \lor p$

3. $p \rightarrow q, p \rightarrow \neg q, p$

4. $p \rightarrow q, \neg p \rightarrow q, \neg q$

5. $(\neg q \& r) \rightarrow \neg p, p \rightarrow r, \neg(q \leftrightarrow (\neg r \& p))$

Symbolize these arguments in SL and evaluate the resulting argument forms for validity. If the argument form is invalid, describe the interpretations under which the premises would be true but the conclusion false.

6. If logic is difficult, then philosophers are peddlers of the obscure. But philosophers peddle obscurity only if the minds of the masses are incapable of rational thought. The minds of the masses, however, are indeed capable of rational thought, so logic is difficult only if you are a monkey's uncle.

7. If our group of stockholders remains resolute, we can stop Digital's hostile takeover attempt. Of course, unless the company can improve its cash flow position from that of last quarter, there will be no way to stop the takeover. Our group will remain resolute only if we all understand the consequences of the acquisition for our own holdings. Our group understands the consequences, but the company can't improve its cash flow position. Therefore, the group of stockholders will not remain resolute.

8. If Congress raises taxes in general and grants some tax relief to the middle class, then inflation will pick up steam. If that happens, however, then, unless taxes remain indexed, Congress won't be able to give the middle class tax relief after all. If Congress doesn't raise taxes in general, it won't be able to give tax relief to the middle class, but at least inflation won't pick up steam. Therefore, unless taxes remain indexed, the middle class will get no tax relief from Congress.

9. If the present philosophy of enforcement of pollution control standards is viable, then we must be able to assign responsibility for violations to specific corporations. We can do this, at present, just in case we monitor each company individually. But we can't adopt such individual monitoring unless we are willing to spend vast sums of money on pollution control. Finally, if we can't assign responsibility to specific corporations, we must develop a new approach to pollution control that is not based on traditional concepts of responsibility and enforcement. Consequently, we must either spend vast sums of money or develop a new approach to pollution control.

▶ **10.** If the Soviets march into Poland, or even denounce free trade unions in the wrong way, the Communist Party's ideal of representing workers will be exposed to the Russian people as a sham. The ideal won't be exposed, moreover, just in case Poland remains within the Soviet orbit. But if the Soviets don't march into Poland, that nation will drift from the Soviet orbit. So either the Communist Party's ideal will be exposed to the Russian people as a sham, or Poland will drift from Soviet control.

11. If Congress approves a liberalized depreciation plan, a corporation will be able to increase expenditures on plant and equipment, but only if it's in a reasonably good cash flow position. If a corporation has a reasonably good cash flow, it can increase such expenditures anyway. The ability of corporations to increase investment expenditures is a necessary condition for the full revitalization of the economy. So the economy will be fully revitalized if Congress approves a liberalization of depreciation.

12. If you want to maximize your job opportunities after graduation, you will major in business. But you'll succeed in the top ranks of industry only if you can write effectively; you'll be able to write effectively, however, only if you major in the liberal arts. So you'll succeed in the top ranks of industry only if you don't want to maximize your job opportunities after graduation.

13. That woman either loves or hates, but never both. If she loves, and her quarry is indifferent, then she hates instead. Here her quarry has been indifferent. Therefore, if she loves, you have played the fool.

14. If God is all powerful, He is able to prevent evil. If He is all good, He is willing to prevent evil. Evil does not exist unless God is either unwilling or unable to prevent it. If God exists, He is both all good and all powerful. Therefore, since evil exists, God doesn't.

▶ **15.** My cat will not sing opera unless all the lights are out. If I am very insistent, then my cat will sing opera; and if I either turn all the lights out or howl at the moon, you can be sure that I am very insistent indeed. I always howl at the moon if I am not very insistent. Therefore, my lights are never turned on, my cat is singing opera, and I am perpetually very insistent.

Use semantic tableaux to solve these problems.

16. A brutal ax murder has been committed and is being investigated by none other than Sherlock Holmes. Holmes feels that he has evidence to support each of these contentions concerning the case: (a) If Mr. Perry committed the murder, Ms. Jackson is innocent. (b) Ms. Jackson is guilty unless the butler didn't deliver his testimony with a clear conscience. (c) If Mr. Perry committed the murder and Ms. Jackson is innocent, then the butler didn't deliver his testimony with a clear conscience. (d) If the butler isn't guilty of the murder, then neither is Mr. Perry. Holmes, stumped by the situation, calls you in as a logical consultant. To get your share of the reward, you must answer these questions: (i) Is Holmes's theory satisfiable? (ii) Is one of Holmes's contentions implied by the other three? (iii) Does Holmes's theory imply anything about the guilt or innocence of Mr. Perry? (iv) Does Holmes's theory imply anything about the guilt or innocence of Ms. Jackson?

17. You have been diagnosed as having a rare psychological disorder manifesting itself in intense bouts of fear and loathing whenever you are faced with a logic problem. Your psychiatrist tells you that (a) if you don't undergo psychoanalysis, you won't recover. But you know, given his rates, that (b) if you do undergo psychoanalysis, you'll be poverty-stricken. Unknown to both of you, the reasoning involved in undergoing psychoanalytic treatment will make your condition worse; if you improve at all, it will be for reasons unrelated to your treatment. So (c) if you undergo psychoanalysis, then you will recover only if you improve spontaneously. Do these facts imply that you will recover only if you improve spontaneously? That you will recover only if you become poverty-stricken? That if you become poverty-stricken, you'll recover?

18. Roger, a hapless accounting major, works on a take-home final exam in a course on U.S. tax law. He attempts to analyze a problem concerning the tax liability of a corporation involved in overseas shipping. Some of the fleet counts as American for tax purposes, but some doesn't. Roger thinks that the definition of *American vessel* is: "A ship is an American vessel if and only if (a) it is either numbered or registered in the United States; or (b) if it is neither numbered nor registered in the U.S., and not registered in any foreign country, then its crew members are all U.S. citizens or are all employees of corporations based in the U.S." Unfortunately for Roger, this definition is incorrect. Show that Roger's definition implies that, if a ship is registered in a foreign country, it's an American vessel.

▶ **19.** Roger, at the last moment, revises his definition to read: "A ship is an American vessel if and only if (a) it is either numbered or registered in the United States, or (b) it is neither numbered nor registered in the U.S. and not registered in any foreign country, but its crew members are all U.S. citizens or are all employees of corporations based in the U.S." Show that this version does not imply that, if a ship is registered in a foreign country, it's an American vessel.

20. Devise semantic tableau rules for Sheffer's stroke. (See Section 2.3.) Use them to show that $p / (p / p)$ is valid and that the following pairs of formulas are equivalent: (a) $\neg p$ and p / p; (b) $p \& q$ and $(p / q) / (p / q)$; (c) $p \vee q$ and $(p / p) / (q / q)$; and (d) $p \rightarrow q$ and $p / (q / q)$.*

21. Devise semantic tableau rules for nondisjunction. (See Section 2.3.) Use them to show that $p \downarrow (p \downarrow p)$ is contradictory and that the following pairs of formulas are equivalent: (a) $\neg p$ and $p \downarrow p$; (b) $p \& q$ and $(p \downarrow p) \downarrow (q \downarrow q)$; (c) $p \vee q$ and $(p \downarrow q) \downarrow (p \downarrow q)$; and (d) $p \rightarrow q$ and $((p \downarrow p) \downarrow q) \downarrow ((p \downarrow p) \downarrow q)$.*

Notes

[1] The Dutch logician E. W. Beth and the Finnish-American logician Jaacko Hintikka independently developed this technique in 1955 by simplifying a logical system that the German logician Gerhard Gentzen (1909–1945) devised in the 1930s. See E. W. Beth, "Semantic Entailment and Formal Derivability," in J. Hintikka (ed.) *Philosophy of Mathematics* (Oxford: Oxford University Press, 1969): 9–41; J. Hintikka, "Form and Content in Quantification Theory, "*Acta Philosophica Fennica* 8 (1955): 7–55; G. Gentzen, "An Investigation into Logical Deduction," in M. Szabo (ed.), *The Collected Papers of Gerhard Gentzen* (Amsterdam: North-Holland, 1969). The method of this chapter is very close to Gentzen's original system, although Gentzen thought of the technique as purely syntactic.

4

NATURAL
DEDUCTION

e can use the methods of truth tables and semantic tableaux to evaluate arguments in theoretically clear and practically efficient ways. But these methods provide little insight into how people construct arguments and, especially, extended arguments and proofs.

In this chapter, we'll develop a system designed to simulate people's construction of arguments. Although the system forces arguments into rather rigid structures, it is natural in the sense that it approaches, in certain respects, the way people argue, particularly in legal, scientific, and philosophical contexts. It comes even closer to the way mathematicians prove theorems.

4.1 NATURAL DEDUCTION SYSTEMS

A *natural deduction system* is a collection of rules of inference.[1] The central notion of a natural deduction system is that of *proof*. Some proofs, called *hypothetical proofs,* begin with *assumptions,* or *hypotheses.* The assumptions serve as premises for the argument found in the proof; the conclusion depends on these assumptions. Such proofs show that the conclusion is true, not outright, but if the assumptions are true. Other proofs, however, contain no assumptions; they show that their conclusions are true outright.

A *proof* in a natural deduction system is a sequence of *lines;* on each line is a formula. Each formula in a proof must either be an assumption or derive, by a rule of inference, from formulas on previously established lines. In our formulation, the formula on the topmost *Show* line of a proof is its conclusion; the proof is a proof *of* that formula *from* the assumptions. Formulas proved from no assumptions are *theorems.* A theorem of a natural deduction system, then, is any formula that can be proved from no hypotheses in the system.

Rules of inference are either *simple* or *complex*. Simple rules allow us to write down formulas having certain shapes in a proof if we've already established formulas of certain kinds in that proof. For example, we'll have a rule that lets us write $\mathcal{A}$ or $\mathcal{B}$ if we've already established $\mathcal{A}$ & $\mathcal{B}$.

Complex rules, in contrast, allow us to write down a formula of a certain shape in a proof if some other proof is completed. One rule, for instance, allows us to write $\neg\mathcal{A}$ if we've proved a contradiction from the assumption $\mathcal{A}$. Another complex rule allows us to assert a conditional formula $\mathcal{A} \rightarrow \mathcal{B}$ if we can, in a subordinate proof, assume $\mathcal{A}$ and derive $\mathcal{B}$. Because our system has some complex rules, proofs sometimes appear within other proofs. A proof appearing within another is *subordinate* to it. In the larger, *superordinate* proof, we use the information in the subordinate proof by way of a complex rule. In the larger proof we sometimes simply take over the conclusion of its smaller partner. At other times, in the larger proof we use the fact that we successfully proved a subordinate conclusion from a given assumption to state an assertion that doesn't itself depend on this assumption. An assertion within a proof established by a subordinate proof is called a *lemma*.

As our talk of "shapes" suggests, natural deduction systems are purely syntactic. It's possible to verify that a proof is successful without any reference to the meanings of the symbols in the proof. Nevertheless, we use the rules of inference we do because, taken together, they allow us to prove from a set of assumptions only those formulas that follow from the set. The rules themselves are syntactic, based on the shapes of formulas, but their justification is semantic.

For the most part, the deduction system of this book has two rules for each connective. One rule tells us how to prove a formula with that connective as main connective. In short, it tells us how to *introduce* formulas of that form into proofs. The other rule tells us how to use the information encoded in a formula having that connective as main connective. That is, it tells us how to *exploit* formulas of that kind in proofs. For this reason, the basic rules of the system fall largely into two groups: *introduction* rules and *exploitation* (or *elimination*) rules. Most connectives have one rule of each sort.

Before discussing these rules in detail, we need to know more about which proofs the system allows. Proofs can appear inside other proofs. The subordinate proofs fulfill various functions. In this chapter, there are three such functions. Accordingly, there are three methods of proof. Two of them are really complex connective-introduction rules. The remaining, "pure" method is that of *direct proof*.

All proofs are sequences of lines that are structured in certain ways. As we'll write them, proofs will have three columns. The middle column consists of a sequence of formulas; some may be preceded by the word *Show*. The left column numbers these formulas. The right column provides justifications for the formulas. Thus, if a formula on a given line derives from previously established formulas by a rule of inference, the right column specifies that rule of inference and references the earlier lines used. If the formula is an assumption, the right column indicates so. Only when the formula derives from an entire subordinate proof is the right column empty.

A direct proof begins with premises—or, if it appears within another proof, formulas deduced from earlier lines—and proceeds to its conclusion. We'll begin a direct proof, once any premises or earlier lines are recorded, by stating what we want to show:

n. Show $\mathscr{A}$

Lines that we hope to establish by doing proofs—that is, theorems and lemmas—will always have this form. The left column contains a line number; the right column is empty. The formula we hope to prove is prefaced by the word *Show* to indicate that we haven't proved it yet; so far, the information recorded on the line is just wishful thinking.

How can we make our wishes reality? Clearly, by proving $\mathscr{A}$. If we can prove $\mathscr{A}$ from what we are given, then we can go back to line n and cross out the *Show* signaling that the following formula was, at that point, only fantasy. The proof allowing us to do so is our justification for canceling the *Show*. That proof follows line n immediately. To show graphically what lines constitute the proof, we'll draw a bracket encompassing those lines to the left of the formulas in the proof. A successful, completed direct proof looks like this:

Direct Proof

n. $\underline{\text{Show}}\ \mathscr{A}$

n + m. $\mathscr{A}$

Obtaining $\mathscr{A}$ allows us to complete the direct proof only if two conditions are satisfied.

1. $\mathscr{A}$, on line n + m, must not already be enclosed in another set of brackets.

2. There must be no uncanceled *Show* statements in the area to be bracketed.

We want to show that $\mathscr{A}$ follows from what has gone before. If $\mathscr{A}$, on line n + m, were already enclosed in a bracket, or if we enclosed any uncanceled *Show* statements by drawing a new bracket to complete the proof, we could not be sure that $\mathscr{A}$ followed from what we were given. We would know only that $\mathscr{A}$ followed from what we were given together with some additional assumptions, for other proof methods introduce assumptions into the proof.

Neither of the following, then, count as legitimate instances of direct proof:

n. ~~Show~~ *p*   n. ~~Show~~ *p*
m. ⌈~~Show~~ *q*   m. ⌈Show *q*
   ⌈ :   :
k. ⌊⌊*p*   k. ⌊*p*
(Wrong)   (Wrong)

On the left, we are trying to use the occurrence of *p* on line k to complete the direct proof of *p*. But *p*, on that line, is already enclosed in another set of brackets. Because the proof of *q* may have begun with an added assumption, we have no guarantee that *p* follows from the formulas above line n alone. On the right, the uncanceled *Show* statement within the bracket may signal the introduction of added assumptions.

Since the bracketed lines constitute a proof that justifies line n, no other justification is needed. The right column for line n can thus remain empty.

4.2 RULES FOR NEGATION AND CONJUNCTION

In this section, we'll develop rules of inference for negation and conjunction. First, however, we'll introduce a simple rule that allows us to construct hypothetical proofs. The *assumption rule* asserts that you may begin a proof by listing premises or assumptions. That is, before the very first *Show* line, you may write premises. The conclusion you derive will depend on these premises. Thus:

Assumption

n. 𝒜   A

Here line n must precede the first *Show* line in the proof.

Suppose that we want to prove *q* hypothetically from *p* → *q* and *p* ∨ *q*. We would begin the proof by writing the premises, and then the *Show* line containing the conclusion:

1. *p* → *q*   A
2. *p* ∨ *q*   A
3. Show *q*

The rule of *conjunction exploitation* (&E) shows how we can use the information encoded in a conjunction. If a conjunction is true, in other words, what follows? Clearly, the truth of both conjuncts. If $\mathscr{A}$ & $\mathscr{B}$ is true, then both $\mathscr{A}$ and $\mathscr{B}$ must be true. The rule of conjunction exploitation thus takes two forms, since the truth of a conjunction implies the truth of both conjuncts:

Conjunction Exploitation (&E)

n.	$\mathscr{A}$ & $\mathscr{B}$	
n + m.	$\mathscr{A}$	&E, n
n.	$\mathscr{A}$ & $\mathscr{B}$	
n + m.	$\mathscr{B}$	&E, n

Hereafter, we'll abbreviate rules having two forms by using parentheses. We can write &E as

n.	$\mathscr{A}$ & $\mathscr{B}$	
n + m.	$\mathscr{A}$ (or $\mathscr{B}$)	&E, n

This rule, often called *simplification,* asserts that, from a conjunction, we can prove either or both conjuncts. The conjuncts may be written on any later line. When applying this rule, we write a conjunct together with the explanation that the line comes by application of &E—conjunction exploitation—to the formula on line n. (The line just above the formula that results from applying this rule separates the formula deduced from its necessary antecedents; it won't appear in actual proofs.) When we apply this rule to a formula, the conjunction must be its main connective. We can move from p & $(q \rightarrow r)$ to $q \rightarrow r$, but we cannot go from $(p$ & $q) \rightarrow r$ to $p \rightarrow r$.

Let's do a simple proof using this rule. Let's assume $((p$ & $q)$ & $\neg r)$ and prove q. We begin with an assumption and a *Show* statement.

1. $(p$ & $q)$ & $\neg r$ A
2. Show q

Now, we can exploit the conjunction to derive the smaller conjunction $(p$ & $q)$. Note that we can't leap inside to derive q directly: The connective to which we apply the rule must always be the main connective of the formula.

1. $(p$ & $q)$ & $\neg r$ A
2. Show q
3. $(p$ & $q)$ &E, 1

The right column tells us that the formula on line 3 comes from line 1 by applying conjunction exploitation. At this stage, we can easily apply that same rule again to obtain q, which is what we need to finish the proof:

$$
\begin{array}{lll}
1. & (p \ \& \ q) \ \& \ \neg r & \text{A} \\
2. & \text{Show } q & \\
3. & \lceil (p \ \& \ q) & \text{\&E, 1} \\
4. & \lfloor q & \text{\&E, 3}
\end{array}
$$

The rule of conjunction introduction (&I) tells us how to prove conjunctions. It's a very simple rule, since there is an obvious strategy for proving a conjunction: Prove each conjunct. If you know that $\mathcal{A}$ is true, and that $\mathcal{B}$ is true, you can conclude that $\mathcal{A}$ & $\mathcal{B}$ is true as well. The rule thus states that from the two formulas $\mathcal{A}$ and $\mathcal{B}$ you can derive $\mathcal{A}$ & $\mathcal{B}$:

Conjunction Introduction (&I)

$$
\begin{array}{lll}
\text{n.} & \mathcal{A} & \\
\text{m.} & \mathcal{B} & \\
\hline
\text{p.} & \mathcal{A} \ \& \ \mathcal{B} & \text{\&I, n, m}
\end{array}
$$

The right column simply indicates that $\mathcal{A}$ & $\mathcal{B}$ comes from applying conjunction introduction to the formulas on lines n and m. The order in which $\mathcal{A}$ and $\mathcal{B}$ appear in the proof makes no difference. We could just as easily have concluded $\mathcal{B}$ & $\mathcal{A}$.

To see how this might be used in a proof, let's show that $(p \ \& \ q)$ allows us to derive $(q \ \& \ p)$. Again we begin by using the assumption rule and writing a *Show* line:

$$
\begin{array}{lll}
1. & p \ \& \ q & \text{A} \\
2. & \text{Show } q \ \& \ p &
\end{array}
$$

Now, we need to show $q \ \& \ p$. To do this, we need to separate the two conjuncts. We can therefore derive them separately, using conjunction exploitation, and put them back together in the other order, using conjunction introduction:

$$
\begin{array}{lll}
1. & p \ \& \ q & \text{A} \\
2. & \text{Show } q \ \& \ p & \\
3. & \lceil p & \text{\&E, 1} \\
4. & \mid q & \text{\&E, 1} \\
5. & \lfloor q \ \& \ p & \text{\&I, 4, 3}
\end{array}
$$

One negation rule serves as both an introduction and an exploitation rule. Because it always introduces or exploits two negation symbols at once, we'll refer to it as $\neg\neg$. It asserts, basically, that two consecutive negation signs "cancel each other out": $\neg\neg\mathscr{A}$ is equivalent to $\mathscr{A}$. First formulated by the Stoics, this rule is often called *double negation*. One can add two consecutive negation signs, or delete them, without affecting truth values.

Negation Introduction/Exploitation ($\neg\neg$)

n.	$\mathscr{A}$	
n + p.	$\neg\neg\mathscr{A}$	$\neg\neg$, n

n.	$\neg\neg\mathscr{A}$	
n + p.	$\mathscr{A}$	$\neg\neg$, n

We can express this rule more concisely by writing a double line between $\mathscr{A}$ and $\neg\neg\mathscr{A}$. A double line indicates that the rule is *invertible:* that one can go from what is above the lines to what is below them, or vice versa. The rule works in both directions. So:

n.	$\mathscr{A}$	$\neg\neg$, m
m.	$\neg\neg\mathscr{A}$	$\neg\neg$, n

To illustrate the use of this rule, let's show that we can derive $\neg\neg p \mathbin{\&} \neg q$ from $\neg\neg(p \mathbin{\&} \neg\neg\neg q)$.

1.	$\neg\neg(p \mathbin{\&} \neg\neg\neg q)$	A
2.	Show $\neg\neg p \mathbin{\&} \neg q$	
3.	$\lceil\, p \mathbin{\&} \neg\neg\neg\neg q$	$\neg\neg$, 1
4.	$\mid\, p$	&E, 3
5.	$\mid\, \neg\neg p$	$\neg\neg$, 4
6.	$\mid\, \neg\neg\neg\neg q$	&E, 3
7.	$\mid\, \neg q$	$\neg\neg$, 6
8.	$\lfloor\, \neg\neg p \mathbin{\&} \neg q$	&I, 5, 7

Another, more powerful negation rule is *indirect proof,* which is essentially another negation introduction rule. The indirect proof rule is complex. It says that we can write $\neg\mathscr{A}$ if we can derive a contradiction from the assumption that $\mathscr{A}$. Indirect proofs always introduce assumptions, called *assumptions for indirect proof (AIPs),* from which they try to prove contradictions. What point is there to that? If an assumption leads to a contradiction, it must be false. Consequently, indirect proofs always establish the negations of their assumptions.

An indirect proof begins with a statement of what we want to prove: a formula prefaced by the word *Show.* This formula has a negation as a main connective. Next, we make an assumption. The assumption is the same as the formula we're trying to establish, but with the main negation omitted. An indirect proof thus begins:

> n.   Show $\neg \mathscr{A}$
> n + 1. $\mathscr{A}$ AIP

To complete the proof, we need to prove a contradiction. The contradiction doesn't have to relate directly to the assumption; we do not, in other words, have to use the assumption $\mathscr{A}$ to prove $\mathscr{A}$ & $\neg \mathscr{A}$, or anything else containing $\mathscr{A}$. Furthermore, we don't have to get our contradiction into a single formula; two formulas, one of which is the negation of the other, suffice. What we want, then, is to prove a formula $\mathscr{B}$ and also its negation $\neg \mathscr{B}$. A completed indirect proof, then, looks like this:

Indirect Proof

> n.   Show $\neg \mathscr{A}$
> n + 1. $\ulcorner \mathscr{A}$ AIP
> $\quad \vdots$
> n + p. $\mid \mathscr{B}$
> n + q. $\llcorner \neg \mathscr{B}$

It makes no difference whether $\mathscr{B}$ or $\neg \mathscr{B}$ is proved first. Notice that, once the proof is complete, we cancel the word *Show* to indicate that we've established what, earlier, we had merely hoped for. Again, neither $\mathscr{B}$ nor $\neg \mathscr{B}$ may already be enclosed in brackets on lines n + p and n + q, and we may enclose no uncanceled *Show* statements when we draw a bracket to complete the proof.

For example, this rule allows us to prove the thesis that a contradiction implies anything.[2] To see this, take a contradiction such as p & $\neg p$, and a completely unrelated formula q. We can show that we can derive q from p and $\neg p$. So we begin with an assumption and a *Show* line:

> 1. p & $\neg p$ A
> 2. Show q

Now, how can we get to q? There seems to be no way to get there by a direct proof, since q bears no relation to our assumption. Our only choice, at this point, is to use indirect proof. This bit of reasoning is not restricted to this example. Throughout this text we'll resort to indirect proof when nothing else suggests itself as a good proof technique.

Of course, as we have stated the rule of indirect proof, we can prove only negated formulas. But our rule for negation tells us that q and $\neg\neg q$ are equivalent. Instead of proving q, then, we can prove $\neg\neg q$, and then use the rule for double negation to obtain q. This strategy leads to the proof:

$$
\begin{array}{lll}
1. & p \;\&\; \neg p & \text{A} \\
2. & \text{Show } q & \\
3. & \lceil \text{Show } \neg\neg q & \\
4. & \quad \lceil \neg q & \text{AIP} \\
5. & \quad \mid p & \&\text{E, 1} \\
6. & \quad \lfloor \neg p & \&\text{E, 1} \\
7. & \lfloor q & \neg\neg, 3 \\
\end{array}
$$

Notice that, in this proof, we used a line that contained the word *Show;* that's how we got from $\neg\neg q$ to q. Yet we can't use such lines all the time. If we could use *Show* lines anytime we wanted, we could easily prove anything. Consider, for instance, this "proof" that pigs fly:

$$
\begin{array}{lll}
1. & \text{Show Pigs fly} & \\
2. & \lceil \neg\neg \text{ Pigs fly} & \neg\neg, 1 \quad \text{(Error)} \\
3. & \lfloor \text{Pigs fly} & \neg\neg, 2 \\
\end{array}
$$

Clearly we've got to restrict the circumstances in which we can use such lines.

We'll say that any line we can use in a proof at a given point is *free* at that point. In sentential logic, every line is free, except (a) lines that begin with an uncanceled *Show;* and (b) lines that are imprisoned within a bracket. Lines beginning with an uncanceled *Show* contain formulas that we haven't yet proved. All we can say is that we hope to prove them. So, the information on those lines is inaccessible to us in the proof. That's why the proof that pigs fly seems silly; we used what we wanted to prove to prove it. Once a *Show* has been canceled, however, we've proved the formula, so we can use it throughout the rest of the proof.

It's only a little harder to see the point of the second restriction. Lines that are enclosed in a completed bracket or "prison" may depend on a particular assumption. Both indirect and conditional proofs introduce assumptions; the assumptions themselves, and the following lines that depend on them, are true only given those assumptions. We can't assume that they are true in general. Above, for example, we introduced a contradiction—$p \;\&\; \neg p$—as an assumption. We proceeded to show that, if this were true, then anything would be true. Here, and in subordinate proofs, we wouldn't want to say outright that the contradiction was true; we would merely pretend that it was true. The contradiction, in other words, would serve only as a dialectical premise. If this proof were part of a larger proof, it would be a terrible mistake to come back and claim that, in general, we could prove $p \;\&\; \neg p$ because it was introduced earlier, within a subordinate proof, as an assumption.

The only free lines, then, are those that are neither prefaced with an uncanceled *Show* nor imprisoned within a bracket (that is, a completed proof). These "proofs," consequently, contain grave errors:

1. S̶h̶o̶w̶ $\neg p$
2. ⌈ S̶h̶o̶w̶ $(p \ \& \ \neg p) \to p$
3. | ⌈ $p \ \& \ \neg p$ AIP
4. | | p &E, 3
5. | ⌊ $\neg p$ &E, 3   (Error)

1. S̶h̶o̶w̶ $p \ \& \ \neg p$
2. ⌈ p &E, 1   (Error)
3. | $\neg p$ &E, 1   (Error)
4. ⌊ $p \ \& \ \neg p$ &I, 2, 3

In summary, formulas preceded by an uncanceled *Show* have not yet reached legal age, and those in a bracket prison are sentenced to life. Both kinds of formulas are inaccessible.

One rule in this section, the assumption rule, is *structural* in that it makes no reference to any particular connectives. We'll now introduce another structural rule to the effect that whatever is true is true. This rule allows you to repeat yourself.

The *reiteration* rule allows us to repeat a formula that appeared earlier in the proof, so long as the line containing that formula is still free.

Reiteration

$$n. \quad \mathscr{A}$$
$$\overline{n + p. \quad \mathscr{A} \qquad R, n}$$

This rule supposes, in effect, that whatever we've shown to be true is still true. Repeating what we've established can never get us into trouble.

To see how this rule works, consider a simple argument: 'It's not true that Congress will cut military spending and refuse to raise taxes, because Congress will raise taxes.' We can symbolize this as

p
$\therefore \neg(q \ \& \ \neg p)$

We begin a proof to demonstrate the validity of the argument form by using the rule of assumption to introduce the premise. Then we try to show the conclusion.

1. p A
2. Show $\neg(q \ \& \ \neg p)$

The first rule of thumb for constructing proofs is this:

> Choose a proof method by looking at the main connective of the formula being proved.

In trying to prove a negation, use indirect proof.

$$
\begin{array}{llc}
1. & p & A \\
2. & \text{Show } \neg(q \,\&\, \neg p) & \\
3. & (q \,\&\, \neg p) & \text{AIP}
\end{array}
$$

We can apply conjunction exploitation to obtain $\neg p$, and then get a contradiction by reiterating the premise:

$$
\begin{array}{llc}
1. & p & A \\
2. & \text{Show } \neg(q \,\&\, \neg p) & \\
3. & \lceil (q \,\&\, \neg p) & \text{AIP} \\
4. & \mid \neg p & \&\ \text{E, 3} \\
5. & \lfloor p & \text{R, 1}
\end{array}
$$

In general, reiteration is useful chiefly for completing indirect and other subordinate proofs.

Problems

Using deduction, show that each individual formula is provable and that the conclusion of each argument form is provable from the premises.

1. $p \,\&\, q; \; \therefore \; q$ 2. $p \,\&\, (q \,\&\, r); \; \therefore \; q$

3. $\neg(p \,\&\, \neg p)$ 4. $\neg(p \,\&\, (q \,\&\, \neg p))$

▶ 5. $\neg((p \,\&\, \neg q) \,\&\, (q \,\&\, \neg p))$ 6. $\neg p \,\&\, \neg q; \; \therefore \; \neg(p \,\&\, q)$

7. $(p \,\&\, q) \,\&\, r; \; \therefore \; p \,\&\, (q \,\&\, r)$ 8. $p \,\&\, \neg q; \neg r \,\&\, s; \; \therefore \; s \,\&\, p$

9. $\neg(p \,\&\, q); q; \; \therefore \; \neg p$ 10. $q; \; \therefore \; \neg(\neg p \,\&\, \neg q)$

11. $p; \; \therefore \; \neg(q \,\&\, \neg(p \,\&\, q))$

12. $\neg((p \,\&\, \neg q) \,\&\, r); \; \therefore \; \neg((p \,\&\, r) \,\&\, \neg q)$

13. $\neg(p \,\&\, \neg q); p; \; \therefore \; q$

14. $\neg(\neg p \,\&\, q); \neg(p \,\&\, q); \; \therefore \; \neg q$

▶ 15. $p \,\&\, q; \neg(q \,\&\, \neg r); \neg(p \,\&\, \neg s); \; \therefore \; r \,\&\, s$

16. $\neg(\neg(\neg(p \,\&\, \neg q) \,\&\, \neg p) \,\&\, \neg p)$

17. $p \& q; \neg(\neg r \& q); \neg(r \& \neg s); \therefore p \& s$

18. $\neg(p \& \neg q); \neg(q \& \neg p); \neg(p \& q); \neg(\neg p \& \neg q); \therefore r*$

19. $\neg(\neg s \& q); \neg(p \& (\neg q \& \neg r)); \neg(r \& \neg s); \therefore \neg(\neg s \& p)*$

▶ 20. $\neg(p \& r); \neg(\neg(p \& q) \& \neg p); r; \therefore \neg s*$

4.3 RULES FOR THE CONDITIONAL AND BICONDITIONAL

The rule of conditional exploitation ($\rightarrow$E) is very simple. Many axiomatic systems, in fact, use only conditional exploitation as a rule of inference. Often called *modus ponens,* this rule sanctions the inference from p and $p \rightarrow q$ to q. It thus stands behind such arguments as 'If you're smart, you'll do well at logic; you're smart; so, you'll do well at logic.'

Conditional Exploitation

n.	$\mathcal{A} \rightarrow \mathcal{B}$	
m.	$\mathcal{A}$	
p.	$\mathcal{B}$	$\rightarrow$E, n, m

To illustrate this rule, let's show that we can derive $p \& q$ from the hypotheses $r \rightarrow q$ and $r \& p$.

1. $r \rightarrow q$	A
2. $r \& p$	A
3. Show $p \& q$	
4. $\lceil r$	&E, 2
5. $\mid p$	&E, 2
6. $\mid q$	$\rightarrow$E, 1, 4
7. $\lfloor p \& q$	&I, 5, 6

Conditional introduction is a complex rule, which constitutes the method of *conditional proof.*[3] A conditional proof, like an indirect proof, may use some premises or earlier lines of a proof, but it doesn't have to. It always establishes a conditional formula, that is, a formula with a conditional as its main connective. It begins with a statement of what we want to prove and proceeds to an assumption, called the *assumption for conditional proof (ACP).* This assumption is always the antecedent of the conditional we're trying to establish. A conditional proof, then, begins as follows:

n.	Show $\mathcal{A} \rightarrow \mathcal{B}$	
n + 1. $\mathcal{A}$		ACP

The method mimics actual arguments for conditional statements in English. To argue for a conclusion such as 'If the fetus is a person, it has a right to life,' we can begin by supposing that the fetus is a person, and seeing what follows. If we can show that it follows from the assumption that the fetus has a right to life, then we have established the original conditional. In summary, to show that if $\mathcal{A}$, then $\mathcal{B}$, we assume $\mathcal{A}$ and try to show that $\mathcal{B}$ follows. A successful conditional proof has this form:

Conditional Proof

$$
\begin{array}{ll}
\text{n.} & \text{Show } \mathcal{A} \rightarrow \mathcal{B} \\
\text{n} + 1. & \mathcal{A} \qquad\qquad \text{ACP} \\
& \vdots \\
\text{n} + \text{p.} & \mathcal{B}
\end{array}
$$

Once again, obtaining the subordinate conclusion $\mathcal{B}$ allows us to cancel the *Show* above it and count the conditional statement as established. A proof proves its topmost *Show* line whenever the *Show* is canceled. At that point, the conditional proof is complete. Drawing a bracket around the proof indicates that it is complete and also allows us to see easily what lines constitute the justification for the conditional statement. Just as in direct and indirect proofs, we may not draw the bracket if $\mathcal{B}$, on line n + p, appears inside another bracket, or if we would be enclosing an uncanceled *Show* statement.

For example, we can use conditional proof to show that $r \rightarrow q$ is derivable from $p \& q$.

$$
\begin{array}{lll}
1. & p \& q & \text{A} \\
2. & \text{Show } r \rightarrow q & \\
3. & r & \text{ACP} \\
4. & q & \&\text{E, 1}
\end{array}
$$

Another example points out that $(p \rightarrow \neg q) \rightarrow (q \rightarrow \neg p)$ is provable:

$$
\begin{array}{lll}
1. & \text{Show } (p \rightarrow \neg q) \rightarrow (q \rightarrow \neg p) & \\
2. & p \rightarrow \neg q & \text{ACP} \\
3. & \text{Show } q \rightarrow \neg p & \\
4. & q & \text{ACP} \\
5. & \text{Show } \neg p & \\
6. & p & \text{AIP} \\
7. & \neg q & \rightarrow\text{E, 2, 6} \\
8. & q & \text{R, 4}
\end{array}
$$

Here, an indirect proof is subordinate to a conditional proof that is, in turn, subordinate to another conditional proof.

The rules for the biconditional are straightforward. First, consider biconditional introduction. Under what circumstances may one introduce a biconditional into a proof? What do we need to do, that is, to establish the truth of a biconditional? Recall that a biconditional is so called because it amounts to two conditionals, a crucial point to remember in devising a proof strategy. Mathematicians, for example, tend to prove biconditionals in two steps. They do the "left-to-right direction" and the "right-to-left direction" separately. In other words, they prove two conditionals to establish the biconditional. Our rule for biconditional introduction similarly requires two conditionals:

Biconditional Introduction

$$n. \quad \mathcal{A} \to \mathcal{B}$$
$$m. \quad \underline{\mathcal{B} \to \mathcal{A}}$$
$$p. \quad \mathcal{A} \leftrightarrow \mathcal{B} \qquad \leftrightarrow\text{I, n, m}$$

The rule for exploiting biconditionals reflects the fact that a biconditional asserts that two sentences have the same truth value. If we know the truth of a biconditional and also the truth of one of its components, we can deduce the truth of the other.

Biconditional Exploitation

$$n. \quad \mathcal{A} \leftrightarrow \mathcal{B}$$
$$m. \quad \underline{\mathcal{A} \qquad \qquad (\text{or } \mathcal{B})}$$
$$p. \quad \mathcal{B} \qquad \qquad (\text{or } \mathcal{A}) \qquad \leftrightarrow\text{E, n, m}$$

Notice that this rule differs from conditional exploitation by allowing us to deduce $\mathcal{A}$ from $\mathcal{B}$ or vice versa. If we have one component, we can derive the other; it makes no difference which appears on which side of the biconditional. The rule of conditional exploitation, however, works only in one direction. If we have the antecedent of the conditional, we can obtain the consequent. But we can't derive the antecedent from the consequent.

To see how these rules work, let's show that $p \leftrightarrow q$ is derivable from $(p \to q) \mathbin{\&} (q \to p)$, and vice versa.

1.	$p \leftrightarrow q$	A
2.	Show $(p \to q) \mathbin{\&} (q \to p)$	
3.	⌈Show $p \to q$	
4.	⌈p	ACP
5.	⌊q	$\leftrightarrow$E, 1, 4
6.	Show $q \to p$	
7.	⌈q	ACP
8.	⌊p	$\leftrightarrow$E, 1, 7
9.	⌊$(p \to q) \mathbin{\&} (q \to p)$	&I, 3, 6

1. $(p \rightarrow q) \,\&\, (q \rightarrow p)$ A
2. Show $p \leftrightarrow q$
3. $\lceil p \rightarrow q$ &E, 1
4. $\,| \; q \rightarrow p$ &E, 1
5. $\lfloor p \leftrightarrow q$ $\leftrightarrow$I, 3, 4

Problems

Theophrastus (371–286 B.C.), a pupil of Aristotle, cited these principles as hypothetical syllogisms. Use natural deduction to demonstrate their validity.

  1. $p \rightarrow q$; $q \rightarrow r$; $\therefore p \rightarrow r$ **2.** $p \rightarrow q$; $q \rightarrow r$; $\therefore \neg r \rightarrow \neg p$

  3. $p \rightarrow q$; $\neg p \rightarrow r$; $\therefore \neg q \rightarrow r$ **4.** $p \rightarrow q$; $\neg p \rightarrow r$; $\therefore \neg r \rightarrow q$

▶ **5.** $p \rightarrow r$; $q \rightarrow \neg r$; $\therefore p \rightarrow \neg q$

Show, using deduction, that these argument forms are valid.

  6. $p \leftrightarrow (q \,\&\, p)$; $\therefore p \rightarrow q$ **7.** $p \rightarrow q$; $\neg(p \rightarrow r)$; $\therefore \neg(q \rightarrow r)$

  8. $p \rightarrow q$; $p \rightarrow r$; $\therefore p \rightarrow (q \,\&\, r)$ **9.** $p \leftrightarrow q$; $p \leftrightarrow r$; $\therefore q \leftrightarrow r$

▶ **10.** $p \rightarrow q$; $\neg q$; $\therefore \neg p$ **11.** $p \leftrightarrow q$; $\neg p$; $\therefore \neg q$

  12. $(p \rightarrow q) \rightarrow p$; $\therefore p$ **13.** $p \rightarrow q$; $\therefore p \leftrightarrow (p \,\&\, q)$

  14. $p \,\&\, \neg q$; $r \rightarrow (r \,\&\, q)$; $\neg r \rightarrow s$; $\therefore s$

▶ **15.** $s \rightarrow (r \,\&\, p)$; $q \rightarrow (\neg r \,\&\, \neg p_1)$; $\therefore (q \,\&\, s) \rightarrow p_2$

  16. $p \rightarrow (q \,\&\, (\neg r \rightarrow \neg p))$; $q \leftrightarrow \neg r$; $\therefore \neg p$

  17. $(p \rightarrow q) \rightarrow (r \rightarrow s)$; $\neg(p \,\&\, q) \rightarrow s$; $\therefore \neg s \rightarrow \neg r$

  18. $(s \,\&\, \neg r) \rightarrow \neg p$; $(q \rightarrow \neg s) \leftrightarrow \neg p$; $\therefore p \leftrightarrow (r \,\&\, (s \,\&\, q))$*

  19. $p \leftrightarrow q$; $\therefore \neg p \leftrightarrow \neg q$*

  20. $(p \leftrightarrow q) \leftrightarrow (p \leftrightarrow r)$; $\therefore (q \leftrightarrow r)$*

4.4 Rules for Disjunction

Like most connectives, disjunction has both an introduction and an exploitation rule. The introduction rule is very simple. It asserts that we may introduce a disjunction into a proof if we have already obtained either disjunct.[4]

Disjunction Introduction (∨I)

n. $\mathscr{A}$ (or $\mathscr{B}$)

n + p. $\mathscr{A} \vee \mathscr{B}$ ∨I, n

To see how this rule works in practice, let's try to prove "the law of the excluded middle": $p \vee \neg p$. We can't prove either disjunct separately—neither p nor $\neg p$ is valid—so we need to use indirect proof. This means we need to prove $\neg\neg(p \vee \neg p)$. Furthermore, introducing the assumption for indirect proof leaves us with very little to work with. We should be able to prove a contradiction, but it's not obvious how we can get anything out of $\neg(p \vee \neg p)$. So, we can begin by trying to show $\neg p$. (After all, anything should follow from a contradiction.)

1. Show $p \vee \neg p$
2. ⎡ Show $\neg\neg(p \vee \neg p)$
3. ⎢⎡ $\neg(p \vee \neg p)$ AIP
4. ⎢⎢ Show $\neg p$
5. ⎢⎢⎡ p AIP
6. ⎢⎢⎢ $p \vee \neg p$ $\vee$I, 5
7. ⎢⎢⎣ $\neg(p \vee \neg p)$ R, 3
8. ⎢⎣ $p \vee \neg p$ $\vee$I, 4
9. ⎣ $p \vee \neg p$ $\neg\neg$, 2

This shows how useful disjunction introduction is, even in proving a fairly simple theorem.

Disjunction exploitation is perhaps the most complicated rule in our entire proof system. How can we exploit the information encoded in a disjunction? That is, if we know a disjunction, how can we use it to obtain some conclusion? A mathematician encounters this same problem in doing a "proof by cases." Often, one can say only that there are several possibilities, and then examine each individually. If the conclusion one is seeking holds in every case, then one can conclude that the conclusion holds in general, since those were the only possibilities. So, if we have a disjunction and ways of getting from each disjunct to a conclusion, then we can obtain that conclusion. This leads us to the rule often called *constructive dilemma:*[5]

Disjunction Exploitation ($\vee$E)

n. $\mathscr{A} \vee \mathscr{B}$
m. $\mathscr{A} \rightarrow \mathscr{C}$
p. $\mathscr{B} \rightarrow \mathscr{C}$
q. $\mathscr{C}$ $\vee$E, n, m, p

Faced with a disjunction, then, we must usually prove our conclusion in each of the two cases the disjunction presents. We've got to "get down to cases."

To see how disjunction exploitation works, let's derive q from $p \vee q$ and $p \rightarrow q$.

1. $p \vee q$ A
2. $p \rightarrow q$ A
3. Show q
4.   Show $q \rightarrow q$
5. q ACP
6. q ∨E, 1, 2, 4

Note that we can prove $q \rightarrow q$ in just one step, since the antecedent and consequent are the same.

Problems

Show that the conclusion formulas of these argument forms can be proved from the premise formulas.

1. $p \vee q$; $\neg p$; $\therefore q$

2. $p \vee q$; $p \rightarrow r$; $q \rightarrow s$; $\therefore r \vee s$

3. $(p \& q) \vee (\neg p \& \neg q)$; $\therefore p \leftrightarrow q$

4. $\neg p \vee \neg q$; $\therefore q \rightarrow \neg p$

▶ 5. $\neg p \vee \neg r$; $\therefore \neg(p \& r)$

6. $p \vee q$; $\neg p \vee \neg q$; $\therefore p \leftrightarrow \neg q$

7. $(p \vee q) \vee r$; $\therefore p \vee (q \vee r)$

8. $p \& (q \vee r)$; $\therefore (p \& q) \vee (p \& r)$

9. $(r \& \neg p) \vee (q \& r)$; $\therefore (p \rightarrow q) \& r$

10. $p \vee q$; $r \vee s$; $\neg(p \vee s)$; $\therefore (q \& r) \vee p_1$

11. p; $\neg s \vee \neg p$; $p \rightarrow r$; $\therefore \neg s \& r$

12. $p \vee q$; $p \rightarrow r$; $\neg r$; $\therefore q$

13. $\neg p \vee q$; $\neg q \vee r$; $\neg r$; $\therefore \neg p$

14. $(p \& \neg q) \rightarrow \neg r$; r; $\therefore \neg p \vee (p \& q)$

▶ 15. $\neg s \vee (s \& p)$; $(s \rightarrow p) \rightarrow r$; $\therefore r$

16. $p \rightarrow \neg q$; $\neg p \vee r$; q; $(q \& r) \rightarrow p$; $\therefore \neg r$

17. $\neg(p \& \neg q) \vee \neg p$; $\therefore p \rightarrow q$

18. $p \& s$; $p \rightarrow (\neg s \vee r)$; $\therefore r$

19. $\neg p \vee q$; $\neg q$; $\neg p \rightarrow r$; $\therefore r$

▶ 20. $p \rightarrow q$; $r \rightarrow p$; $\therefore \neg r \vee q$

21. $r \rightarrow p$; $\neg r \rightarrow q$; $q \rightarrow s$; $\therefore p \vee s$

22. $p \to \neg q; r; r \to (q \lor \neg s); \therefore s \to \neg p$

23. $p \to (\neg r \lor s); p \to \neg s; \therefore p \to \neg r$

24. $p \to \neg q; \neg p \to \neg r; r \lor \neg s; \therefore \neg q \lor \neg s$

25. $p \& \neg s; r \to s; p \to (q \lor r); \therefore q$

26. $p \& q; r \& \neg s; q \to (p \to p_1); p_1 \to (r \to (s \lor q_1)); \therefore q_1{}^*$

27. $p \& q; p \to (s \lor r); \neg (r \& q); \therefore s^*$

28. $p \lor (q \lor s); s_1 \& \neg s_2; \neg(\neg s_1 \lor s_2) \to \neg p; (s \to r) \& \neg r; \therefore q^*$

29. $p \& (\neg q \& \neg p_1); p \to (s \to r); s \to (r \leftrightarrow (p_1 \lor q)); \therefore \neg s^*$

▶ **30.** $s \to p; (s \& p) \to q; r \to s_1; r \lor s; \therefore q \lor s_1{}^*$

31. $p \lor (r \lor q); (r \to s_1) \& (q \to s_2); (s_1 \lor s_2) \to (p \lor q); \neg p; \therefore q^*$

32. $p \to (s \& r); (r \lor \neg s) \to (q \& q_1); q_1 \leftrightarrow q_2; \therefore p \to q_2{}^*$

33. $(p \& q) \lor (q \& r); \therefore \neg q \to s^*$

34. $(p \lor q) \& r; q \to s; \therefore \neg p \to (r \to s)^*$

▶ **35.** $\neg p \lor (q \& r); (r \lor \neg q) \to (s \& q_1); (q_1 \& q_2) \lor \neg(q_1 \lor q_2); \therefore \neg p \lor q_2{}^*$

36. $\neg(\neg p \& q) \& (p \leftrightarrow \neg q); \therefore p \leftrightarrow (q \to r)^*$

4.5 Derivable Rules

The series of proof methods and rules we've adopted allow us to prove the validity of any valid argument form in sentential logic. That is to say, the system is *complete:* Every valid argument form can be proved valid in the system. This guarantees that we have enough rules. Our proof system is also *sound* in the sense that every argument form we can show to be valid is in fact valid. Our rules, in other words, never lead us astray. To speak in terms of formulas rather than argument forms: The system is complete in that every valid formula is a theorem and also sound in that every theorem is valid. The provable formulas are exactly the valid formulas. So, this system matches precisely our semantics for the sentential connectives. Every aspect of the meanings of the connectives we've captured in some rule, proof method, or combination of these. Establishing the soundness and completeness of this chapter's natural deduction system requires a fairly sophisticated proof in our metalanguage; such a proof is possible, but we won't attempt it here.

To cover all of sentential logic, therefore, we need no more rules or proof techniques. Nevertheless, this section will present some added rules and methods. Everything we can prove with them is still valid. But they are all *derivable* rules, since they force us to accept nothing new about the logical connectives. They are shortcuts; they abbreviate series of proof lines that we could write in terms of our basic rules. Though they are theoretically unnecessary, then, these rules save a great deal of time and effort.

First, we can develop a new proof method, or, really, an amendment to a basic proof method. An indirect proof shows that something is false by showing that the assumption that it is true leads to a contradiction. As we've introduced it, this means that the conclusion of an indirect proof is always a negation. But it's easy to extend this method to any formula. If we want to show that a formula, $\mathscr{A}$, is true, we can assume that $\mathscr{A}$ is false by assuming $\neg \mathscr{A}$ and showing that a contradiction follows. So, our extended method of indirect proof will work as follows:

Indirect Proof

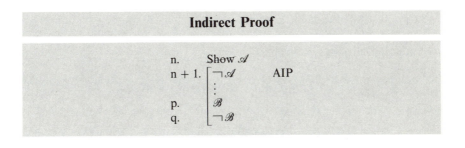

Using this method eliminates at least one application of negation exploitation.

The first derivable rules pertain to negated formulas. Negations are somewhat difficult to exploit in our basic system, since we can eliminate negations only two at a time. It's very useful, therefore, to simplify formulas that begin with a negation. These derivable rules all give equivalents for negated formulas. Because they do give equivalents, they are all invertible; they work in either direction.

The first two are often called *DeMorgan's Laws.*[6]

Negation-Conjunction ($\neg$ &)

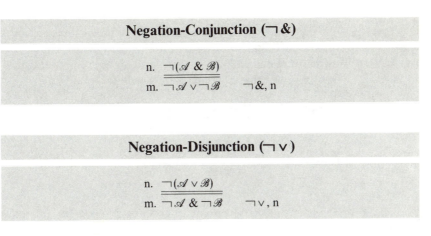

Negation-Conditional (¬→)

n. $\underline{\underline{\neg(\mathscr{A} \to \mathscr{B})}}$

m. $\mathscr{A} \,\&\, \neg\mathscr{B}$ ⬩ ¬→, n

Negation-Biconditional (¬↔)

n. $\underline{\underline{\neg(\mathscr{A} \leftrightarrow \mathscr{B})}}$

m. $\neg\mathscr{A} \leftrightarrow \mathscr{B})$ (or $\mathscr{A} \leftrightarrow \neg\mathscr{B}$) ⬩ ¬↔, n

---

Closely related to these rules are several others that define a connective in terms of other connectives. The first allows us to transform disjunctions into conditionals, and vice versa:

Conditional-Disjunction (→∨)

n. $\underline{\underline{\mathscr{A} \to \mathscr{B}}}$

m. $\neg\mathscr{A} \vee \mathscr{B}$ ⬩ →∨, n

The next allows us to characterize the biconditional in terms of the conditional:

Conditional-Biconditional (→↔)

n. $\underline{\underline{\mathscr{A} \leftrightarrow \mathscr{B}}}$

m. $\mathscr{A} \to \mathscr{B}$ (or $\mathscr{B} \to \mathscr{A}$) ⬩ →↔, n

---

The next four rules assert that the order and grouping of subformulas is irrelevant in continued conjunctions and continued disjunctions. The first rule is that the order of conjuncts makes no difference: $\mathscr{A} \,\&\, \mathscr{B}$ is equivalent to $\mathscr{B} \,\&\, \mathscr{A}$. It thus indicates that conjunction is *commutative*.

Commutativity of Conjunction (&C)

n. $\underline{\underline{\mathscr{A} \,\&\, \mathscr{B}}}$

m. $\mathscr{B} \,\&\, \mathscr{A}$ ⬩ &C, n

The second rule is that the grouping of conjuncts makes no difference or, in other words, that conjunction is *associative*.

Associativity of Conjunction (&A)

n. $\underline{(\mathscr{A} \mathbin{\&} \mathscr{B}) \mathbin{\&} \mathscr{C}}$

m. $\mathscr{A} \mathbin{\&} (\mathscr{B} \mathbin{\&} \mathscr{C})$ &A, n

The third and fourth rules assert that disjunction is also commutative and associative.

Commutativity of Disjunction ($\vee$C)

n. $\underline{\mathscr{A} \vee \mathscr{B}}$

m. $\mathscr{B} \vee \mathscr{A}$ $\vee$C, n

Associativity of Disjunction ($\vee$A)

n. $\underline{(\mathscr{A} \vee \mathscr{B}) \vee \mathscr{C}}$

m. $\mathscr{A} \vee (\mathscr{B} \vee \mathscr{C})$ $\vee$A, n

Finally, four rules abbreviate commonly used proof steps. The first is a variation of conditional exploitation; the second, a variation of biconditional exploitation; the third, a variation of disjunction exploitation. All these variations allow negations to function readily, without detours. The fourth expresses the principle that anything follows from a contradiction.[7]

Conditional Exploitation* ($\rightarrow$E*)

n. $\mathscr{A} \rightarrow \mathscr{B}$

m. $\underline{\neg \mathscr{B}}$

p. $\neg \mathscr{A}$ $\rightarrow$E*, n, m

The abbreviated rule of conditional exploitation is sometimes called *modus tollens*.

Biconditional Exploitation* ($\leftrightarrow$E*)

n. $\mathscr{A} \leftrightarrow \mathscr{B}$

m. $\underline{\neg \mathscr{A}}$ (or $\neg \mathscr{B}$)

p. $\neg \mathscr{B}$ (or $\neg \mathscr{A}$) $\leftrightarrow$E*, n, m

Disjunction Exploitation* (∨E*)

n. $\mathcal{A} \vee \mathcal{B}$
m. $\neg \mathcal{A}$ (or $\neg \mathcal{B}$)
p. $\mathcal{B}$ (or $\mathcal{A}$)   ∨E*, n, m

This abbreviated rule of disjunction exploitation is occasionally called *disjunctive syllogism*.

Contradiction (!)

n. $\mathcal{A}$
m. $\neg \mathcal{A}$
p. $\mathcal{B}$!, n, m

In addition to these derivable rules, we'll adopt a principle of *replacement*. In general, we can apply rules only to formulas with the appropriate main connectives. For instance, we can apply conjunction exploitation only to conjunctions: formulas with & as their main connectives. But invertible rules are justified by the equivalence of the formulas they link. And if we replace a subformula of any formula with an equivalent subformula, we obtain a formula equivalent to the original. If p and $\neg\neg p$ are equivalent, for instance, then so are $p \rightarrow q$ and $\neg\neg p \rightarrow q$. Consequently, we can apply invertible rules to subformulas as well as formulas. So we can use $\neg\neg$ to move from $\neg\neg p$ to p, but also from $\neg\neg p \rightarrow q$ to $p \rightarrow q$. The derivable rules are often tremendous time-savers, as attempts to show that they are derivable from the basic rules will demonstrate.

We'll close this chapter by summarizing strategy hints. Overall proof strategies derive, most significantly, from what we are trying to prove, and, secondarily, from what we already have. This table contains some of the most important strategies. In all cases, a direct proof is easiest when it can be achieved. These strategies are useful, however, when it's not obvious how to prove the conclusion directly.

Proof Strategies

To Get	Try
$\neg \mathcal{A}$	Using indirect proof
$\mathcal{A}$ & $\mathcal{B}$	Proving $\mathcal{A}$ and $\mathcal{B}$ separately
$\mathcal{A} \vee \mathcal{B}$	(a) Using indirect proof, or (b) proving $\mathcal{A}$ or $\mathcal{B}$ separately
$\mathcal{A} \rightarrow \mathcal{B}$	Using conditional proof
$\mathcal{A} \leftrightarrow \mathcal{B}$	Proving the two conditionals $\mathcal{A} \rightarrow \mathcal{B}$ and $\mathcal{B} \rightarrow \mathcal{A}$

To Exploit	Try
$\neg\mathscr{A}$	(a) Using it with other lines that have $\mathscr{A}$ as a part, or (b) using a derivable rule
$\mathscr{A}$ & $\mathscr{B}$	Using &E to get $\mathscr{A}$ and $\mathscr{B}$ individually
$\mathscr{A} \vee \mathscr{B}$	(a) Getting the negation of one disjunct and using $\vee$E* to get the other, or (b) using $\vee$E by taking each case separately
$\mathscr{A} \rightarrow \mathscr{B}$	(a) Getting $\mathscr{A}$ and then reaching $\mathscr{B}$ by $\rightarrow$E, or (b) getting $\neg\mathscr{B}$ and then reaching $\neg\mathscr{A}$ by $\rightarrow$E*
$\mathscr{A} \leftrightarrow \mathscr{B}$	(a) Getting either component and then reaching the other by $\leftrightarrow$E, or (b) getting the negation of either component and then the negation of the other by $\leftrightarrow$E*

These strategies indicate how to construct proofs of various kinds. They are a helpful guide in a wide variety of situations. Sometimes, however, the obvious ploys may not work. When this happens, there are two "safety valves": strategies that work well when pressure is high.

1. When in doubt, use indirect proof. Anything provable can be proved with an indirect proof.

2. If it's not clear what to try to prove in an indirect proof, choose a sentence letter and try to prove it. The assumption for indirect proof should lead to a contradiction, so absolutely anything should follow. No matter what letter you select, therefore, you should be able to prove it.

Problems

Construct a deduction to show that each of these arguments is valid.

1. If you are ambitious, you'll never achieve all your goals. But life has meaning only if you have ambition. Thus, if you achieve all your goals, life has no meaning.

2. God is that, the greater than which cannot be conceived. If the idea of God exists in our understanding, but God does not exist in reality, then something is conceivable as greater than God. If the idea of God exists in our understanding, therefore, God exists in reality.

3. God is omnipotent if and only if He can do everything. If He can't make a stone so heavy that He can't lift it, then He can't do everything. But if He can make a stone so heavy that He can't lift it, He can't do everything. Therefore, either God is not omnipotent, or God does not exist.

4. If the objects of mathematics are material things, then mathematics can't consist entirely of necessary truths. Mathematical objects are immaterial only if the mind has access to a realm beyond the reach of the senses. Mathematics does consist of necessary truths, although the mind has no

access to any realm beyond the reach of the senses. Therefore the objects of mathematics are neither material nor immaterial.

▸ 5. If the president pursues arms limitations talks, then, if he gets the foreign policy mechanism working more harmoniously, the European left will acquiesce to the placement of additional nuclear weapons in Europe. But the European left will never acquiesce to that. So either the president won't get the foreign policy mechanism working more harmoniously, or he won't pursue arms limitations talks.

6. If we either introduce a new product line or give an existing line a new advertising image, then we'll be taking a risk, and we may lose market share. If we don't introduce a new product line, we won't have to make large expenditures on advertising. So, if we don't take risks, we won't have to make large expenditures on advertising.

7. If we can avoid terrorism only by taking strong retaliatory measures, then we have no choice but to risk innocent lives. But if we don't take strong retaliatory measures, we'll certainly fall prey to attacks by terrorists. Nevertheless, we refuse to risk innocent lives. Consequently, terrorists will find us, more and more, an appealing target.

8. If God is all powerful, He is able to prevent evil. If He is all good, He is willing to prevent evil. Evil does not exist unless He is both unwilling and unable to prevent it. If God exists, He is both all good and all powerful. Therefore, since evil exists, God does not.

9. My cat does not sing opera unless all the lights are out. If I am very insistent, then my cat sings opera; but if I either turn out all the lights or howl at the moon, I am very insistent indeed. I always howl at the moon if I am not very insistent. Therefore, my lights are out, I am very insistent, and my cat is singing opera.

10. If we continue to run a large trade deficit, then the government will yield to calls for protectionism. We won't continue to run a large deficit only if our economy slows down or foreign economies recover. So, if foreign economies don't recover, then the government will resist calls for protectionism only if our economy slows down.

11. If companies continue to invest money here, then the government will sustain its policies. If they don't invest here, those suffering will be even worse off than they are now. But if the government sustains its policies, those suffering will be worse off. Thus, no matter what happens, those suffering will be worse off.

12. We cannot both maintain high educational standards and accept almost every high school graduate unless we fail large numbers of students when (and only when) many students do poorly. We will continue to maintain high standards; furthermore, we will placate the legislature and admit

almost all high school graduates. Of course, we can't both placate the legislature and fail large numbers of students. Therefore, not many students will do poorly.

Construct deductions to demonstrate the validity of these argument forms.

13. $p \leftrightarrow q;\ \neg q;\ \therefore\ \neg p$ **14.** $p \leftrightarrow q;\ \therefore\ p \to q$

15. $p \leftrightarrow q;\ \therefore\ q \to p$ **16.** $p\ \&\ q;\ \therefore\ q\ \&\ p$

17. $p \lor q;\ \therefore\ q \lor p$

Use deduction to solve each of these problems.

18. Holmes and Watson question three suspects: Peters, Quine, and Russell. Hearing that their responses conflict, Holmes declares, "If Peters and Quine are telling the truth, then Russell is lying." Watson seemingly assents, saying, "Indeed, at least one of them is telling us a falsehood." Irritated, Holmes insists, "That's not all, my dear Watson! We know that Russell is the trickster, if the other two are telling us the truth!" Show that Holmes's irritation is unjustified by showing that his original statement is equivalent to Watson's.

19. Jones, feeling upset about the insecurity of the Social Security system, sighs that he faces a dilemma: "If taxes aren't raised, I'll have no money when I'm old. If taxes are raised, I'll have no money now." Smith, ever the even-tempered one, reasons that neither of Jones's contentions is true. Jones answers, "Aha! You've contradicted yourself!" Show that Smith's assertion that both Jones's claims are false is indeed contradictory.

20. On the way to the barber shop (adapted from Lewis Carroll), you are trying to decide which of three barbers—Allen, Baker, and Carr—will be in today. You know Allen has been sick and so reason that (a) if Allen is out of the shop, his good friend Baker must be out with him. But, since they never leave the shop untended, (b) if Carr is out of the shop, then, if Allen is out with him, Baker must be in. Show that (a) and (b) imply (c) that not all three are out; (d) that Allen and Carr are not both out; and (e) that, if Carr and Baker are in, so is Allen.

Show that each of our derivable rules is in fact derivable from the basic rules by using only basic rules to prove that these argument forms are valid.

21. $\neg(p\ \&\ q);\ \therefore\ \neg p \lor \neg q$*   **22.** $\neg p \lor \neg q;\ \therefore\ \neg(p\ \&\ q)$*

23. $\neg(p \lor q);\ \therefore\ \neg p\ \&\ \neg q$*   **24.** $\neg p\ \&\ \neg q;\ \therefore\ \neg(p \lor q)$*

25. $\neg(p \to q);\ \therefore\ p\ \&\ \neg q$*   **26.** $p\ \&\ \neg q;\ \therefore\ \neg(p \to q)$*

27. $p \leftrightarrow \neg q;\ \therefore\ \neg(p \leftrightarrow q)$*   ▶ **28.** $\neg(p \leftrightarrow q);\ \therefore\ p \leftrightarrow \neg q$*

29. $p \to q;\ \therefore\ \neg p \lor q$*   **30.** $\neg p \lor q;\ \therefore\ p \to q$*

Use deduction to show that the following argument forms are valid.

31. $\neg(p \to \neg q); r \to (\neg p \lor \neg q); (r \lor s) \leftrightarrow s_1; \therefore s_1 \leftrightarrow s^*$

32. $\neg(\neg p \,\&\, q); (p \lor r) \to \neg(q_1 \,\&\, s); q \leftrightarrow s; \therefore q_1 \to \neg(q \lor s)^*$

33. $p \to (q \lor r); (\neg q \,\&\, q_1) \lor (s \to p); \neg(\neg r \to \neg p); \therefore \neg s \lor q^*$

34. $p \to (q \to r); (\neg q \leftrightarrow s) \to \neg p; p \lor q_1; \therefore (\neg r \,\&\, s) \to q_1{}^*$

35. $(p \leftrightarrow q) \leftrightarrow r; \neg(p \leftrightarrow \neg r); (q \lor \neg s) \to s_1; \therefore s_1{}^*$

36. $\neg s \to \neg s_2; (s \,\&\, s_1) \to (p \leftrightarrow q); \neg(\neg p \lor q); \therefore s_1 \to \neg s_2{}^*$

37. $(q \leftrightarrow \neg p) \to \neg r; (\neg q \,\&\, s) \lor (p \,\&\, q_1); (s \lor q_1) \to r; \therefore p \to q^*$

▶ **38.** $(p \,\&\, \neg r) \leftrightarrow (s \lor \neg q); s_1 \,\&\, ((\neg s \,\&\, \neg r) \to p); ((s_1 \to q) \lor (s_1 \to r)) \,\&\, ((((p \,\&\, q) \,\&\, s) \to r); \therefore q \,\&\, r^*$

Notes

[1] Natural deduction systems are relatively new; Gerhard Gentzen, a German logician, and Stanislaw Jaskowski, a Polish student of Jan Lukasiewicz, independently proposed the first natural deduction systems in 1934. The system of this book owes a great deal, as well, to innovations by the American logicians Willard van Orman Quine, Frederic B. Fitch, Donald Kalish, and Richard Montague.

[2] This thesis was apparently first defended by the fourteenth-century logician Pseudo-Scot.

[3] Although Aristotle took this method of proof for granted, the Stoics first formulated it explicitly. Having been ignored and forgotten for centuries, it reemerged in the "Port Royal Logic," *La Logique ou l'Art de Penser (Logic, or, the Art of Thinking)* written by the French philosophers Antoine Arnauld (1612–1694) and Pierre Nicole (1625–1695) in 1662.

[4] This rule was first formulated by Robert Kilwardby, an English Dominican who served as Archbishop of Canterbury from 1272 to 1277.

[5] This rule seems to have been known by the Stoics.

[6] They are so-called after the nineteenth-century British logician Augustus DeMorgan (1806–1871), but William of Ockham (1285–1349), a fourteenth-century English philosopher, noted them roughly 500 years before DeMorgan.

[7] The rules →E* and ∨E* (*modus tollens* and disjunctive syllogism) both occur in the works of Ockham. (The Stoics developed a rule that resembled disjunctive syllogism, but they read the disjunction as exclusive.) The rule of contradiction occurs first in Pseudo-Scot. Most of our derived rules, then, have been known since medieval times.

CLASSICAL
QUANTIFICATIONAL
LOGIC

CHAPTER

QUANTIFIERS

Using sentential logic—taking sentences as basic units of analysis—we can handle arguments with any number of premises and sentences of any length and any degree of complexity. Nevertheless, sentential logic can narrow our horizons. Consider even a simple example:

(1) All cows are mammals.
     All mammals are animals.
     ∴ All cows are animals.

This argument is surely valid. Yet we cannot use sentential logic to explain why. We have no choice but to construe (1) as

(2) p
     q
     $\therefore r$

which is plainly not valid. The validity of arguments such as (1) depends on the structure within sentences that in sentential logic we are forced to call "atomic." No theory that fails to analyze such "atomic" sentences can hope to account for such reasoning.

In 1879, the two logicians Gottlob Frege and Charles Sanders Peirce, working independently, developed a way of extending sentential logic to handle arguments such as (1). They introduced symbols representing *determiners,* such as *all, some, no, every, any,* and so on. Frege and Peirce used two symbols: the *universal quantifier,* which we will write as ∀ (Peirce used Π, and the *existential quantifier,* ∃ (in Peirce, Σ). The universal quantifier corresponds roughly to the English *all, every,* and *each;* the existential quantifier, to the English *some, a,* and *an.*

116

5.1 Constants and Quantifiers

Sentential logic is limited precisely because sentences are its basic, unanalyzed units of explanation. To expand this logic to include a broader range of arguments, we must look inside "atomic" sentences to see how they are put together. In general, they consist of a main or subject noun phrase and a main verb phrase.

We can gain some insight by examining the character of verb phrases. Consider a few examples:

(3) a. is a man
  b. knows some people who live in Oklahoma City
  c. sleeps very soundly
  d. kicked the ball into the end zone
  e. thought that Yosemite would be more fun to visit
  f. gave Fred a copy of the letter

All verb phrases are *general terms* in the sense that they are true or false of individual objects. We can pick any object we like; it will either be a man or not be a man. It will either sleep very soundly or not sleep very soundly. It will be true either that it gave Fred a copy of the letter or that it didn't. Alone, verb phrases and other general terms are not true or false. But we can think of them as yielding a truth value when combined with the name of an object. Objects of which the verb phrase or general term are true *satisfy* it; the verb phrase or general term *applies to* them. Equivalently, we can think of verb phrases and other general terms as classifying objects into two categories: those of which they are true, and those of which they are false. The first category—the set of objects of which the verb phrase or general term is true— is called its *extension*.

Verb phrases combine with noun phrases to form sentences. Since verb phrases are expressions that are true or false of particular objects, noun phrases must specify an object, or a group of objects, and say something about the application of the verb phrase to them. In the sentence 'Several people gave Fred a copy of the letter,' for example, the noun phrase 'Several people' specifies the set of people and says that the verb phrase applies to several objects in that set. Noun phrases take several forms. In this chapter, we'll concentrate on two of them.

First, noun phrases may simply pick out a single object by naming it. A sentence with a proper name as its subject is true if the verb phrase is true of the object named, and false otherwise. Each of these sentences results from combining a proper name with a verb phrase from (3):

(4) a. Socrates is a man.
  b. Maria knows some people who live in Oklahoma City.
  c. Mr. Hendley sleeps very soundly.
  d. Bahr kicked the ball into the end zone.

   e. Penelope thought that Yosemite would be more fun to visit.

   f. Nate gave Fred a copy of the letter.

In our formal language, we thus need at least two kinds of symbol to represent sentences such as (4)a. First, we'll use lowercase letters from the beginning of the alphabet, with or without numerical subscripts, to represent proper names and pronouns; we'll call these symbolic names *individual constants* or, more simply, *constants.* In this and the next several chapters, we'll use constants to symbolize only those names that *denote,* that is, refer to something that actually exists.

Second, we'll use uppercase letters, with or without numerical subscripts, as *predicate constants,* or more simply, *predicates.* Each predicate is assigned a number as a superscript; predicates are called n-*ary* if they are assigned the number *n.* Every predicate yields a truth value when combined with a certain number of objects. The assigned number indicates of how many objects at once the predicate is true or false. Usually, in writing predicates, we'll omit the superscripts: We can tell what number a predicate is assigned by looking at the number of constants or other terms that follow it.

Recall that a general term is true or false of individual objects. We will therefore symbolize simple general terms such as *sleeps* with *singulary* predicates, that is, predicates to which 1 is assigned. These predicates are true or false of single objects and produce sentences when combined with a single proper name. Other predicates yield sentences only when combined with two names; they are true or false of two objects taken together. Such predicates are assigned the number 2, and are called *binary.* They are useful for symbolizing, among other things, transitive verbs. *Respect,* for example, applies not to objects taken individually, but to objects taken in pairs. It requires a direct object. We can ask whether Robin respects Julia, but not simply whether Robin respects.

   Letting *a* symbolize 'Socrates,' and *M* symbolize 'man' (or, equivalently, 'is a man'), we can symbolize 'Socrates is a man' by

(5) *Ma*

which is merely a symbolic version of (4)a. *Ma* is a formula of quantificational logic. One way of building formulas, then, is to combine individual constants with predicates.

A second sort of noun phrase consists of a determiner, such as *every* or *some,* together with a common noun, such as *man* or *truck.* The common noun may be modified by adjectives, adjectival phrases, prepositional phrases, or relative clauses. Whatever its grammatical structure, however, the modified noun constitutes a general term.

Noun phrases of this more complex sort, when combined with the verb phrases in (3), yield the following sentences:

(6) a. One reporter who covered the match is a man.

   b. A few friends know some people who live in Oklahoma City.

   c. Every endomorph sleeps very soundly.
   d. Several prospects kicked the ball into the end zone.
   e. A taxi driver thought that Yosemite would be more fun
      to visit.
   f. Nobody gave Fred a copy of the letter.

Nobody, in (6)f, is a special case; the word itself contains both a determiner and a general term, and is equivalent to 'no person.'

We can gain some insight into the structure of the sentences in (6) by examining some related sentences that contain subject noun phrases that, while complex, consist only of determiners and the rather colorless general terms *thing* and *object.*

We'll begin by considering this sentence:

(7) Something is peculiar.

To symbolize this sentence, we can't use any individual constant for 'something.' We don't want to say that 'peculiar' applies to any object in particular. The sentence says just that 'peculiar' applies to some object. Quantification theory allows us to represent this with *variables.* We will say, in effect, "'x is peculiar' is true, for some object x." Variables, unlike constants, do not denote particular objects. Instead, they *range over* a *domain,* a set of objects that they can take as *values.*

To express the idea that 'x is peculiar' is true for some x, we write the symbolic equivalent of

(8) (for some x)(x is peculiar)

which is

(9) ∃xPx

We may read (9) as meaning the following:

   (10) a. For some x, x is peculiar.
        b. Some x is such that x is peculiar.
        c. There is an x such that x is peculiar.
        d. An x is such that x is peculiar.

In better English, these become:

   (11) a. Something is peculiar.
        b. There is something peculiar.
        c. An object is peculiar.

In English, the words *thing* and *object* serve some of the purposes that variables such as x serve in quantification theory. Variables link quantifiers to the predicates they accompany.

The strategy of quantificational logic, then, requires two new sorts of symbol. First, we must introduce *individual variables,* or, more simply, *variables.* Individual variables are so called because they range over individual, particular objects. Variables will be lowercase letters from the end of the alphabet, with or without numerical subscripts. As with sentence letters, individual constants, and predicate letters, subscripts allow us to have as many variables as we need. (The superscripts on predicates serve a very different purpose.) Variables in quantificational logic act, for the most part, much as variables for numbers such as n or x act in arithmetic or algebra. They denote no objects in particular; instead, they range over a set of objects. Second, the language of quantification theory must include quantifiers.

As another simple example, suppose we want to say, recalling the title of a popular song, that everything is beautiful. To say that Pittsburgh is beautiful, or that Ingrid Bergmann is beautiful, we can introduce the predicate B and the individual constants p and b and write Bp and Bb. To say that everything is beautiful, however, we need to say that 'x is beautiful' is true for every object x. The universal quantifier does just this. We can use the quantifier as a prefix to Bx, writing the symbolic equivalent of

(12) (for every x)(x is beautiful)

or

(13) $\forall x Bx$

(13) says

(14) a. For every x, x is beautiful.
  b. For all x, x is beautiful.
  c. For each x, x is beautiful.
  d. For any x, x is beautiful.
  e. Every x is such that x is beautiful.
  f. All x are such that x is beautiful.
  g. Each x is such that x is beautiful.
  h. Any x is such that x is beautiful.

or, in more acceptable English,

(15) a. Everything is beautiful.
  b. All things are beautiful.
  c. Each object is beautiful.
  d. Any object is beautiful.

All, every, each and (usually) *any* generally receive the same symbolization. In English, these words differ subtly but significantly in meaning. Note, for example, that though 'All things are beautiful' seems to mean just what 'Everything is beautiful' does, 'Anything is beautiful' sounds strange. Quantification theory cannot capture all the differences between these determiners, but it can capture a very important class of them, as a later section of this chapter will show. Note also that all the sentences in (15) sound vague: Every *what* is beautiful? What is the range of the variable? Usually, the context of use indicates what the variable ranges over.

Variables have no meanings independent of their symbolic context. In (9) and (13), P and B represent 'peculiar' and 'beautiful,' respectively. We cannot interchange them without changing the translation manual relating English sentences to their symbolic representations. We can interchange variables, in contrast, with very few restrictions, without altering meaning. We can symbolize 'Something is peculiar' just as well by

(16) $\exists y P y$

as by (9), and we can represent 'Everything is beautiful' just as well by

(17) $\forall z B z$

as by (13). We can do so because the variables x, y, and z denote no particular objects but range over a domain.

5.2 CATEGORICAL SENTENCE FORMS

Sentences such as 'Something is peculiar' or 'Everything is beautiful' take us only so far. Most of the time, we want to say something about, for example, some people or every frog, not about just something or everything. We must be able to handle sentences with subject noun phrases that contain more complicated general terms. If we can do this, it will be easy to represent any sentence having one of four classic *categorical forms* in quantification theory.

Universal affirmative sentences have the structure

(18) All F G

We might want to represent, for instance,

(19) All frogs swim.

We already know that Sa can represent, say, 'Albert swims,' and that $\forall x S x$ can represent 'Everything swims.' To represent 'All frogs swim,' we must use quantification theory to focus on the determiner *all*. *All* combines with a general term, in this case *frogs,* to form a noun phrase; that noun phrase in turn combines with another general term, the verb phrase *swims*. We can see *all,* then, as relating two general terms. The general terms, in this case, are simple, so we can symbolize them as singulary predicates F and S. We can symbolize *all* with the universal quantifier. To link the quantifier and the predicates, we can use a variable, say, x. We need to search, then, for the proper logical relationship among $\forall x$, Fx, and Sx; we need to say that every x such that x is a frog swims.

In essence, the theory points out that 'All frogs swim' says that if an object is a frog, it swims. So, the relationship we are looking for is the conditional; we want to say something like

(20) For all x, if x is a frog, then x swims.

which we can represent, using quantifiers and sentential connectives, as

(21) $\forall x(Fx \rightarrow Sx)$

This reasoning applies to all universal affirmative sentences. So we can represent anything having the form (18) as

(22) $\forall x(Fx \rightarrow Gx)$

Second, *particular affirmative* sentences have the structure

(23) Some *F G*

For example, let's try to represent

(24) Some people are bothersome.

We can represent 'Edna is bothersome' as *Be*, and 'Some things are bothersome' as $\exists x Bx$. (24) seems to require that we express a relationship between the general terms *people* and *bothersome* in a way corresponding to the meaning of the determiner *some*. We want to say that some object is a person and is bothersome. Conjunction expresses the right relationship: To say that some people are bothersome is to say that, for some object x, x is a person, and x is bothersome. So we can write (24) as

(25) $\exists x(Px \& Bx)$

In general, particular affirmative sentences translate into quantification theory as

(26) $\exists x(Fx \& Gx)$

Conjunction, together with the existential quantifier, succeeds in representing sentences having the form of (23).

Finally, *universal negative* sentences have the form

(27) No *F G*

They have two equivalent and equally natural symbolizations in the language of quantificational logic. Suppose we want to symbolize 'No man is an island' in its literal sense. We might think that we need a new quantifier to symbolize the determiner *no*. But we can symbolize the relation between general terms that *no* expresses by using either the existential or the universal quantifier, together with negation. Because 'No man is an island' is a direct denial of the sentence 'Some man is an island,' we can represent it as the negation of a particular affirmative sentence form. If we let *F* and *G* represent 'man' and 'island,' respectively, this strategy yields

(28) $\neg \exists x(Fx \& Gx)$

which we can read

(29) a. It is not the case that, for some x, x is a man and x is an island.
 b. There is no x such that x is a man and x is an island.

or, more naturally,

(30) a. It is not true that some man is an island.
b. There is no man who is an island.
c. There is nothing that is both a man and an island.
d. No man is an island.

But 'No man is an island' also bears some similarity to universal affirmative sentence forms, being equivalent to "If an object is a man, it is not an island.' It says about every man, in other words, that he is not an island. So, adopting the same dictionary as before, we can write

(31) $\forall x(Fx \rightarrow \neg Gx)$

which says

(32) a. For all x, if x is a man, then x is not an island.
b. Every x is such that, if x is a man, then x is not an island.

or, in better English,

(33) a. No man is an island.

(28) and (31) both represent universal negative sentences. As we might hope, they are equivalent.

Any categorical sentence form, therefore, has a representation in quantification theory. The logic of this chapter has the power to cover the entire realm of syllogistic reasoning. In fact, as the next few sections demonstrate, it has the power to capture an extremely wide range of English arguments.

Problems

Symbolize the following in quantificational logic.

1. All men are born good (Confucius).

2. All that I know, I learned after I was thirty (Georges Clemenceau).

3. Children are always cruel (Samuel Johnson).

4. All who remember, doubt (Theodore Roethke).

▶ **5.** All big men are dreamers (Woodrow Wilson).

6. All are not friends that speak us fair (James Clarke).

7. All's well that ends well (William Shakespeare).

8. No sound is dissonant which tells of life (Samuel Taylor Coleridge).

9. There is no detail that is too small (George Allen).

▶ **10.** Alas! It is delusion all (George Gordon, Lord Byron).

11. It is always the secure who are humble (G. K. Chesterton).

12. A thing of beauty is a joy forever (John Keats).

5.3 Polyadic Predicates

General terms are true or false of objects. *Man, woman, animal,* and *mortal,* in addition to intransitive verbs such as *swim* and *live,* all apply or fail to apply to objects considered one by one. Most frequently, the objects we have in mind are physical. But this is not essential; *prime number, hypotenuse,* and *integral* all apply to mathematical objects. Because general terms are true or false of single objects, we can symbolize them using *monadic* or *singulary* predicates.

Some English expressions, as we have seen, are true or false of objects taken in pairs, or even triples, quadruples, and so on. Predicates symbolizing them are *polyadic.* Consider, for example, the English verb *distrust.* It makes no sense to ask whether *distrust* applies to an object considered alone. Does Socrates distrust? Does this book distrust? These questions seem incoherent. (We can, of course, ask whether Socrates is distrustful.) But we can ask whether Socrates distrusts Alcibiades (yes) or whether this book distrusts its author (no, books don't distrust anything). So *distrust* applies, or fails to apply, to objects taken in pairs. The verb needs not only a subject but also a direct object. The same holds of most transitive verbs. Predicates that are true or false of pairs of things are not only polyadic but, more specifically, *dyadic* (or *binary*).

To represent general terms symbolically in quantification theory, we use predicates. Thus *Ma* might symbolize 'Alonzo is a man,' and *Cj* might represent 'Joan lives in California.' The same strategy works for polyadic predicates. For instance, how do we symbolize

(34) Hanno admires Bob Dobbs

Because it asserts that 'admires' applies to the pair of objects consisting of Hanno and Bob, this sentence has the structure

(35) Admires (Hanno, Bob Dobbs)

We can thus symbolize it as

(36) *Mhb*

Similarly, we can represent

(37) Joanie loves Chachi

as

(38) *Ljc*

Given a dyadic predicate, say, *D*, symbolizing 'distrusts,' we can construct a formula of quantification theory in several ways. First, we can combine the predicate with constants, obtaining formulas such as *Dab* and *Dcc*. The first constant following the predicate marks the *subject* position; the second marks the *object* position. If we write '*x* distrusts *y*' as *Dxy*, then *x* is the distruster, and *y* is the distrusted. Notice that the same constant may appear more than once; this allows us to symbolize sentences such as 'Alan distrusts himself.'

Second, we may prefix a quantifier to the predicate and combine it with variables. We might do this in only one place, filling the other with a constant. For example, to symbolize

(39) Sam distrusts everyone

given a domain of persons, we can write

(40) ∀*xDsx*

And, to translate

(41) Everyone distrusts Sam

we can write

(42) ∀*xDxs*

Notice that the only difference between these formulas is the order of the variables and constants. This order reflects the distinction in English between subject and object. Since *Dxy* symbolizes '*x* distrusts *y*,' the first position is always occupied by the subject of distrust; the second, by its object. So, in ∀*xDsx*, *s* is in the subject position. In ∀*xDxs*, *s* is in the object position.

We may construct a formula from a predicate by prefixing a quantifier for each place. There may be two variables in the formula; each may link the predicate to either a universal or an existential quantifier. So, using combinations of quantifiers and variables with *D* might yield any of the following formulas:

(43) a. ∃*x*∃*yDxy*
    b. ∃*y*∃*xDxy*
    c. ∀*x*∀*yDxy*
    d. ∀*y*∀*xDxy*
    e. ∃*x*∀*yDxy*
    f. ∃*y*∀*xDxy*
    g. ∀*x*∃*yDxy*
    h. ∀*y*∃*xDxy*
    i. ∃*xDxx*
    j. ∀*xDxx*

To see what these formulas mean, let's examine them one by one.

$\exists x \exists y Dxy$ says that for some x and some y, x distrusts y. This is an ugly way of saying that something distrusts something, or, if we take our domain to consist only of people, that somebody distrusts somebody. $\exists y \exists x Dxy$ says that for some y and some x, x distrusts y. In English, we can express this idea by saying that somebody is distrusted by somebody. We can capture the effect of reversing the variables by using the passive voice, which reverses subject and object in English. Of course, 'Somebody distrusts somebody' and 'somebody is distrusted by somebody' are equivalent. So are $\exists x \exists y Dxy$ and $\exists y \exists x Dxy$. In general, reversing the order of adjacent existential quantifiers produces an equivalent formula.

$\forall x \forall y Dxy$ says that, for all x and for all y, x distrusts y. If we again restrict our attention to people, this means that everybody distrusts everybody. $\forall y \forall x Dxy$ reverses the order of the quantifiers, saying that, for all y and all x, x distrusts y. Again, we can use the passive voice to reverse subject and object in English; $\forall y \forall x Dxy$ represents the English 'Everybody is distrusted by everybody,' which is equivalent to 'Everybody distrusts everybody.' Reversing adjacent universal quantifiers yields an equivalent formula.

Combinations of existential and universal quantifiers, however, do not allow such switches. $\exists x \forall y Dxy$ says that there is an x such that, for all y, x distrusts y. This corresponds to the English 'Somebody distrusts everybody.' $\exists y \forall x Dxy$ says that there is a y such that, for all x, x distrusts y; that amounts to the English 'Somebody is distrusted by everybody.' $\forall x \exists y Dxy$ says that, for all x, there is a y such that x distrusts y. It thus represents 'Everybody distrusts somebody.' $\forall y \exists x Dxy$ says that, for all y, there is an x such that x distrusts y. This corresponds to 'Everybody is distrusted by somebody.' Finally, $\exists x Dxx$ says that some x is such that x distrusts x, corresponding to the English 'Somebody distrusts himself/herself,' while $\forall x Dxx$ says that every x is such that x distrusts x, corresponding to 'Everybody distrusts himself/herself.'

We can display these correspondences in a table (again assuming that we are speaking only of people):

(44)

	Formula	English Sentence
a.	$\exists x \exists y Dxy$	Somebody distrusts somebody.
b.	$\exists y \exists x Dxy$	Somebody is distrusted by somebody.
c.	$\exists x \forall y Dxy$	Somebody distrusts everybody.
d.	$\exists y \forall x Dxy$	Somebody is distrusted by everybody.
e.	$\forall x \exists y Dxy$	Everybody distrusts somebody.
f.	$\forall y \exists x Dxy$	Everybody is distrusted by somebody.
g.	$\forall x \forall y Dxy$	Everybody distrusts everybody.
h.	$\forall y \forall x Dxy$	Everybody is distrusted by everybody.
i.	$\exists x Dxx$	Somebody distrusts himself.
j.	$\forall x Dxx$	Everybody distrusts himself.

In this table, no formula that contains both universal and existential quantifiers is equivalent to any other. 'Somebody distrusts everybody' and

'Everybody is distrusted by somebody' differ in meaning. 'Somebody distrusts everybody' suggests that some one person (for instance, Sam) distrusts everybody. 'Everybody is distrusted by somebody,' in contrast, suggests that everybody is the object of somebody else's distrust. The latter sentence, that is, asserts that Fred distrusts you, Greta distrusts me, and so on. The former asserts that one individual distrusts you, me, and everyone.

Much the same is true of 'Everybody distrusts somebody' and 'Somebody is distrusted by everybody.' The former means that everybody distrusts somebody or other; you may distrust Pat, I may distrust Lou, and so on. The latter, however, means that one person is the object of everyone's distrust; you, I, and everyone else distrust one individual, say, Zeke.

In quantification theory, these differences in meaning are reflected by the different positions of quantifiers. When an existential quantifier appears to the left of a universal, the formula means that one object stands in a particular relation to every object. When an existential appears to the right of a universal, the formula means that, for each object, some object or another stands in some relation to it.

English often distinguishes these senses by placing the existential first, for the former, or the universal first, for the latter. So, the order of the English determiners frequently matches the order of the quantifiers. There are exceptions—*each* and *any,* for example, almost always go to the extreme left—and ambiguities, as the sentence 'Everybody loves somebody' shows. The linguistic rules governing quantifier scope in English are highly complex and controversial. Consequently, we will often have to rely on our own intuitions about meaning to determine, in a given context, the correct order of the quantifiers.

Problems

Symbolize these sentences in quantification theory.

1. All roads lead to Rome.

2. Nobody knows all the trouble I've seen.

3. There is a name that is greater than every name.

4. No number is greater than every number.

▶ 5. All finite things reveal infinitude (Theodore Roethke).

6. Every country can produce good men (Gotthold Lessing).

7. A dull axe never loves grindstones (Henry Ward Beecher).

8. He that has no patience has nothing at all (Italian proverb).

9. There is a singer everyone has heard (Robert Frost).

▶ 10. To do nothing is in every man's power (Samuel Johnson).

5.4 THE LANGUAGE QL

So far we've spoken of quantification theory in informal terms. Before we go further, however, it is important to define the language of the theory more precisely. In the next section, we'll discuss some guidelines for representing English sentences as quantificational formulas. These guidelines naturally depend on the notion of what a formula of quantification theory is.

Recall that any logical language consists of a syntax and a semantics. The semantics of quantification theory is complex; we'll postpone treatment of it until the next chapter. The syntax comprises a vocabulary of symbols and a set of formation rules for combining the symbols to form formulas. The vocabulary of our quantification language, QL, includes the following:

Vocabulary

Sentence letter constants: p, q, r, s, with or without numerical subscripts

n-ary predicate constants: A^n, B^n, . . . , Z^n, with or without numerical subscripts

Individual constants: a, b, c, . . . , o, with or without numerical subscripts

Individual variables: t, u, v, w, x, y, z, with or without numerical subscripts

Sentential connectives: $\neg$, $\rightarrow$, $\&$, $\vee$, $\leftrightarrow$

Quantifiers: $\forall$, $\exists$

Grouping indicators: (,)

This is the vocabulary of SL, supplemented with predicates, constants, variables, and quantifiers.

The rules for combining these symbols to construct formulas are more complex than in sentential logic. To understand them, we need to understand substitution. Suppose that we are given an expression $\mathcal{A}$. We can construct a new expression, which we can call, abstractly, $\mathcal{A}[c/d]$ ($\mathcal{A}$, with c substituted for d), by replacing every occurrence of the constant d with an occurrence of the constant c. Similarly, we can form $\mathcal{A}[x/y]$ by replacing every occurrence of the variable y with an occurrence of the variable x. We can also substitute constants for variables, and variables for constants as well. To take an example, suppose $\mathcal{A}$ is Fab. Then $\mathcal{A}[c/a]$ is Fcb. Or, suppose $\mathcal{A}$ is $\forall xFxa$; then $\mathcal{A}[y/a]$ is $\forall xFxy$. The substitution operation is defined for all expressions, whether or not they are formulas. But applying it to a formula may yield a string of symbols that is no longer a formula.

Formation Rules

Any sentence letter constant is a formula.

An *n*-ary predicate followed by *n* constants is a formula.

If $\mathscr{A}$ is a formula, $\neg\mathscr{A}$ is a formula.

If $\mathscr{A}$ and $\mathscr{B}$ are formulas, then $(\mathscr{A} \rightarrow \mathscr{B})$, $(\mathscr{A} \mathbin{\&} \mathscr{B})$, $(\mathscr{A} \vee \mathscr{B})$, and $(\mathscr{A} \leftrightarrow \mathscr{B})$ are formulas.

If $\mathscr{A}$ is a formula with a constant c, and v is a variable that does not appear in $\mathscr{A}$, then $\exists v\mathscr{A}[v/c]$ and $\forall v\mathscr{A}[v/c]$ are formulas.

Every formula can be constructed by a finite number of applications of these rules.

The formation rules characterize the formulas of the language of quantification theory, QL. Every formula of sentential logic is also a formula of QL. In addition, QL allows us to link quantified formulas with sentential connectives and to combine sentence letters with the quantificational apparatus. When we form $\forall v\mathscr{A}[v/c]$ or $\exists v\mathscr{A}[v/c]$ in accordance with these rules, we'll say that the *scope* of $\forall v$ or $\exists v$ is all of $\forall v\mathscr{A}[v/c]$ or $\exists v\mathscr{A}[v/c]$.

As in the case of sentential logic, we'll adopt a few simplifying conventions to make formulas more readable. First, we'll generally drop the superscript that indicates whether a given predicate is singular, binary, and so on. Second, as in SL, we'll abbreviate a formula by deleting the outside parenthesis pair. So, the first formula in each of these pairs abbreviates the second:

(45) a. $\exists x(Fx \rightarrow Gx) \rightarrow \forall x\exists yHxy$
  $(\exists x(Fx \rightarrow Gx) \rightarrow \forall x\exists yHxy)$
  b. $\forall z\forall w\exists t(Fzt \mathbin{\&} Gwz) \leftrightarrow p$
  $(\forall z\forall w\exists t(Fzt \mathbin{\&} Gwz) \leftrightarrow p)$

Third, because we'll usually drop predicate superscripts, we'll avoid using the same predicate letter as both a monadic and a polyadic predicate within the same formula. For instance

(46) $\exists x\forall y(Fxy \rightarrow Fa)$

is correctly formed; however, using F to represent two English expressions at once tends to be confusing. This formula is more properly written out as $\exists x\forall y(F^2xy \rightarrow F^1a)$, which makes it clear that there are two different predicates.

Several points about formulas of QL deserve mention. Note that only variables may appear with quantifiers. $\exists aFa$ is not a formula; neither is $\forall p(p \rightarrow q)$ nor $\exists FFa$. Individual constants and variables take objects as values. In QL we can quantify over objects, speaking about all objects of a certain kind, some objects of that kind, and so on. We cannot do the same with sentences or predicates. Because it allows quantification over individuals alone, QL is a system of *first-order* quantification, sometimes called *first-order logic*. Other logical theories, called *higher-order logics,* do allow quantification over sentences and predicates, but at the price of substantial complication.

Problems

Evaluate each of the following as (a) a formula of QL, (b) a conventional abbreviation of a formula of QL, or (c) neither of the above.

1. F^1x **2.** F^1a

3. $Fx \rightarrow Fy$ **4.** $Fa \rightarrow Fc$

▶ **5.** $(Fx \rightarrow Fy)$ **6.** $(F^1a \rightarrow F^1b)$

7. $\exists xFx \rightarrow Fx$ **8.** $\exists x(Fx \rightarrow Fx)$

9. $\exists xF^1x \rightarrow F^1a$ ▶ **10.** $\exists x(F^1x \rightarrow F^1a)$

11. $(\exists xFx \rightarrow Fa)$ **12.** $\forall x\forall yFxy \rightarrow Fyx$

13. $\forall x\forall yF^2xy \rightarrow F^2ab$ **14.** $\forall x\forall yFxy \rightarrow Fa$

▶ **15.** $\forall x\forall y(Fxy \rightarrow Fyx)$ **16.** $\forall x\forall yFxy \rightarrow \forall x\forall yFyx$

17. $(\forall x\forall yF^2xy \rightarrow \forall x\forall yF^2yx)$ **18.** $\forall x\forall yFxy \rightarrow Fy$

19. $\forall xFx \rightarrow \exists xFx$ ▶ **20.** $\forall x(Fx \rightarrow \exists xFx)$

21. $(\forall xF^1x \rightarrow \exists xF^1x)$ **22.** $\exists x\forall yGy$

23. $\forall x\forall y\forall b(Fxy \vee Fyb)$ **24.** $\forall x\exists F(Fx \rightarrow Fa)$

▶ **25.** G^1y **26.** G^2b

27. $\forall xG^1x$ **28.** $\forall xGxy$

29. $\forall yG^1y \rightarrow G^1z$ ▶ **30.** $\forall y(G^1y \rightarrow G^1z)$

31. $\forall yGy \rightarrow Gy$ **32.** $\forall y(G^1y \leftrightarrow G^1y)$

33. $\forall xF^2xy \rightarrow \forall yG^2yx$ **34.** $\forall x(Fxy \rightarrow \forall yGyx)$

▶ **35.** $\forall x\forall y(Fxy \rightarrow Gyx)$ **36.** $\forall y(\forall xFxy \rightarrow Gyx)$

37. $\forall x\forall y(\exists zFyz \leftrightarrow (Gy \mathrel{\&} Hzx))$ **38.** $\forall x(\forall y\exists zF^2yz \leftrightarrow (G^1y \mathrel{\&} H^2zx))$

39. $\forall x\forall y\exists z(Fyz \leftrightarrow (Gy \mathrel{\&} Hzy))$ ▶ **40.** $(\forall x\forall y\exists z(Fyz \mathrel{\&} Gx) \leftrightarrow Hzy)$

Taking each expression below as $\mathscr{A}$, write (a) $\mathscr{A}[c/d]$, (b) $\mathscr{A}[d/c]$, (c) $\mathscr{A}[x/c]$, (d) $\mathscr{A}[d/y]$, and (e) $\mathscr{A}[y/x]$ and say whether the result in each case is a formula. (Count abbreviations of formulas as formulas.)

41. Hcd **42.** Hcc

43. Hcx **44.** Hxy

▶ **45.** $\forall x Fx \leftrightarrow Gc$

46. $\forall x (Fx \leftrightarrow Gc)$

47. $\forall x Fxc \leftrightarrow \exists x Fdx$

48. $\forall x Fxc \leftrightarrow \exists y Fdy$

49. $\forall x (Fxc \leftrightarrow \exists y Fdy)$

▶ **50.** $Fxc \leftrightarrow \exists y Fdy$

51. $\exists x Fxc \ \& \ \forall x Fxd$

52. $Fxc \ \& \ \forall x Fxd$

5.5 SYMBOLIZATION

With the addition of polyadic predicates, quantification theory has the power to express and evaluate a very large group of sentences and arguments. This section contains a guide to representing English sentences in the theory.

To translate even a simple sentence, we must distinguish its grammatical subject from its grammatical predicate. Because the word *predicate* has a different meaning in logic, in this book grammatical subjects are called subject (or main) noun phrases, and grammatical predicates are called main verb phrases. It is easy to translate simple sentences such as 'All men are mortal,' 'Some computers are not reliable,' and 'Nobody admires everybody' into Q. But noun and verb phrases may become far more complex. After listing some of the ways in which this can happen, we'll explore how quantification theory can incorporate them.

Noun Phrases

Some noun phrases are easy to handle in Q. Proper names translate as individual constants; common nouns, such as *woman* and *airplane,* translate as monadic predicates. But here the simplicity ends.

Determiners

The determiners *all, each, any,* and *every* generally translate as universal quantifiers, while *some* and *a(n)* generally translate as existentials. But even these rules have exceptions. First, *a* and *an* have a *generic* use, in which they refer to typical members of a kind. Thus

(47) A whale is a mammal

does not mean that some whales are mammals, but that all whales are. In these cases, *a* and *an* correspond roughly to universal quantifiers.

Second, *a, an, any,* and *some* all interact with conditionals when they are part of the antecedent.

(48) If you steal something, you'll get into trouble

can be translated straightforwardly as a conditional with a quantified antecedent (where *a* represents *you*):

(49) $\exists x Sax \rightarrow Ta$

But, when the consequent contains a word that refers back to something in the antecedent, this strategy doesn't work. We could try to translate

(50) If you steal something, you'll pay for it

as

(51) $\exists x Sax \rightarrow Pax$

but this is not a formula; *Pax* cannot come from our formation rules. Changing the parentheses so that the quantifier has scope over the entire formula in itself does not help.

(52) $\exists x(Sax \rightarrow Pax)$

is a formula, but it says the wrong thing. Because of the nature of the conditional, it is equivalent to

(53) $\exists x(\neg Sax \lor Pax)$

But this formula says that there is an object that either you don't steal or you pay for. And this fact is true so long as there is an object you don't steal. But the same does not hold of (50), whose truth value is not determined by whether or not there are things you don't steal.

To represent (50), we must use a universal quantifier with the entire formula as its scope:

(54) $\forall x(Sax \rightarrow Pax)$

This formula says that everything you steal, you pay for, which is equivalent to (50). So, in certain cases in which they appear in the antecedent of a conditional, with reference back to the antecedent in the consequent, the determiners *a, an,* and *some* correspond to universal quantifiers in Q.

Notice that 'If you steal something, you'll pay for it' is equivalent to 'If you steal anything, you'll pay for it.' *Any* translates as a universal quantifier, but with the widest possible scope. *Each* usually takes wide scope among quantifiers; *any,* however, demands wide scope over connectives as well. *Every* and *all* make no such demand. For this reason, the above sentences are not equivalent to 'If you steal everything, you'll pay for it' and 'If you steal each thing, you'll pay for it.' It also explains why *any* often seems similar to *some* or *a.* 'John didn't see any deer' is equivalent to 'John didn't see a deer,' not to 'John didn't see every deer,' because *any* represents a universal quantifier to the left of the negation sign ($\forall x \neg (Sjx \& Dx)$).

Everything, anything, something, and the like all act like the corresponding determiners; *thing* functions, more or less, as a variable. *Everybody, anybody, somebody, everyone, anyone,* and *someone* all act like *every person, any person,* and so on. They generally force the use of a quantifier together with the monadic predicate *P* (for *person*) linked to the remainder of the formula in the appropriate way.

No, and the related *nobody, nothing,* and so on all translate into QL in the ways suggested by representations of categorical sentence forms. They correspond to negations of existential quantifiers or to universal quantifiers applying to negations.

Only, though it is not really a determiner, functions much like *all*, except that it reverses the order of the relevant expressions. 'Only *F G*' amounts to 'All *G F*.'

English contains many other determiners that QL can't translate. *Many, several, a few, few, most, infinitely many,* and *much,* for example, elude the powers of QL. Sentences containing them cannot be translated into quantification theory.

We haven't yet discussed the semantics of QL. But an important part of interpreting formulas of QL is assigning them a *universe of discourse* or *domain.* This is a set of objects. The universe of discourse is a set containing the objects to which the formulas refer. The quantifiers and variables *range over* the domain, in the sense that we interpret the universal quantifier as meaning that something is true for all elements of the domain and the existential quantifier as meaning that something is true for some element of the domain.

We can often simplify symbolization by assigning an appropriate universe of discourse. If, within the context of an argument, we are speaking of nothing but people, then we can limit the domain to the set of people. If we do, then a universal quantifier has the effect of 'for all x in the set of people,' that is, 'for all people x.' The universal quantifier, in such a case, represents the English expressions *anybody* and *everybody;* we need not use a predicate meaning *person.*

Adjectives

Adjectives, words such as *good, red, friendly,* and *logical,* modify nouns. With a few exceptions, they translate into QL as monadic predicates, linked to the predicates representing the nouns they modify by conjunctions. Thus

(55) All friendly cats purr

becomes

(56) $\forall x((Fx \,\&\, Cx) \rightarrow Px)$

and

(57) Some artists are unhappy people

becomes

(58) $\exists x(Ax \,\&\, (Ux \,\&\, Px))$

Notice that these sentences are basically categorical sentence forms.

Adjectival phrases, consisting of an adjective modified by an adverb, for example, function in the same way. They must be treated as a single unit. Thus

(59) John is a very wealthy logician

translates into QL as

(60) *Wj & Lj*

where *W* represents 'very wealthy.'

Most adjectives and adjectival phrases thus translate as conjunctions. The set of colorless gases is the set of things that are both colorless and gases. It is thus the intersection of the set of colorless things and the set of gases. For this reason, we'll call adjectives that work in the standard way *intersective*.

Certain adjectives, however, do not translate into QL directly. They have meanings that relate in some way to the nouns they modify. So, such adjectives and their nouns must be translated as a single unit. Luckily there is an easy test for identifying these adjectives, which we'll call *nonintersective*. Wealthy logicians are both wealthy and logicians; red Chevrolets are both red and Chevrolets. But alleged criminals are not alleged and criminals. Good pianists are not simply pianists and good. Former Congressmen are not former and Congressmen; large mice are not large and mice. 'Good pianist' means something like 'good as a pianist'; 'large mouse,' something like 'large for a mouse.' This is why the following arguments fail:

(61) a. Every pianist is a lover.
  ∴ Every good pianist is a good lover.
  b. All mice are animals.
  ∴ All large mice are large animals.

To avoid trouble, then, we must translate nonintersective adjectives modifying nouns as single monadic predicates. Note, however, that there is an important difference among nonintersective adjectives. The set of large mice is a subset of the set of mice; the set of good pianists is similarly a subset of the set of pianists. Nonintersective adjectives such as *large* and *good* are therefore called *subsective*. Other nonintersective adjectives, such as *fake* and *former*, are not even subsective. The set of fake diamonds is not a subset of the set of diamonds at all. Likewise, the set of former congressmen is not a subset of the set of congressmen. Nonintersective, nonsubsective adjectives must be translated, together with the nouns they modify, as a single unit. Nonintersective but subsective adjectives may be translated as conjunctions, but with an important difference from intersective adjectives. We can render 'large mouse' as *Lx & Mx*, where *M* represents 'mouse,' but only if we construe *L* as translating not 'large' but 'large for a mouse.'

Relative Clauses

Relative clauses are English expressions formed from sentences. They begin, generally, with *that* or a word starting with *wh-*, such as *who, which, when,* or *where,* though these words are often omitted. Relative clauses frequently act like adjectives, modifying nouns or noun phrases. Thus 'that I used to attend,' 'who once denounced Richard Nixon,' and '(when) I've placed my hopes in something' behave as relative clauses in the following sentences:

(62) a. A school that I used to attend has been closed.
    b. Senator McCarthy, who once denounced Richard Nixon, is now retired.
    c. Every time (when) I've placed my hopes in something I've been disappointed.

Like intersective adjectives, relative clauses are conjoined to the nouns they modify. 'That I used to attend,' for example, derives from the open sentence 'I used to attend x.' Similarly, 'who once denounced Richard Nixon' derives from 'x once denounced Richard Nixon,' and '(when) I've placed my hopes in something' derives from 'I've placed my hopes in something at (time) x.' Using the obvious representations, we can symbolize these as Aix, Dxn, and $\exists y Hiyx$. Conjoining them in the appropriate manner to representations of their nouns, we obtain:

(63) a. $\exists x((Sx \,\&\, Aix) \,\&\, Cx)$
    (There is an x such that x is a school, I used to attend x, and x is closed.)
    b. $Dmn \,\&\, Rm$
    (Senator McCarthy once denounced Richard Nixon, and McCarthy is now retired.)
    c. $\forall x((Tx \,\&\, \exists y Hiyx) \rightarrow Dix)$
    (For every x, if x is a time and, for some y, I've placed my hopes in y at x, then I've been disappointed at x.)

Relative clauses, in general, translate quite easily. Only one minor wrinkle ruins their simplicity. Some relative clauses restrict the group of things the noun phrase they modify applies to. If I tell you that everyone I know prefers Mexican to Chinese food, then I am speaking, not of everyone, but just of everyone I know. (62a) and (62c) contain such *restrictive* relative clauses. Other clauses, however, make almost parenthetical comments about their nouns or noun phrases. (62b) contains such an *appositive* relative clause.

Most relative clauses in actual discourse are restrictive. To tell whether a given clause is restrictive or appositive, we can ask whether the clause helps to specify a topic occurring in the sentence or provides additional information concerning an already determinate topic. English does offer two linguistic hints. First, *that* often signals that a clause is restrictive; *which,* with some exceptions (for example, the phrases *in which* and *with which*), often signals an appositive. Relative clauses often begin with other *wh*-words, however, or with no special word at all. In these cases, there are no signals. Furthermore, the use of *that* and *which* is not firmly established; these words are unreliable guides. Second, and more reliably, commas often do, and always can, set off appositive clauses from the rest of the sentence. Restrictives, by contrast, reject commas in this role. So virtually all relative clauses set off by commas are appositive. For those not set off by commas, we can use a simple test; try inserting commas. If the result sounds acceptable, the clause is probably appositive. Otherwise, it is restrictive.

Restrictives and appositives, in symbolic representations, both connect to the remainder of the formula by conjunction. Most of the time, therefore, it makes no difference to the translation whether a given clause is restrictive or appositive. When universal quantifiers are involved, however, and the clause modifies the subject noun phrase, it does matter. Consider these sentences:

> (64) a. All the Democratic candidates for president, who are already campaigning, support labor unions.
> b. All the Democratic candidates for president who are already campaigning support labor unions.

The only difference between them is the pair of commas setting off the relative clause in (64a). In that sentence, the clause is clearly appositive. It asserts that all the Democratic presidential candidates support labor unions, and remarks, on the side, as it were, that all those candidates are already campaigning. (64b), in contrast, does not claim that all the Democratic candidates support labor unions; it asserts only that all those who are already campaigning do so. (64b) is thus a weaker contention than (64a).

To translate these sentences, we first translate the relative clause. 'Who are already campaigning' derives from 'x is(are) already campaigning,' which we can write as Cx. Adopting the obvious representations (and letting Lx correspond to 'x supports labor unions'), then, we can symbolize the sentences in (64) as

> (65) a. $\forall x(Dx \rightarrow Lx)$ & $\forall x(Dx \rightarrow Cx)$ (or, equivalently,
> $\forall x(Dx \rightarrow (Lx \ \& \ Cx)))$
> b. $\forall x((Dx \ \& \ Cx) \rightarrow Lx)$

Note that the restrictive clause is conjoined, in effect, to the rest of the subject; the appositive clause, to the rest of the entire sentence.

Prepositional Phrases

Prepositions are rather ordinary English words such as *in, to, of, about, up, over, from,* and so on. They combine with noun phrases to form prepositional phrases, which act as either adjectives or adverbs: 'up a creek,' 'from Pennsylvania,' and 'in the middle of Three Chopt Road.' We'll discuss those acting as adverbs, which we can translate together with the verbs or adjectives they modify as single units, in a few pages. First, we'll talk about prepositional phrases modifying nouns, which have separate translations.

In prepositional phrases that function more or less as adjectives, prepositions relate two noun phrases. They thus translate into QL as dyadic predicates. The representatives of prepositional phrases themselves connect to the symbolizations of the noun phrases they modify by conjunction.

Consider these examples:

(66) a. Everyone from Pittsburgh loves the Steelers.
b. If I don't meet you, I'll be in some jail.

(66a) contains the prepositional phrase 'from Pittsburgh.' Since 'from' translates into a dyadic predicate, say F (and since 'the Steelers' here functions as the proper name of a team), (66a) becomes

(67) $\forall x((Px \ \& \ Fxb) \rightarrow Lxs)$

(66b) contains the prepositional phrase 'in some jail,' which itself contains a determiner. The conjunction of prepositional phrase to noun phrase, then, occurs within the scope of a quantifier:

(68) $\neg Hab \rightarrow \exists x(Jx \ \& \ Iax)$

(68), in which a and b symbolize 'I' and 'you,' respectively, symbolizes (66b). Prepositional phrases modifying nouns thus translate readily into quantification theory.

Verb Phrases

So far we've discussed how noun phrases and their modifiers translate into quantification theory. Since sentences consist of a subject noun phrase and a verb phrase, however, we also need to explain how to symbolize verb phrases and their modifiers in QL.

In any verb phrase, of course, there is a verb. Verbs fall into several categories, depending on their ability to take certain kinds of objects. Some verbs are *intransitive;* they cannot take objects at all. *Fall, walk,* and *die* are all intransitive. *Transitive* verbs take noun phrases as direct objects. Examples are *throw, win,* and *send.* Some of these, such as *give,* also take noun phrases as indirect objects. Other verbs take sentences, or grammatical constructions closely related to sentences, as objects. *Believe, know,* and *persuade* are such *clausally complemented* verbs. The logic of verbs taking sentential complements remains the subject of much debate. Here, therefore, we'll consider only transitive and intransitive verbs.

Note that many verbs fall into more than one category. *Eat,* for example, can have a noun phrase object (in, for example, 'We eat spaghetti every Wednesday night'), but does not need one ('Let's eat out'). *Believe* can take a sentence ('I believe that God exists') or a noun phrase ('I believed him').

Intransitive verbs translate into QL as monadic predicates. 'John walks,' for instance, becomes Wj; 'Everyone who doesn't own a car walks' becomes

(69) $\forall x((Px \ \& \ \neg \exists y(Cy \ \& \ Oxy)) \rightarrow Wx)$

Transitive verbs translate into QL as polyadic predicates. Usually, they become dyadic predicates; Lmf represents 'Mary loves Fred,' and so on. Occasionally, however, a verb relates more than two noun phrases. 'Mike gave John *War and Peace,*' for example, translates as *Gmja.* In general, predicates of more than two places prove very useful in symbolizing sentences with indirect objects or adverbial modifiers of certain kinds. In this context, however, the general issue of adverbs arises.

Adverbial Modifiers

Adverbs, such as *quickly, well, anytime,* and *somewhere,* modify verbs. They specify how, when, or where a certain condition holds or a certain activity occurs. Unfortunately, most adverbs have no direct symbolizations in quantificational logic. QL must represent them, together with the verbs they modify, as it does nonintersective adjectives; expressions such as 'walks slowly' or 'plays well' become predicates such as W or P. These adverbs are subsective: Anyone who is walking slowly is walking. To express this, we can write Wx & Sx, where W symbolizes 'walks' and S symbolizes 'walks slowly.' But beware of nonsubsective adverbs such as *allegedly:* That John allegedly stole the money does not entail that he stole the money.

Some adverbs, however, translate into QL differently. We'll call *always, anytime, whenever, wherever, anywhere, sometime,* and so on *adverbs of quantification.* Consider the sentence 'I like Alfred sometimes.' 'I like Alfred,' normally, would become *Lia.* So how do we represent '*Lia,* sometimes'? *Some* is a determiner. So, the sentence is saying, in effect, that, for some times x, 'I like Alfred' is true at x. Instead of *Lia,* then, we need *Liax,* meaning 'I like Alfred at x.' ' I like Alfred sometimes' thus becomes

(70) $\exists x(Tx$ & *Liax*)

Similarly, consider 'Everywhere I look there are timeshare resorts.' This, in essence, amounts to 'For every x, if x is a place and I look at x then there are timeshare resorts at x,' or, in symbolic notation

(71) $\forall x((Px$ & *Lix*) $\rightarrow \exists y(Ry$ & *Ayx*))

Some adverbs of quantification, however, have no correlates in QL. *Frequently,* which amounts roughly to 'at many times,' and *rarely* or *seldom,* which amount roughly to 'at few times,' could be translated into QL only if we had a way of symbolizing *many* and *few* in QL. Since quantification theory represents only a few determiners, it can represent only a few adverbs of quantification.

Prepositional phrases, as we've seen, can modify nouns. They can also modify verbs. 'John ran down the street' and 'We're singing in the rain' contain prepositional phrases functioning adverbially. Just as adverbs, in most cases, do not translate into QL except as parts of verb phrases that become predicates, so prepositions linking noun phrases to verbs or verb phrases translate together with the modified verb or verb phrase. They do not become dyadic predicates in their own right, as they do when modifying noun phrases. Nevertheless, because prepositional phrases contain noun phrases, their symbolic representations are more interesting than those of adverbs.

Think about a sentence such as 'Laura lives on East 72nd Street.' This becomes, when symbolized, *Lle*, where *e* represents 'East 72nd Street.' 'Lives on' translates as a single dyadic predicate. Note that we cannot apply the strategy appropriate to adjectival prepositional phrases; the above sentence is not equivalent to 'Laura lives and is on East 72nd Street.'

A more complex example is 'Richard has worked in every division of Reynolds Metals Company,' which contains an adverbial prepositional phrase that itself contains a determiner. Here, 'work' is intransitive. It would usually translate into a monadic predicate. But *in* is a preposition that, in effect, can combine with the verb; 'work in,' as a unit, is transitive. Instead of using the simple open sentence '*x* works,' therefore, we can use '*x* works in *y*,' a dyadic predicate, to obtain

(72) $\forall x(Dxc \rightarrow Wdx)$

Notice that in this sentence *of* does not add another place to the predicate; 'of Reynolds Metals Company' functions adjectivally, modifying 'division.'

Connectives

Quantification theory includes sentential logic. Sentential connectives can link quantified sentences; they can even inhabit noun and verb phrases. Recall that noun and verb phrases can be joined by *and, or,* and *if not*. In Chapter 2, we recommended a policy of splitting such phrases. Connectives linking noun or verb phrases usually can be transformed into connectives linking sentences. Thus, 'Abraham Lincoln and Calvin Coolidge were Republican presidents' amounts to 'Abraham Lincoln was a Republican president, and Calvin Coolidge was a Republican president.' Similarly, 'Fred likes hot dogs and hamburgers' amounts to 'Fred likes hot dogs, and Fred likes hamburgers.'

In quantification theory this advice becomes more important in many cases. 'All lions and tigers are cats' is equivalent to 'All lions are cats, and all tigers are cats.' But the conjoined noun phrase can tempt us into a translation

(73) $\forall x((Lx \& Tx) \rightarrow Cx)$

which says that everything that is both a lion and a tiger is a cat. But nothing is both a lion and a tiger. So, this symbolization is true whether lions are cats or not. Separating sentences results in the formula

(74) $\forall x(Lx \rightarrow Cx) \& \forall x(Tx \rightarrow Cx)$

which captures the meaning of the original. When existential quantifiers are involved, or when the connectives are in the verb phrase, splitting makes little difference. But, in subject noun phrases, it is vital.

As in the case of sentential logic, however, we must take care to split only those sentences for which the process preserves meaning. 'Harry loves pork and beans' may not be equivalent to 'Harry loves pork, and Harry loves beans'; he may love the combination without liking the individual components, or vice versa. 'Mary and Susan own the entire company' probably does not mean that Mary owns the entire company and that Susan does too, but that they own the entire company between them. The best we can do in QL is to treat 'pork and beans' and 'Mary and Susan' as units denoted by a constant. So, we could symbolize 'Harry loves pork and beans' as Lhb, where b represents 'pork and beans.'

Another problem pertains to connectives such as *if, only if,* and so on. They cannot join two noun phrases or two verb phrases, but they can appear in sentences in ways that cannot be reduced to simple sentential connection. Consider:

(75) A formula is contingent only if it's not valid.

We might think of this as a sentence with 'a formula' as subject noun phrase and 'is contingent only if it's not valid' as main verb phrase. The determiner *a* is clearly functioning generically, so it translates as a universal quantifier. The common noun *formula* appears as a monadic predicate. (75) thus looks like a complex version of a universal affirmative sentence form. Its symbolization begins with $\forall x(Fx \rightarrow$. The main verb phrase contains two connectives, *only if* and *not;* the adjectives *contingent* and *valid* appear as monadic predicates. *It* acts much like a variable. So the symbolization of (75) turns out to be

(76) $\forall x(Fx \rightarrow (Cx \rightarrow \neg Vx))$

Alternatively, we might think of (75) as containing a connective, *only if,* joining together two sentences, 'a formula is contingent' and 'it's not valid.' When we take this approach, (75) resembles (50), amounting to 'If a formula is contingent, it's not valid.' This we might be tempted to translate as

(77) $\exists x(Fx \ \& \ Cx) \rightarrow \neg Vx$

but, in (77), the final occurrence of x is not in the scope of the existential quantifier. So, (77) is not even a formula. Once again, the solution is to use a universal quantifier with the entire formula as its scope:

(78) $\forall x((Fx \ \& \ Cx) \rightarrow \neg Vx)$

This, too, is an acceptable translation of (75). Fortunately, (78) is equivalent to (76). So, the ways of construing the sentence's structure yield equivalent results.

Naturally, connectives can also join entire sentences that have no troublesome links between them: 'If we don't hang together, we'll surely all hang separately,' 'Some political parties die out after a short time, but others last for centuries,' and 'Unless everyone leaves, I'll refuse to come out' all work as we might expect from our knowledge of sentential logic.

Finally, quantification theory contains not only connectives but sentence letters. It might seem that we can translate any sentence into QL by using just predicates, constants, variables, and quantifiers. But a few, very simple sentences—'It is raining' and 'It's three o'clock,' for example—resist this analysis. *It,* in these sentences, does not stand for an object, so it would be very odd to translate 'It is raining' as, say, *Ri.* To see why, we can ask, "What is raining?" The question doesn't make very good sense. Even here, however, we have an alternative: 'It is raining' seems to assert that it's raining at a certain time and in a certain location; the time and location are determined by context. So perhaps, after all, we can render 'It is raining' as *Ri,* where *i* represents, not *it,* but a space-time location.

Problems

Translate the following sentences into QL, exposing as much structure as possible. If any translate only with difficulty, explain why.

1. Only the shallow know themselves (Oscar Wilde).

2. If any would not work, neither should he eat (II Thessalonians 3:10).

3. . . . and now nothing will be restrained from them, which they have imagined to do (Genesis 11:6).

4. Poets are the unacknowledged legislators of the world (Percy Bysshe Shelley).

▶ 5. . . . we are dust and dreams (A. E. Housman).

6. Everything that man esteems endures a moment or a day (William Butler Yeats).

7. To be beloved is all I need, and whom I love, I love indeed (Samuel Taylor Coleridge).

8. Hope is a delusion; no hand can grasp a wave or a shadow (Victor Hugo).

9. . . . the things which are seen are temporal; but the things which are not seen are eternal (II Corinthians 4:18).

▶ 10. Some people with great virtues are disagreeable while others with great vices are delightful (La Rochefoucauld).

11. Nothing which is true or beautiful or good makes complete sense in any immediate context of history (Reinhold Neibuhr).

12. They also live who swerve and vanish in the river (Archibald MacLeish).

13. Nothing is done. Everything in the world remains to be done or done over (Lincoln Steffens).

14. So then neither is he that planteth any thing, neither he that watereth; but God that giveth the increase (I Corinthians 3:7).

▶ 15. Loafing needs no explanation and is its own excuse (Christopher Morley).

16. Any mental activity is easy if it need not take reality into account (Marcel Proust).

17. . . . it is not poetry, if it make no appeal to our passions or our imagination (Samuel Taylor Coleridge).

18. When a man is wrong and won't admit it, he always gets angry (Thomas Haliburton).

19. My only books were women's looks, and folly's all they've taught me (Thomas Moore).

▶ 20. All things fall and are built again (William Butler Yeats).

21. All men have aimed at, found, and lost (William Butler Yeats).

22. Great is the hand that holds dominion over man by a scribbled name (Dylan Thomas).

23. He that stays in the valley shall never get over the hill (Jonn Ray).

24. To whom nothing is given, of him nothing can be required (Henry Fielding).

▶ 25. There has never been any thirty-hour week for men who had anything to do (Charles F. Kettering).

26. Work is a grand cure of all the maladies that ever beset mankind (Thomas Carlyle).

27. Every man without passions has within him no principle of action, no motive to act (Claude-Adrien Helvétius).

28. You can't have a better tomorrow if you are thinking about yesterday all the time (Charles F. Kettering).

29. Nothing will ever be attempted if all possible objections must be first overcome (Jules W. Lederer).

▶ 30. All man's friend, no man's friend (John Wodroephe).

31. Nobody ever did anything very foolish except from some strong principle (William Lamb).

32. We receive only what we give (Samuel Taylor Coleridge).

33. If you build a castle in the air, you won't need a mortgage (Philip Lazarus).

34. Nothing is more boring than a man with a career (Aleksandr Solzhenitsyn).

▶ 35. . . . the Bears were good Bears, who did nobody any harm, and never suspected that anyone would harm them (Robert Southey). (Use a predicate *B* for 'is one of the Bears.')

QUANTIFIED
TABLEAUX

The method of semantic tableaux extends easily to incorporate quantifiers. As in sentential logic, semantic tableaux provide a very powerful and efficient means of determining whether argument forms are valid. Nevertheless, tableaux fall short of being a decision procedure for validity and related semantic concepts in quantification theory. Within the bounds of sentential logic, tableaux constitute a decision procedure for determining validity. They remain such a technique in quantificational logic if all the predicates in the language are monadic. When we admit polyadic predicates, however, this decision procedure breaks down. The method of semantic tableaux is still mechanical, but some tableaux don't terminate after a finite time. As we'll see near the end of this chapter, applying the tableau rules sometimes results in infinitely long tableaux.

This fact might tempt us to search for another method that would constitute a decision procedure for quantificational logic. In 1936, however, the American logician Alonzo Church proved that quantification theory is *undecidable*, that is, that there is no decision procedure for quantificational validity.[1] Any method for demonstrating validity, or satisfiability, or contradictoriness, and so on within this logic must at some point yield infinite procedures or require the use of nonmechanical insights. Quantification theory is thus so powerful that no effective procedure can capture it.

6.1 QUANTIFIER TABLEAU RULES

Semantic tableaux for quantificational logic use all the sentential tableau rules, together with four new rules for the quantifiers. Each quantifier has two associated rules: One deals with its occurrences on the left side of tableau branches, while the other deals with its occurrences on the right side.

The quantifier rules all rely on the notion of an instance. Remember that the left side of a tableau represents truth, while the right represents falsity. "Left" rules thus tell us what we can infer from the supposition that a quantified formula is true; "right" rules tell us what we can infer from the assumption that such a formula is false.

Existential Left (∃L)

$$\checkmark \quad \exists v \mathscr{A}$$
$$\mathscr{A}[c/v]$$

Here c must be a constant new to the tableau.

This rule asserts that an existential formula on the left may be replaced by one of its instances, provided that the constant that substitutes for the variable is new to the tableau. To see why the rule has this form, assume that an existential sentence—say, 'Someone in this room is a spy'—is true.[2] This assumption, by itself, tells us nothing about who the spy is. Nevertheless, it's important to introduce some way of referring to the spy, so that we can record additional information about the same person. In English, we would generally just use 'the spy' in this role, as we've in fact been doing in the last two sentences. In logic, we achieve the same effect by introducing a name for the spy—as we might also in English by saying, "call the spy *Karla*"—in the guise of a constant that hasn't appeared anywhere on the tableau before. We must use a new constant precisely because we can't say who the spy is. It's hardly fair to say to Fred, "Someone in this room is a spy. Let's call him *Fred.*" Similarly, it would be outrageous for a mathematician to say, "So, there's a point on the interval at which the derivative is zero. Let's call this point '9.3.'" In the absence of any information about which objects make the existential sentence true, we need a new constant to avoid making any illicit identifications.

For example, suppose that we want to learn whether the formula $\exists x(Fx \,\&\, \neg Fx)$—'Something is both F and not F'—is satisfiable or contradictory. As in sentential logic, we place the formula on the left of a tableau, assuming that it is true, and see whether a contradiction results.

$$
\begin{array}{c|c}
\checkmark \quad \exists x(Fx \,\&\, \neg Fx) & \\
\checkmark \quad Fa \,\&\, \neg Fa & \\
Fa & \\
\checkmark \quad \neg Fa & \\
& Fa \\
\hline
& \text{Cl}
\end{array}
$$

The tableau closes, so the formula is contradictory; nothing can be both F and not F at the same time.

Existential Right (∃R)

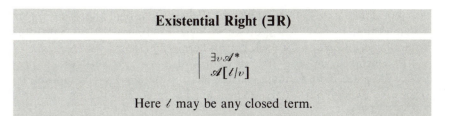

$$\exists v \mathscr{A}^*$$
$$\mathscr{A}[t/v]$$

Here t may be any closed term.

This rule asserts that we may dispatch an existential formula on the right by writing, also on the right, any instance of that formula. Later, we'll have closed terms of various kinds; now, however, constants are our only closed terms. The constant we use in ∃R need not be new, though it may be; any constant whatever will do. Furthermore, we may apply this rule more than once. The asterisk is a *temporary dispatch mark* indicating that, though we have already applied ∃R to this formula, we may come back and apply it again. To see why this rule is repeatable, suppose that we know it's false that someone in this room is a spy. Then we can infer that it's false that Al is a spy, that Beth is a spy, that Carl is a spy, that Dorothy is a spy, and so on for each person in the room.

For example, suppose that we want to find out whether the formula $\exists x(Fx \vee \neg Fx)$ is valid. As usual, we place it on the right side of a tableau, assuming that it can be false.

$$
\begin{array}{c|l}
 & \exists x(Fx \vee \neg Fx)^* \\
 & Fa \vee \neg Fa \quad \checkmark \\
 & Fa \\
 & \neg Fa \quad \checkmark \\
Fa & \\
\hline
 & \text{Cl}
\end{array}
$$

Since the tableau closes, the formula is valid. Notice that the instance here contained a new constant. It had to, not because of the rule, but because no constants had yet appeared on the branch.

Universal Left (∀L)

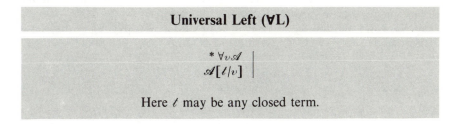

$$^*\forall v \mathscr{A}$$
$$\mathscr{A}[t/v]$$

Here t may be any closed term.

This rule asserts that a universal formula on the left allows us to write any instance of it on the left. Again, the constant we use doesn't have to be new. In fact, any instance of the formula will do. Furthermore, this rule is also repeatable. If we know that it's true that everyone here knows logic, then it follows that I know logic, you know logic, Fred knows logic, Samantha knows logic, and so on for each person here.

For instance, suppose that we want to discover whether $\forall x Fx$ implies $\exists x Fx$. We place these formulas on the left and right sides of a tableau, respectively, and find out whether there is any interpretation making $\forall x Fx$ true but $\exists x Fx$ false.

$$
\begin{array}{c|c}
* \forall x Fx & \\
& \exists x Fx* \\
& Fa \\
Fa & \\
\hline
& \text{Cl}
\end{array}
$$

As this tableau shows, there is no such interpretation, so $\forall x Fx$ does indeed imply $\exists x Fx$. Notice that neither quantifier rule used in this tableau required the use of new constants. There was no problem, then, about using a single constant for both instances, no matter in what order we applied the rules.

Universal Right (∀R)

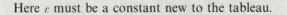

$$
\begin{array}{c|}
\forall v \mathscr{A} \quad \checkmark \\
\mathscr{A}[c/v]
\end{array}
$$

Here c must be a constant new to the tableau.

This rule asserts that, given a universal formula on the right, we can write an instance of it on the right, using a constant that hasn't appeared before on the tableau. Suppose that it's false that everyone in the department has a PhD. It follows that at least one person in the department doesn't have a PhD. We can't infer anything about who these persons are, but we want to reason about them, so we need some way of referring to them. To correspond roughly to the English "the person or persons without a PhD," we introduce a new constant. If the constant were not new, we would be illicitly assuming something about the identity of those without PhDs.

For example, suppose that we want to know whether $\forall x(Fx \rightarrow Fx)$ is valid.

$$
\begin{array}{c|c}
& \forall x(Fx \rightarrow Fx) \quad \checkmark \\
& Fa \rightarrow Fa \qquad \checkmark \\
Fa & \\
& Fa \\
\hline
& \text{Cl}
\end{array}
$$

This tableau tells us that it is. Our search for an interpretation making the formula false results in contradiction.

6.2 STRATEGIES

In sentential logic, any way of applying tableau rules produces the same result. Nevertheless, some ways are more efficient in reaching that result than others. We thus adopted two strategies for simplifying tableaux. The first was: *Close branches as soon as possible.* Once some formula appears on both sides of a tableau branch, applying further rules to the branch can make no difference; the branch—or branches, if further applications split the original—will still have a formula that appears on both sides and so will close. The second strategy was: *Avoid splitting tableaux as long as possible.* Once a branch splits, applying rules to the formulas above the split forces us to write the result on each of the resulting branches.

We will continue to use these strategies in quantificational logic. Furthermore, two other strategies help to simplify tableaux. Here is the third strategy:

> Introduce the constants we need as quickly as possible by applying ∃L and ∀R before applying ∃R and ∀L. That is, apply the quantifier rules introducing new constants as soon as possible.

Observing this principle allows us to minimize the number of constants in a tableau.

Let's consider an example. Suppose that we want to evaluate the argument form

$\exists x(Fx \ \& \ Gx)$
$\forall x(Gx \rightarrow Hx)$
$\therefore \ \exists x(Fx \ \& \ Hx)$

This is a form of arguments such as 'Some rulers are dictators; all dictators have absolute power; so, some rulers have absolute power.' Following the third strategy results in the first tableau, while introducing the new constant after applying the other quantifier rules results in the longer, second tableau.

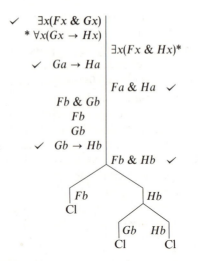

The fourth strategy is:

Use constants already on the tableau whenever possible.

Two quantifier rules, ∃R and ∀L, don't require a new constant. They accept any constant, whether or not it's already on the tableau. Consequently, when applying these rules, we use any constants that are already available. We introduce new ones only if no constants are available on the tableau.

To see why, again consider an example. The argument 'Some conservative Southerners are Democrats; therefore some Democrats are conservative' has this form:

$$\exists x((Fx \,\&\, Gx) \,\&\, Hx)$$
$$\therefore \exists x(Hx \,\&\, Fx)$$

Following the fourth strategy produces the first tableau, while ignoring it results in the second tableau.

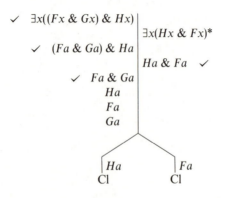

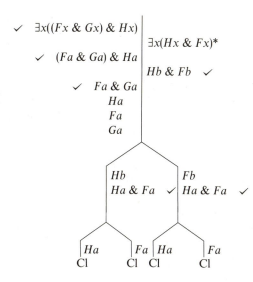

Notice that the second tableau closes only because we returned to instantiate the existential formula on the right a second time, using the original constant.

This example points out an important fact: Some tableaux close only if we use the right constants in applying quantifier rules. The last tableau we saw, for instance, would not have closed if we had given up too early. Consequently, it's important to:

> Continue instantiating as far as possible using the constants on the branch.

∃R and ∀L don't require new constants and can be repeated. Indeed, they must be repeated, using the constants available on the tableau, until (a) the tableau closes, (b) the available constants are exhausted, or (c) it seems clear that the tableau isn't going to close. The last case is possible precisely because tableaux do not constitute a decision procedure for quantificational logic. If applying the rules for tableau construction results in an infinite tableau in a particular case, then the tableau will never close, but the opportunities for further instantiation will never be exhausted. Sometimes it will become obvious that a repeating pattern has developed, but there is no general test for this.

Let's first look at a tableau that closes only through the return to a previously instantiated formula. In the last chapter, we saw that there is a great difference between the quantifier strings ∀x∃y and ∃y∀x. The former says that each thing stands in some relation to something or other, allowing the possibility that *a* stands in relation to *b*, *c* to *d*, and so on. The latter, in contrast, says that there is a single object to which everything stands in some relation. It should be clear that, if this is true, then the former is true as well. That is, if everything relates to some one thing, then everything relates to something or other. Thus, if God created everything, then everything was created (by something or other). So, switching the order of quantifiers in a string by moving universals to the left preserves truth. Normally, however, the reverse is not true; switching existentials to the left results in a stronger assertion that doesn't follow from the original. In a moment we'll consider a fallacious argument of this form that some people think St. Thomas Aquinas advocated: The inference from 'Everything has a cause' to 'Something is the cause of everything.'

For now, however, we'll examine an unusual case. Sometimes moving an existential quantifier to the left does preserve truth. Consider the argument form

$\forall x \exists y (Fx \ \& \ Gy)$
$\therefore \ \exists y \forall x (Fx \ \& \ Gy)$

Switching the quantifiers results in a valid argument form here only because the variables *x* and *y* have no relation in the formula; no atomic portion of it contains both variables. In fact, both premise and conclusion are equivalent to the much clearer formula

$\forall x Fx \ \& \ \exists y Gy$

We could begin a tableau to evaluate the above argument form in the following way, observing our strategy rules:

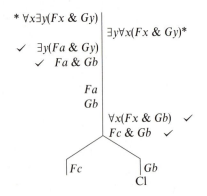

At this point, the fainthearted might be tempted to stop and declare the argument form invalid. But there is still a universal formula on the left that has been dispatched only temporarily; we have taken an instance using the

constant *a*, but not using *b* or *c*. Similarly, we've instantiated the existential formula on the right using *b*, without trying *a* or *c*. It seemed wise for us to use *b* here, because doing so gave us a portion of the formula, *Gb*, that matched a formula on the other side of the branch. Thinking about possible matches may lead us to look for something that might match the newly introduced *Fc*. The best hope for doing so seems to reside in instantiating the universal formula on the left, again using *c*. This instantiation allows us to close the tableau:

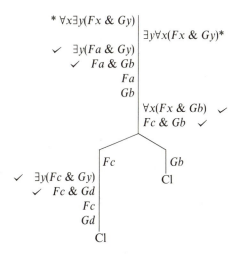

To see an example of the second possibility, let's consider a case where we exhaust the opportunities for instantiation without closing the tableau. A relation is *reflexive* if every object stands in relation to itself. A relation is *symmetric* if, whenever it holds between objects in one direction, it holds between them in the other direction. A relation is *transitive* if, whenever it holds between *x* and *y* and between *y* and *z*, it also holds between *x* and *z*. This table summarizes these definitions and provides a few examples, using relations between numbers.

R is	Iff	Examples
Reflexive	$\forall x R x x$	$=, \leq, \geq$
Symmetric	$\forall x \forall y (R x y \rightarrow R y x)$	$=$
Transitive	$\forall x \forall y \forall z ((R x y \ \& \ R y z) \rightarrow R x z)$	$=, <, \leq, >, \geq$

Suppose we want to find out whether all symmetric and transitive relations are reflexive. We can assume that *R* is symmetric and transitive and construct a tableau to determine whether that fact implies that *R* is reflexive.

Following our strategy principles, we can produce the tableau

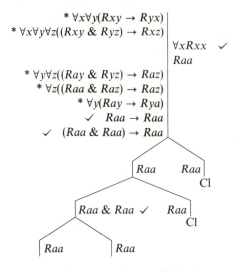

Notice that we have exhausted the available instances. Our strategy principles tell us that it's pointless to instantiate with constants not already available on the branch if we don't have to. So introducing a constant, *b*, would do us no good here. Once we finish instantiating with *a* throughout, we are finished. The argument form is invalid.

Incidentally, this tableau points out that instantiating a string of universal quantifiers on the left, or a string of existential quantifiers on the right, is a dull matter. We can substitute any constant for any of the variables. Our rules instruct us to take instances, dropping one quantifier at a time. It's far more convenient, however, to drop several quantifiers and make several substitutions at once. We'll allow ourselves, therefore, to apply the rules ∀L and ∃R in tandem, to move from formulas to instances of their instances, or instances of instances of their instances, and so on. So, supposing that we have the formula $\forall x \forall y (Rxy \rightarrow Ryx)$ on the left, we can move immediately to the formulas $Raa \rightarrow Raa$, $Rab \rightarrow Rba$, $Rba \rightarrow Rab$, $Rbb \rightarrow Rbb$, and so on.

Similarly, it can be convenient to treat several consecutive existential quantifiers on the left, or universal quantifiers on the right, at the same time. Each application of these rules, however, requires a new constant. We can take the quantifiers together as a shortcut, then, only if we use a new constant for each quantifier. Thus, assuming that *a* is the only constant appearing earlier in the tableau, we can move from the formula $\exists x \exists y \exists z (Fxy \ \& \ Fxz)$ to *Fbc* & *Fbd*.

Finally, some tableaux never close, but neither do they allow us to deplete their stores of available constants. These infinite, nonterminating tableaux are the source of the undecidability of quantificational logic. They also present a serious practical problem. There is no way of telling when a particular tableau construction will be nonterminating. In many simple cases, nonetheless, it's easy to tell that the tableau will never close; it traps us in a loop that continually forces us to introduce new constants.

Consider the argument mentioned earlier, which Aquinas allegedly advanced, from 'Everything has a cause' to 'Something is the cause of everything.' We can symbolize the argument as

$\forall x \exists y C y x$
$\therefore \exists y \forall x C y x$

Constructing a tableau for this argument quickly leads us into a loop in which we introduce new constants, take new instances, introduce new constants, and so on in a never-ending cycle.

$$
\begin{array}{c|c}
* \; \forall x \exists y C y x & \\
& \exists y \forall x C y x * \\
\checkmark \;\; \exists y C y a & \\
C b a & \\
& \forall x C b x \quad \checkmark \\
& C b c \\
\checkmark \;\; \exists y C y c & \\
C d c & \\
& \forall x C d x \quad \checkmark \\
& C d e \\
& \vdots
\end{array}
$$

Problems

Evaluate these arguments for validity by using tableaux.

1. All computers are logical. Everything that is illogical is irrational. Therefore, some irrational things are not computers.

2. Something is rotten in the state of Denmark. Something in my refrigerator is rotten. Thus, something in my refrigerator is in the state of Denmark.

3. All flying horses are quick and clever; all flying horses live forever. And sad but true, all horses die. It follows that no horses fly.

4. Bani-Sadr hates everyone who hates the Ayatollah. The Ayatollah hates everyone. So, the Ayatollah and Bani-Sadr hate each other.

▶ 5. Any moral person respects the dignity of everyone. Julie is a moral person. Thus, some who are moral respect Julie's dignity.

6. Some elected politicians who fail to consider the wishes of their constituents soon find themselves voted out of office. All politicians who win election by very large margins fail to consider the wishes of their constituents. Thus, some politicians who win election by very large margins soon find themselves voted out of office.

7. Everyone who has a parent born with the taint of original sin is also tainted by original sin. Mary was Jesus' parent. Mary's parents were born with the taint of original sin. Thus, Jesus was tainted by original sin.

8. I like everyone who likes everyone I like. Thus, I don't dislike everyone.

9. All who are Christians love all their neighbors if they love even themselves. In a profound sense, we are all neighbors. Every Christian loves himself or herself. So, all Christians love each other.

▶10. The Longhorns can beat anybody who can beat everybody the Longhorns can. Therefore the Longhorns can beat themselves.

Evaluate these sentences as valid, contingent, or contradictory.

11. There exists an object that is either physical or nonphysical.

12. There is a barber who shaves all and only those who do not shave themselves.

13. There are people who love each other only if they both love some one person.

14. Nobody's seen all the trouble I've seen.

▶15. There is something such that, if it's an object of art, anything is.

16. God created everything, but nothing created God.

17. Somebody respects everyone unless he or she respects only somebody who has self-respect.

18. If, for any gun, there is a faster gun, then no gun is faster than itself.

19. A gun is faster than every gun only if it's faster than itself.

20. There's a gun that's faster than every gun only if some gun is faster than itself.

The philosopher Gottfried Leibniz defined a good person as one who loves everyone (as much as reason allows). If we accept this definition

$$\forall x(Gx \leftrightarrow \forall yLxy)$$

then which of these statements follow?

21. All good people love somebody or other.

22. All good people love themselves.

23. Someone is loved by all good people.

24. Somebody loves all good people.

▶25. Everyone good is loved by somebody or other.

26. All good people love all good people.

27. Everybody is loved by somebody or other.

28. If there are any good people, then everybody is loved by somebody.

The natives on the island of Amok practice incest extensively. In fact, incest is so widespread there that, if an Amok native is a parent of natives x and y, then there is a native with both x and y as parents. An anthropologist contends that if an Amok native is a parent of natives x and y, then either x is a parent of y or y is a parent of x.

29. Does the anthropologist's thesis follow from the actual fact about Amok?*

▶ **30.** Does the fact about Amok follow from the anthropologist's thesis?*

Use tableaux to evaluate the following argument forms for validity.

31. $\exists x(Fx \;\&\; \neg Gx)$
$\forall x(Fx \rightarrow Hx)$
$\therefore \exists x(Hx \;\&\; \neg Gx)$

32. $\exists x(Fx \;\&\; Gx)$
$\forall x(Gx \rightarrow \neg Hx)$
$\therefore \exists x \neg Hx$

33. $\forall x(Fx \rightarrow \neg Gx)$
$\forall x(Gx \rightarrow \neg Hx)$
$\therefore \forall x(Fx \rightarrow Hx)$

34. $\forall x(Fx \rightarrow \exists y(Gy \;\&\; Hy))$
$\therefore \forall x \neg Gx \rightarrow \neg \exists z Fz$

▶ **35.** $\exists x(Gx \;\&\; \forall y Fxy)$
$\forall x \forall y(Fxy \rightarrow Fyx)$
$\therefore \exists x(Gx \;\&\; \forall y Fyx)$

36. $\exists x \forall y(Fxy \rightarrow Gyx)$
$\therefore \exists x \exists y(Fxy \;\&\; Fyx)$

37. $\forall x(Fx \rightarrow \forall y(Gy \rightarrow Hxy))$
$\exists x(Gx \;\&\; \forall z \neg Hxz)$
$\therefore \neg \forall x Fx$

38. $\forall x \forall y(Gxy \rightarrow Fxy)$
$\therefore \forall x(Fxx \rightarrow \exists y(Fxy \;\&\; Gyx))$

39. $\forall x \forall y(Fxy \rightarrow (Fyx \rightarrow \exists z Hxyz))$
$\forall x Fax$
$\therefore \exists x \exists y Hyyx$

40. $\exists x(Fxb \;\&\; \forall z(Gz \rightarrow Fzx))$
$\forall x \forall y(Hxy \leftrightarrow ((Gx \;\&\; Gy) \;\&\; Fxy))$
$\therefore \exists x Hxx \rightarrow \exists y \forall z \forall w((Hwz \vee Hzw) \rightarrow (Fzy \;\&\; Fyb))$*

6.3 INTERPRETATIONS

As in sentential logic, an open tableau not only *indicates* that there is an interpretation of the formulas at the top of the tableau making those on the left true and those on the right false, but also *specifies* such an interpretation. Before discussing how tableaux specify interpretations, however, we need to develop a precise notion of what an interpretation in quantificational logic is. Recall that in sentential logic an interpretation is simply an assignment of truth values to atomic formulas, i.e., to sentence letters. Quantificational logic includes sentential logic, so interpretations within it will incorporate such truth value assignments. But quantificational interpretations are more complex.

> **DEFINITION.** An *interpretation, M,* of a set, *S,* of formulas of quantificational logic consists of a nonempty set *D* (*M*'s *domain,* or *universe of discourse*) and a function φ assigning (a) truth values to sentence letters in *S,* (b) elements of *D* to constants in *S,* and (c) sets of *n*-tuples of elements of *D* to *n*-ary predicates in *S.*

An interpretation (sometimes called a *structure* or *model*) thus has two components. The first is a set that specifies what objects the formulas in question are talking about. The quantifiers *range over* this set, in the sense that we construe 'for all x' and 'for some y' as meaning 'for all x in D' and 'for some y in D' or, in other words, 'for all elements of D' and 'for some element of D.'

The second component of an interpretation, *M,* is an *interpretation function.* This function, in effect, assigns meaning to the constants, predicates, and sentence letters in the formulas we're interpreting. It assigns truth values to sentence letters, telling us whether the sentences they represent are true or false. It assigns elements of the domain to constants, telling us which objects they stand for. Finally, the function assigns sets of *n*-tuples of objects to *n*-ary predicates. Consider a unary or monadic predicate, R, which informally means 'red.' The interpretation function assigns R a set of 1-tuples. It tells us which objects satisfy R; it tells us, in other words, which objects are red. The function assigns to a binary or dyadic predicate such as L, meaning 'loves,' a set of ordered pairs. The function tells us, then, who loves who. If $\varphi(L) = \{\langle \text{Bob, Carol}\rangle, \langle \text{Carol, Ted}\rangle, \langle \text{Ted, Alice}\rangle, \langle \text{Alice, Bob}\rangle\}$, then Bob loves Carol, Carol loves Ted, Ted loves Alice, and Alice loves Bob.

This table summarizes how the interpretation function works.

Symbol	Interpretation
Sentence letter	Truth value
Constant	Object in the domain
n-ary predicate	Set of *n*-tuples of objects in the domain

The definition of an interpretation performs only part of the task that we need to accomplish. We want to produce interpretations that make various formulas true and others false. To do so, we need to know how to evaluate the truth value of a formula on an interpretation.

It's easy to judge the truth value of a sentence letter on an interpretation; we just see what value the interpretation function assigns to it. The interpretation, M, consisting of D and φ makes a sentence letter p true just in case φ assigns truth to p:

> p is true on M iff $\varphi(p) = T$.

Other atomic formulas, consisting of an n-ary predicate followed by n constants, are also easy to evaluate. The sentence 'Bob loves Carol' is true if and only if Bob loves Carol. The sentence is true, in other words, just in case the interpretation we assign to the predicate 'loves' includes the pair $\langle$Bob, Carol$\rangle$. In general, then, an atomic formula of the form $\mathscr{R}a_1 \ldots a_n$ is true on an interpretation, M, just in case the set the interpretation function assigns to $\mathscr{R}$ includes the n-tuple consisting of the objects that $a_1, \ldots, a_n$ stand for:

> $\mathscr{R}a_1 \ldots a_n$ is true on M iff $\langle \varphi(a_1), \ldots, \varphi(a_n) \rangle$ belongs to $\varphi(\mathscr{R})$.

Assessing the truth values of formulas with sentential connectives as main connectives is also easy, provided that we know the truth values of the components. We can proceed exactly as in sentential logic:

> $\neg \mathscr{A}$ is true on M iff $\mathscr{A}$ is false on M.
> $(\mathscr{A}\ \&\ \mathscr{B})$ is true on M iff $\mathscr{A}$ and $\mathscr{B}$ are both true on M.
> $(\mathscr{A} \vee \mathscr{B})$ is true on M iff either $\mathscr{A}$ or $\mathscr{B}$ is true on M.
> $(\mathscr{A} \rightarrow \mathscr{B})$ is true on M iff $\mathscr{A}$ is false on M or $\mathscr{B}$ is true on M.
> $(\mathscr{A} \leftrightarrow \mathscr{B})$ is true on M iff $\mathscr{A}$ and $\mathscr{B}$ have the same truth value on M.

The real task is defining the truth value of quantified formulas. Consider an existentially quantified formula, say, $\exists x Fx$. This formula will be true if the predicate F is true of some object in the domain. We can think of ourselves as considering an instance of the formula, such as Fa, and asking whether a could name something that would make Fa true. Similarly, consider a universally quantified formula, say, $\forall x(Fx \rightarrow Gx)$. We can think of an instance of this formula, such as $Fa \rightarrow Ga$; this instance should be true, no matter what object in the domain a stands for, if the original formula is true. So, to judge the truth value of a quantified formula on an interpretation M, we can look at the truth values of an instance of that formula on interpretations which are just like M, except that they may assign different objects from the domain to the constant substituted for the quantified variable.

∃*v𝒜* is true on *M* iff 𝒜[*c*/*x*] is true for some constant *c* on an intepretation *M'* differing from *M* at most in assigning a different element of the domain to *c*.
∀*v𝒜* is true on *M* iff 𝒜[*c*/*x*] is true for every constant *c* on every interpretation *M'* differing from *M* at most in assigning a different element of the domain to *c*.

Another way of thinking about the truth values of quantified formulas is this: A universally quantified formula ∀*v𝒜* is true just in case 𝒜 would be true no matter what object in the universe of discourse *v* would stand for. An existentially quantified formula ∃*v𝒜* would similarly be true if there's an object in the domain such that, if *v* stood for it, 𝒜 would be true.

To see how this works in practice, let's consider the formulas ∀*xFnx* and ∀*x*∃*y*(*Fxy* & *Gyx*). Are these formulas true or false on the following interpretation?

D = {New York, Los Angeles, Chicago}
φ(*n*) = New York
φ(*F*) = {⟨New York, Los Angeles⟩, ⟨New York, Chicago⟩,
　⟨Los Angeles, Chicago⟩}
φ(*G*) = {⟨Los Angeles, Chicago⟩, ⟨Los Angeles, New York⟩,
　⟨Chicago, New York⟩}

The domain here consists of the three largest U.S. cities. We can think of *F* as meaning 'has a larger population than' and *G* as meaning 'is west of.' Informally, then, the first formula we're analyzing means that New York is more populous than each city in the set. Is this true? The instance *Fna* of this formula should be true no matter which city *a* stands for. So, we should find, in the interpretation of *F*, all possible pairs of the form ⟨New York, ——⟩. Do we? Included in *φ*(*F*) are ⟨New York, Chicago⟩ and ⟨New York, Los Angeles⟩, but not ⟨New York, New York⟩. The formula is therefore false; New York is not larger than itself.

The second formula says that, for each of the three largest U.S. cities, there is another that is both smaller than and west of it. This should obviously be false; no city in the set is west of Los Angeles. If we analyze the formula according to our definitions, we can see that the instance ∃*y*(*Fay* & *Gya*) should be true no matter which city *a* stands for. So, let's suppose *a* stands for Los Angeles. We should be able to find another city which, as a referent for *b*, would make *Fab* & *Gba* true. Can we? First, let's contemplate the *Fab* part. Does the interpretation of *F* contain a pair of the form ⟨Los Angeles, ——⟩? Yes: *φ*(*F*) contains the pair ⟨Los Angeles, Chicago⟩. Furthermore, this is the only pair of that form in *φ*(*F*). We now move to the *Gba* part: Does *φ*(*G*) contain the inverse pair ⟨Chicago, Los Angeles⟩? The answer is no. So, we can't find another city that would stand in the proper relations to

Los Angeles. But the universal formula would be true only if we could do this for every city in the universe of discourse, so the formula as a whole is false on *M*.

Notice, incidentally, that we couldn't find a city with the proper relations to Chicago either, because no city in the set is smaller than Chicago; $\varphi(F)$ contains no pair of the form ⟨Chicago, ——⟩. We could, however, find a city with the proper relations to New York. We find both ⟨New York, Chicago⟩ and ⟨New York, Los Angeles⟩ in $\varphi(F)$, and the inverses of both pairs, ⟨Chicago, New York⟩ and ⟨Los Angeles, New York⟩, occupy $\varphi(G)$. This reflects the fact Chicago and Los Angeles are both smaller than and west of New York. It also indicates that, if the initial quantifier of the formula were existential rather than universal, the formula would be true on *M*.

Problems

Each of the following sets of exercises specifies an interpretation and then lists ten or more formulas. What is the truth value of these formulas on the specified interpretation?

Let $D = \{a\}$, $\varphi(a) = a$, and $\varphi(F) = \{⟨a⟩\}$.

1. Fa	**2.** $\neg Fa$	**3.** $Fa \,\&\, Fa$
4. $Fa \vee \neg Fa$	▶ **5.** $Fa \to Fa$	**6.** $\neg Fa \to Fa$
7. $\exists x Fx$	**8.** $\forall x Fx$	**9.** $\forall x(Fx \to \exists y Fy)$

▶ **10.** $\forall x(\neg Fx \to \neg \exists y Fy)$

Let $D = \{a\}$, $\varphi(a) = a$, $\varphi(F) = \emptyset$, and $\varphi(G) = \{⟨a⟩\}$.

11. Fa	**12.** Ga	**13.** $Fa \to Ga$
14. $Ga \to Fa$	▶ **15.** $\exists x Fx$	**16.** $\forall x Fx$
17. $\exists x Gx$	**18.** $\forall x Gx$	**19.** $\forall x(Fx \to Gx)$

▶ **20.** $\forall x((Fx \vee Gx) \to Gx)$

Let $D = \{a\}$, $\varphi(a) = a$, $\varphi(F) = \{⟨a⟩\}$, and $\varphi(R) = \{⟨a, a⟩\}$.

21. Raa	**22.** $\forall x(Fx \to Rxx)$	**23.** $\forall x Rxx$
24. $\forall x Rax$	▶ **25.** $\forall x(Rax \,\&\, Rxa)$	
26. $\forall x \forall y(Rxy \to Ryx)$	**27.** $\forall x(Fx \to \exists y(Fy \,\&\, Rxy))$	
28. $\forall x \forall y \forall z((Rxy \,\&\, Ryz) \to Rxz)$	**29.** $\forall x \forall y(Rxy \to \exists z(Rxz \,\&\, Rzy))$	

▶ **30.** $\exists x(Fx \,\&\, Rxx) \leftrightarrow \forall x(Fx \,\&\, Rxx)$

Let $D = \{a, b\}$, $\varphi(a) = a$, $\varphi(b) = b$, and $\varphi(F) = \{⟨b⟩\}$.

31. Fa	**32.** Fb	**33.** $\exists x Fx$

34. $\forall xFx$ ▶ **35.** $\exists x \neg Fx$ **36.** $\forall x \neg Fx$

37. $\exists xFx \,\&\, \exists x \neg Fx$ **38.** $\exists x(Fx \,\&\, \neg Fx)$ **39.** $\exists x(Fx \rightarrow \forall yFy)$

▶ **40.** $\forall x(Fx \rightarrow \forall yFy)$

Let $D = \{a, b\}$, $\varphi(a) = a$, $\varphi(b) = b$, and $\varphi(R) = \{\langle a, a \rangle, \langle b, a \rangle\}$.

41. Rab **42.** Rba **43.** Rbb **44.** $\exists xRax$

▶ **45.** $\exists xRbx$ **46.** $\exists xRxa$ **47.** $\exists xRxb$ **48.** $\forall xRxx$

49. $\forall x\forall y(Rxy \rightarrow Ryx)$ ▶ **50.** $\forall x\exists yRxy$

Let $D = \{a, b\}$, $\varphi(a) = a$, $\varphi(b) = b$, $\varphi(F) = \{\langle b \rangle\}$, and $\varphi(R) = \{\langle b, b \rangle, \langle b, a \rangle\}$.

51. $\exists x(Fx \,\&\, Rxx)$ **52.** $\exists x(Fx \,\&\, Rxa)$

53. $\forall x(Fx \rightarrow Rxx)$ **54.** $\forall x(Fx \rightarrow Rax)$

▶ **55.** $\forall x\forall y(Rxy \rightarrow Ryx)$ **56.** $\forall x\forall y(Rxy \rightarrow Rxx)$

57. $\forall x\forall y(Rxy \rightarrow Ryy)$ **58.** $\forall x\forall y\forall z((Rxy \,\&\, Ryz) \rightarrow Rxz)$

59. $\forall x\forall y\forall z((Rxy \,\&\, Ryz) \rightarrow \neg Rxz)$

▶ **60.** $\forall x\forall y((Fx \,\&\, Fy) \rightarrow (Rxy \,\&\, Ryx))$

Let $D = \{a, b\}$, $\varphi(a) = a$, $\varphi(b) = b$, $\varphi(R) = \{\langle a, a \rangle, \langle b, a \rangle\}$, $\varphi(S) = \{\langle b, b \rangle, \langle a, b \rangle\}$, and $\varphi(F) = \{\langle a \rangle\}$.

61. $\forall x(Fx \rightarrow Rxx)$ **62.** $\forall x\exists yRxy$

63. $\forall x\exists ySxy$ **64.** $\exists x\forall yRxy$

▶ **65.** $\exists x\forall ySxy$ **66.** $\exists x\forall ySyx$

67. $\exists x\forall yRyx$ **68.** $\forall x\forall y(Rxy \lor Sxy)$

69. $\forall x\forall y(Rxy \rightarrow \neg Sxy)$ ▶ **70.** $\forall x\forall y(Rxy \rightarrow (Fx \lor Fy))$

Let $D = \{a, b\}$, $\varphi(a) = a$, $\varphi(b) = b$, $\varphi(S) = \emptyset$ and $\varphi(R) = \{\langle a, a \rangle, \langle b, b \rangle\}$.

71. $\exists xRxx$ **72.** $\exists xSxx$

73. $\forall xRxa$ **74.** $\exists xRxa$

▶ **75.** $\forall x(Rxb \rightarrow Sxb)$ **76.** $\forall x\forall y(Rxy \rightarrow Sxy)$

77. $\forall x\forall y(Sxy \rightarrow Rxy)$ **78.** $\forall x\forall y((Rxy \,\&\, Sxy) \rightarrow Syx)$

79. $\forall x\forall y((Rxy \,\&\, \neg Sxy) \rightarrow Ryx)$

▶ **80.** $\forall x\forall y((Rxy \,\&\, Ryx) \rightarrow \exists z(Rxz \lor Sxz))$

Let $D = \{a, b, c\}$, $\varphi(a) = a$, $\varphi(b) = b$, $\varphi(F) = \{\langle b \rangle, \langle c \rangle\}$, and $\varphi(R) = \{\langle a, c \rangle, \langle b, c \rangle, \langle c, c \rangle\}$.

81. $\exists xRxx$ **82.** $\forall x(Fx \rightarrow \exists yRxy)$

83. $\exists x(Fx \;\&\; \forall yRxy)$

84. $\exists x(Fx \;\&\; \forall yRyx)$

▸ **85.** $\forall x(Rxx \rightarrow Fx)$

86. $\forall x\exists yRxy$

87. $\exists x\forall yRyx$

88. $(Fa \lor Fb) \rightarrow (Raa \lor Rab)$

89. $\forall y(\exists xRxy \rightarrow \exists zRyz)$

▸ **90.** $\forall y(\exists zRyz \rightarrow \exists xRxy)$

Let $D = \{a, b, c\}, \varphi(a) = a, \varphi(b) = b,$ and $\varphi(R) = \{\langle a, a\rangle, \langle a, b\rangle, \langle b, c\rangle, \langle a, c\rangle,$ $\langle c, b\rangle, \langle b, a\rangle\}.$

91. $\exists xRxx$

92. $\forall x\forall yRxy$

93. $\forall x\exists yRxy$

94. $\forall x\forall y(Rxy \rightarrow Ryx)$

▸ **95.** $\forall x(Rxx \rightarrow \exists yRyx)$

96. $\forall x(\exists yRxy \rightarrow Rxa)$

97. $\forall x(Rxa \rightarrow Rbx)$

98. $\forall x\forall y\forall z((Rxy \;\&\; Ryz) \rightarrow Rxz)$

99. $\forall x\forall y(Rxy \rightarrow \exists z(Rxz \;\&\; Ryz))$

▸ **100.** $\forall x\forall y((Rxy \;\&\; Ryx) \rightarrow \exists z(Rxz \;\&\; Rzz))$

Let $D = \{0, 1, 2, \ldots, 10\}$, with $<$ and $\leq$ having their usual interpretations.

101. $\forall x \; x < x$

102. $\forall x \; x \leq x$

103. $\forall x\forall y(x < y \rightarrow \neg y < x)$

104. $\exists x\exists y(x \leq y \;\&\; y \leq x)$

▸ **105.** $\exists x\exists y(x < y \;\&\; y \leq x)$

106. $\forall x\forall y\forall z((x < y \;\&\; y < z) \rightarrow x < z)$

107. $\forall x\forall y\forall z((x \leq y \;\&\; y \leq z) \rightarrow x \leq z)$

108. $\forall x\forall y\forall z((x < y \;\&\; y \leq z) \rightarrow x < z)$

109. $\forall x\forall y(x < y \rightarrow \exists z(x < z \;\&\; z < y))$

▸ **110.** $\forall x\forall y(x \leq y \rightarrow \exists z(x \leq z \;\&\; z \leq y))$

111. $\forall x\exists y \; x < y$ **112.** $\forall x\exists y \; x \leq y$ **113.** $\forall x\exists y \; y < x$

114. $\exists y\forall x \; x < y$ ▸ **115.** $\exists y\forall x \; y \leq x$

Let $D = \{0, 1, 2, \ldots.\}$, with $<$ and $\leq$ having their usual interpretations.

116. $\forall x \; x < x$

117. $\forall x \; x \leq x$

118. $\forall x\forall y(x < y \rightarrow \neg y < x)$

119. $\exists x\exists y(x \leq y \;\&\; y \leq x)$

▸ **120.** $\exists x\exists y(x < y \;\&\; y \leq x)$

121. $\forall x\forall y\forall z((x < y \;\&\; y < z) \rightarrow x < z)$

122. $\forall x\forall y\forall z((x \leq y \;\&\; y \leq z) \rightarrow x \leq z)$

123. $\forall x\forall y\forall z((x < y \;\&\; y \leq z) \rightarrow x < z)$

124. $\forall x \forall y(x < y \rightarrow \exists z(x < z \ \& \ z < y))$

▶ **125.** $\forall x \forall y(x \leq y \rightarrow \exists z(x \leq z \ \& \ z \leq y))$

126. $\forall x \exists y \ x < y$ **127.** $\forall x \exists y \ x \leq y$ **128.** $\forall x \exists y \ y < x$

129. $\exists y \forall x \ x < y$ ▶ **130.** $\exists y \forall x \ y \leq x$

Let $D = \{\ldots, -2, -1, 0, 1, 2, \ldots\}$, with $<$ and $\leq$ having their usual interpretations.

131. $\forall x \ x < x$ **132.** $\forall x \ x \leq x$

133. $\forall x \forall y(x < y \rightarrow \neg y < x)$ **134.** $\exists x \exists y(x \leq y \ \& \ y \leq x)$

▶ **135.** $\exists x \exists y(x < y \ \& \ y \leq x)$

136. $\forall x \forall y \forall z((x < y \ \& \ y < z) \rightarrow x < z)$

137. $\forall x \forall y \forall z((x \leq y \ \& \ y \leq z) \rightarrow x \leq z)$

138. $\forall x \forall y \forall z((x < y \ \& \ y \leq z) \rightarrow x < z)$

139. $\forall x \forall y(x < y \rightarrow \exists z(x < z \ \& \ z < y))$

▶ **140.** $\forall x \forall y(x \leq y \rightarrow \exists z(x \leq z \ \& \ z \leq y))$

141. $\forall x \exists y \ x < y$ **142.** $\forall x \exists y \ x \leq y$ **143.** $\forall x \exists y \ y < x$

144. $\exists y \forall x \ x < y$ ▶ **145.** $\exists y \forall x \ y \leq x$

Let $D =$ the set of real numbers, with $<$ and $\leq$ having their usual interpretations.

146. $\forall x \ x < x$ **147.** $\forall x \ x \leq x$

148. $\forall x \forall y(x < y \rightarrow \neg y < x)$ **149.** $\exists x \exists y(x \leq y \ \& \ y \leq x)$

▶ **150.** $\exists x \exists y(x < y \ \& \ y \leq x)$

151. $\forall x \forall y \forall z((x < y \ \& \ y < z) \rightarrow x < z)$

152. $\forall x \forall y \forall z((x \leq y \ \& \ y \leq z) \rightarrow x \leq z)$

153. $\forall x \forall y \forall z((x < y \ \& \ y \leq z) \rightarrow x < z)$

154. $\forall x \forall y(x < y \rightarrow \exists z(x < z \ \& \ z < y))$

▶ **155.** $\forall x \forall y(x \leq y \rightarrow \exists z(x \leq z \ \& \ z \leq y))$

156. $\forall x \exists y \ x < y$ **157.** $\forall x \exists y \ x \leq y$ **158.** $\forall x \exists y \ y < x$

159. $\exists y \forall x \ x < y$ ▶ **160.** $\exists y \forall x \ y \leq x$

Let $D =$ the set of Federal League teams in 1915, with P interpreted as 'had at least as high a winning percentage as'; F, as 'finished first'; B, as 'finished at least one game behind'; G, as 'won at least as many games as'; and L, as 'lost at least as many games as.'

The 1915 Federal League Standings

Team	Wins	Losses	Pct.	Games Behind
Chicago	86	66	.566	—
Saint Louis	87	67	.565	—
Pittsburgh	86	67	.562	.5
Kansas City	81	72	.529	5.5
Newark	80	72	.526	6
Buffalo	74	78	.487	12
Brooklyn	70	82	.461	16
Baltimore	47	107	.305	40

161. $\forall x(Fx \rightarrow \forall yPxy)$ **162.** $\forall x(Fx \rightarrow \forall yGxy)$ **163.** $\forall x(Fx \rightarrow \forall yLyx)$

164. $\forall x(\forall yPxy \rightarrow Fx)$ ▶ **165.** $\forall x(\forall yGxy \rightarrow Fx)$ **166.** $\forall x(\forall yLyx \rightarrow Fx)$

167. $\forall x\forall y(Gxy \rightarrow Lyx)$ **168.** $\exists x\exists y(Bxy \& Lxy)$ **169.** $\exists x\exists y(Pxy \& Gyx)$

170. $\forall x\forall y(Pxy \rightarrow Lyx)$ **171.** $\forall x(\exists yBxy \rightarrow \neg\forall zLxz)$

172. $\forall x\forall y(Pxy \rightarrow Gxy)$ **173.** $\forall x\forall y\forall z((Pxy \& Lyx) \rightarrow Gxy)$

174. $\forall x\forall y(Bxy \rightarrow Lxy)$ ▶ **175.** $\forall x\forall y(Bxy \rightarrow Gxy)$

176. $\forall x\forall y(Bxy \rightarrow Pyx)$ **177.** $\forall x\forall y(Lxy \rightarrow Bxy)$

178. $\forall x\forall y(Gyx \rightarrow Bxy)$ **179.** $\forall x\forall y(Pyx \rightarrow Bxy)$

▶ **180.** $\forall x\exists y(Gyx \& Pyx)$

6.4 Constructing Interpretations from Tableaux

We begin a semantic tableau by placing one or more formulas on a particular side. We then proceed to search for an interpretation of these formulas making those appearing on the left true and those appearing on the right false. If a tableau closes, no interpretation meets these conditions. If it remains open, however, some interpretation meets the initial conditions by assigning the proper truth values to the formulas at the top. An open tableau not only indicates that there is such an interpretation but also spells out such an interpretation. Now that we have a precise concept of interpretation in quantificational logic, we can explain how it does so.

Remember that, in sentential logic, each open branch specified an interpretation meeting the tableau's initial conditions. We could read these interpretations from the tableau branches simply by seeing which sentence letters appeared on which sides of the branches.

In quantificational logic, things are only slightly more complicated. Once again, each open branch specifies an appropriate interpretation. Each open branch specifies both a domain and an interpretation function.

To make things easier, let's assume that objects a, b, c, and so on are named by the constants *a*, *b*, *c*, etc. This way, we needn't bother about defining the interpretations of constants. We can simply presume that, in an interpretation $M = \langle D, \varphi \rangle$, $\varphi(a) = a$, $\varphi(b) = b$, $\varphi(c) = c$, and so on. We can also refer to the objects in the domain as *a*, *b*, *c*, etc.

We will also assume at the outset that the tableau is *finished* in the sense that all possible instantiations of temporarily dispatched formulas using the available constants (that is, those that we've had to introduce to satisfy the quantifier rules) have been performed. We need to begin with a tableau in which we have applied all the usual tableau rules and have applied all repeatable rules as many times as possible with the constants already on the tableau. The only undispatched formulas should be atomic. (An infinite tableau, of course, will never be finished. But even an infinite tableau specifies an infinite interpretation according to the rules to follow.)

The domain associated with an open branch of a finished tableau will be the set of objects named by the constants appearing in any undispatched formula on the branch. So, if the undispatched formulas on a branch are *Fa*, *Gba*, and *Haa*, the branch specifies the domain $\{a, b\}$. Similarly, if a branch retains as its undispatched formulas *Fba*, *Fcb*, *Fdc*, and *Fad*, the associated domain will be $\{a, b, c, d\}$. As far as the universe of discourse is concerned, it makes no difference on which side of the branch these formulas appear.

If any sentence letters on the branch are undispatched, then we will adopt our rule from sentential logic for the interpretation function. We will interpret undispatched sentence letters on the left as true and those on the right as false.

Finally, our interpretation function allows us to ignore all other undispatched atomic formulas appearing on the right side of the branch. We will consider only the left side. Whenever something of the form $\mathcal{R}a_1 \ldots a_n$ appears on the left, the *n*-tuple $\langle a_1, \ldots, a_n \rangle$ will appear in the interpretation of $\mathcal{R}$. The set $\varphi(\mathcal{R})$ will thus consist of all *n*-tuples $\langle a_1, \ldots, a_n \rangle$ such that $\mathcal{R}a_1 \ldots a_n$ appears on the left side of the branch. So, suppose that the branch has *Fab*, *Fbb*, and *Ga* on the left and *Gb* and *Faa* on the right. We ignore the formulas on the right. The interpretation of *F* must include the pairs $\langle a, b \rangle$ and $\langle b, b \rangle$, while that of *G* must include $\langle a \rangle$. So the branch specifies that $\varphi(F) = \{\langle a, b \rangle, \langle b, b \rangle\}$ and that $\varphi(G) = \{\langle a \rangle\}$. The domain, of course, will be $\{a, b\}$, the set containing just objects a and b.

To summarize: In a finished, open tableau, each open branch specifies an interpretation assigning truth and falsehood to the formulas at its top, according to whether those formulas appear on its left or right side. The branch specifies an interpretation according to these rules:

Domain

$D = \{a_1, \ldots, a_n, \ldots\}$, where $a_1, \ldots, a_n, \ldots$ are the constants appearing in undispatched atomic formulas on the branch.

Interpretations of Sentence Letters

$\varphi(\rho) = $ T, if ρ appears on the left; F, if ρ appears on the right.

Interpretations of n-ary Predicates

$\varphi(\mathcal{R}) = \{\langle a_1, \ldots, a_n \rangle, \langle \ell_1, \ldots, \ell_n \rangle, \langle c_1, \ldots, c_n \rangle, \ldots\}$, where $\mathcal{R}a_1, \ldots, a_n, \mathcal{R}\ell_1, \ldots, \ell_n, \mathcal{R}c_1, \ldots, c_n$, etc., appear on the left of the branch.

Sometimes, a sentence letter or predicate that figures in the set of formulas at the top of the tableau will not appear on a particular branch in a place where it makes a difference to the interpretation; it may not appear in an undispatched formula at all. If a sentence letter or predicate does not appear in any undispatched formula, then the interpretation can assign it anything. If a predicate appears only on the right, in undispatched atomic formulas of the form $\mathcal{F}a_1 \ldots a_n$, then it can have any interpretation that does not assign $\mathcal{F}$ an n-tuple $\langle a_1, \ldots, a_n \rangle$. For the sake of simplicity, we can assign all such predicates the null set, $\emptyset$, as an interpretation.

These rules for deriving an interpretation from a tableau branch are easier to apply than to explain. So, let's take a simple example. Does 'Any F is G' imply 'Some F are G'? No.

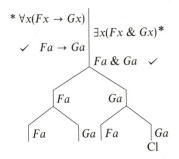

This tableau has three open branches. The leftmost has only one constant, a, so it specifies the domain $\{a\}$. Neither predicate appears in an undispatched formula on the left of the branch, so we can assign both the null set: $\varphi(F) = \varphi(G) = \emptyset$. The middle open branch specifies the same interpretation; again it specifies the domain $\{a\}$, and neither predicate appears in an undispatched formula on the left. The rightmost open branch contains only the constant a, so it too specifies the domain $\{a\}$. But it does have Ga on the left. So, it specifies that $\varphi(G) = \{\langle a \rangle\}$, while $\varphi(F) = \emptyset$. The three open branches thus specify two interpretations, making 'Any F are G' true but 'Some F are G' false: (1) $D = \{a\}$, $\varphi(F) = \varphi(G) = \emptyset$; (2) $D = \{a\}$, $\varphi(F) = \emptyset$, $\varphi(G) = \{\langle a \rangle\}$. These interpretations have in common an assignment of the null set to F, the subject term of the sentences concerned. 'Any violators will be prosecuted' does not imply 'Some violators will be prosecuted,' because there may be no violators.

As a similar example, consider the argument form 'All G are H; no F are G; so, no H are F.' Is this valid? A tableau quickly demonstrates that it isn't:

$$
\begin{array}{c}
* \ \forall x(Gx \to Hx) \\
\checkmark \quad \neg \exists x(Fx \ \& \ Gx) \\
\neg \exists x(Hx \ \& \ Fx) \quad \checkmark \\
\checkmark \quad \exists x(Hx \ \& \ Fx) \\
\exists x(Fx \ \& \ Gx)* \\
\checkmark \quad Ha \ \& \ Fa \\
Ha \\
Fa \\
\checkmark \quad Ga \to Ha \\
Fa \ \& \ Ga \quad \checkmark \\
Fa \qquad Ga \\
\text{Cl} \\
Ga \qquad Ha
\end{array}
$$

Two branches remain open. Both, however, have only the constant a, and both have just Ha and Fa appearing undispatched on the left. They thus specify the same interpretation. The domain, clearly, is the set containing only a. The interpretations of both H and F must contain $\langle a \rangle$, while the interpretation of G can remain empty. So the interpretation is this: $D = \{a\}$; $\varphi(F) = \varphi(H) = \{\langle a \rangle\}$; $\varphi(G) = \emptyset$. It's easy to see that this interpretation does make both 'All G are H' and 'No F are G' true (since there are no G's), but 'No H are F' false (since a is both an F and an H).

As a final example, we can use a tableau to determine whether the set of formulas

$$\{\forall x \forall y(Fx \to Gxy), \exists x Fx, \exists x \forall y \neg Gxy\}$$

is satisfiable. Placing all on the left of the tableau, we obtain a single open branch, telling us that the formulas can all be true together. Notice that we must introduce two constants, a and b, and then instantiate the universal formulas we obtain in all possible ways involving these two constants.

The open branch uses the constants a and b, so its associated domain is $\{a, b\}$. Appearing on the left are the undispatched formulas Fa, Gaa, and Gab. So, the interpretation of F must include $\langle a \rangle$, and that of G must include both $\langle a, a \rangle$ and $\langle a, b \rangle$. The interpretation the branch specifies is thus: $D = \{a, b\}$; $\varphi(F) = \{\langle a \rangle\}$; $\varphi(G) = \{\langle a, a \rangle, \langle a, b \rangle\}$. Notice that a is an F, so

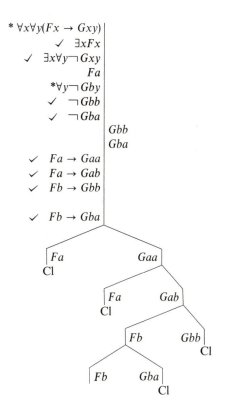

$\exists x Fx$ is true. Further, b has no G-relation to anything, so $\exists x \forall y \neg Gxy$ is true as well. And a is G-related to both b and itself, so it's G-related to everything in the domain; thus $\forall x \forall y (Fx \rightarrow Gxy)$ is true.

Problems

For each of the following formulas, use a tableau to specify an interpretation making it true.

1. $Fa \vee \exists x Fx$
2. $Fa \rightarrow \exists x Fx$
3. $Fa \rightarrow \forall x Fx$

4. $Fa \,\&\, \exists x Rxa$
▶ 5. $Fa \rightarrow \exists y Ray$
6. $\exists x Fx$

7. $\exists x Rxx$
8. $\exists x (Fx \rightarrow \neg Rxb)$

9. $\forall x (Fx \,\&\, (Rxx \rightarrow \forall y Ryx))$
▶ 10. $\forall x \neg Fxx$

11. $\forall x (Fxc \rightarrow Gx)$
12. $\exists x (Fxa \rightarrow Fxx)$

13. $\exists x (Fxa \,\&\, Fxx)$
14. $\neg \exists x (Fx \,\&\, Gx)$

15. $\neg \forall x (Fx \,\&\, Gx)$
16. $\neg \forall x (Fx \rightarrow Gx)$

17. $\forall x(Fx \rightarrow \neg Gx)$ **18.** $\forall x(Fx \leftrightarrow Gx)$

19. $\forall x((Fx \,\&\, Gx) \rightarrow (Hx \vee Jx))$ ▶ **20.** $\exists x \forall y Rxy$

21. $\exists x(\exists y Rxy \,\&\, \forall z Rzx)$ **22.** $\exists x \exists y(Rxy \,\&\, \neg Ryx)$

23. $\exists x \forall z(Rxx \rightarrow \neg Rxz)$ **24.** $\forall x(\exists y(Pyx \,\&\, Ty) \rightarrow Tx)$

▶ **25.** $\forall x(\neg Rxx \rightarrow \forall y(Rxy \rightarrow \neg Ryx))$

For each of the following argument forms, construct a tableau to determine whether it is valid. If it isn't, specify an interpretation making its premise formulas true and its conclusion formula false.

26. $\forall x(Fx \rightarrow Gx)$ **27.** $\exists x Fx$
    $\forall x(Fx \rightarrow Hx)$ $\exists x Gx$
    $\therefore \forall x(Gx \rightarrow Hx)$ $\therefore \exists x(Fx \,\&\, Gx)$

28. $\forall x(Fx \rightarrow \neg Gx)$ **29.** $\forall x(Fx \rightarrow Gx)$
    $\therefore \exists x(Fx \,\&\, \neg Gx)$ $\therefore \forall x(Gx \rightarrow Fx)$

▶ **30.** $\forall x(Fx \rightarrow Gx)$ **31.** $\forall x(\neg Fx \rightarrow \neg Gx)$
    $\therefore \forall x(\neg Fx \rightarrow \neg Gx)$ $\therefore \forall x(Fx \rightarrow Gx)$

32. $Fa \,\&\, \neg Ga$ **33.** $Fa \,\&\, Ga$
    $\therefore \neg \forall x(Gx \rightarrow Fx)$ $\forall x((Fa \,\&\, (Gx \,\&\, Hx)) \rightarrow Ja)$
                                 $\therefore Ja$

34. $\neg \forall x Fx$ ▶ **35.** $\forall x(Cx \rightarrow Bx)$ **36.** $\forall x(Bx \rightarrow Fx)$
    $\therefore \forall x \neg Fx$ $\exists x(Ax \,\&\, \neg Cx)$ $\forall x(Cx \rightarrow Fx)$
                         $\therefore \exists x(Ax \,\&\, \neg Bx)$ $\therefore \forall x(Cx \rightarrow Bx)$

37. $\exists x(Cx \,\&\, \neg Fx)$ **38.** $\neg \exists x(Fx \,\&\, Gx)$ **39.** $\forall x(Fx \rightarrow Gx)$
    $\exists x(Mx \,\&\, \neg Fx)$ $\forall x(Gx \rightarrow Hx)$ $\neg \exists x(Hx \,\&\, Fx)$
    $\therefore \exists x(Cx \,\&\, Mx)$ $\therefore \forall x(Fx \rightarrow \neg Hx)$ $\therefore \neg \exists x(Hx \,\&\, Gx)$

▶ **40.** $\forall x(Fx \rightarrow \neg Gx)$
    $\neg \exists x(Gx \,\&\, Hx)$
    $\therefore \forall x(Fx \rightarrow Hx)$

41. $\forall x \forall y(Rxy \rightarrow \neg \exists z(Rxz \,\&\, Ryz))$
    $\therefore \forall x(\exists y(Rxy \,\&\, Ryx) \rightarrow Fx)$

42. $\forall x \forall y(Fxy \rightarrow \forall z(Fyz \rightarrow Fxz))$
    $\therefore \forall x \forall y(Fxy \rightarrow \neg Fyx)$

43. $\forall x((Fx \vee Gx) \rightarrow \neg Hx)$ **44.** $\forall x(Fx \rightarrow (Gx \vee Hx))$
    $\forall x(Rx \rightarrow \neg Hx)$ $\forall x(Fx \rightarrow Gx)$
    $\therefore \forall x(Rx \rightarrow (Fx \vee Gx))$ $\therefore \forall x(Fx \rightarrow \neg Hx)$

45. $\forall x Fax$ **46.** $\exists x(Fx \,\&\, (Gx \rightarrow Hx))$
    $\forall x(\exists y \neg Fxy \rightarrow \neg Fxx)$ $\exists x(Fx \,\&\, Gx)$
    $\therefore Faa$ $\therefore \exists x(Fx \,\&\, Hx)$

47. $\exists x(Fx \,\&\, (Gx \rightarrow Rxx))$
   $\forall x \forall y(Rxy \rightarrow \neg Ryx)$
   $\forall x(Gx \rightarrow Hx)$
   $\therefore \exists x(Fx \,\&\, \neg Hx)$

48. $\neg \forall x Lax$
   $\therefore \neg \exists z(Laz \,\&\, \neg Lzz)$

49. $\forall x(Rxy \rightarrow \neg Ryx)$
   $\forall x \forall y \forall z((Rxy \,\&\, Ryz) \rightarrow \neg Rxz)$
   $\therefore \forall x \forall y(Rxy \lor Ryx)$

▶ **50.** $\forall x(Fx \rightarrow \exists y(Fy \,\&\, Ryx))$
   $\therefore \neg \exists x(Fx \,\&\, Rxx)*$

Notes

[1] "A Note on the Entscheidungsproblem," *Journal of Symbolic Logic* 1 (1936): 40–41. (Correction on 101–2.)

[2] 'Someone in this room is a spy' asserts that there is at least one spy in the room. We assume in the text, for the sake of simplicity, that there is only a single spy. If there are more than one, then what is said about 'the spy' applies to any of the spies; we cannot distinguish them with the information at hand.

7

Q<small>UANTIFIED</small> N<small>ATURAL</small> D<small>EDUCTION</small>

T he system of natural deduction we developed in Chapter 4 for sentential logic extends easily to quantificational logic. All the rules of sentential deduction apply in quantification theory. But, to deal with quantifiers, we add three new simple rules and a complex rule that has the form of a new method of proof. The deduction system that emerges shares the virtues of its sentential cousin. The system is sound, because every provable formula is valid, and every conclusion that can be proved from a set of premises is implied by the premises. The system is also complete, for every valid formula of quantification theory can be proved in the system, and every valid argument can be shown to be valid within it. Furthermore, the system mirrors closely the processes of reasoning that people use in a wide variety of contexts. It reflects reasoning in mathematics and related disciplines particularly well.

7.1 D<small>EDUCTION</small> R<small>ULES FOR</small> Q<small>UANTIFIERS</small>

The deduction rules needed for quantificational logic are very straightforward. Recall that $\mathscr{A}[c/x]$ is the result of substituting c for every occurrence of x throughout the formula $\mathscr{A}$. If $\forall v \mathscr{A}$ and $\exists v \mathscr{A}$ are formulas, then $\mathscr{A}[c/v]$ is called an *instance* of them. Conversely, $\forall v \mathscr{A}$ and $\exists v \mathscr{A}$ are *generics* of $\mathscr{A}[c/v]$. In later chapters we'll add other singular terms that function, in many respects, like constants; we'll call them, and constants, *closed terms*. Where t is any closed term, we'll say that $\mathscr{A}[t/v]$ is an *instance* of $\forall v \mathscr{A}$ and $\exists v \mathscr{A}$.

Each connective has two rules. We use one rule to introduce the connective into proofs, while the other allows us to exploit the presence of the connective. The existential quantifier similarly comes with two rules: an introduction rule and an exploitation rule. The introduction rule, in essence, allows us to move to an existentially quantified formula from any instance of that formula.[1] The rule, often called *existential generalization,* thus takes the form:

Existential Introduction ($\exists$I)

$$
\begin{array}{ll}
\text{n.} & \mathscr{A}[\ell/v] \\
\text{n + p.} & \overline{\exists v \mathscr{A}} \qquad \exists\text{I, n}
\end{array}
$$

Here ℓ may be any closed term.

Existential introduction allows us to infer an existentially quantified formula from any instance of it. It sanctions the step from an instance to its corresponding existential generic. Suppose that our universe of discourse consists entirely of people. If Jones, for example, is a spy, then we may conclude that someone is a spy. If Susan suspects Harry, then Susan suspects someone; of course, it's also true that somebody suspects Harry and that somebody suspects somebody. Finally, if Frank doesn't trust himself to work around large sums of money, then Frank doesn't trust somebody to work around large sums of money. Additionally, somebody doesn't trust Frank to work around large sums, and someone doesn't trust himself to do so. Each of the following is an acceptable application of existential introduction.

Premise	Conclusion
Fa	$\exists x Fx$
Gab	$\exists x Gax$
Gab	$\exists x Gxb$
Hcc	$\exists x Hcx$
Hcc	$\exists x Hxc$
Hcc	$\exists x Hxx$
$\exists x Fxa$	$\exists y \exists x Fxy$
$\forall x Fx \rightarrow Gb$	$\exists z(\forall x Fx \rightarrow Gz)$

In each case, the premise is an instance of the conclusion.

The rule of existential exploitation allows us to move from an existentially quantified formula to an instance of it.[2] It is almost exactly the reverse, then, of the existential introduction rule. But it does impose a restriction: The instance must involve a constant new to the proof. The rule says that we may (a) drop an existential quantifier serving as a main connective in a formula and (b) substitute for the quantified variable a constant that hasn't appeared earlier in the proof. The constant must have appeared nowhere in the deduction, not even in a *Show* line. (Actually, no harm would result if we were to use, for ∃E, constants that appear earlier only on already bracketed lines. But, to minimize confusion, we'll always use completely new constants.)

Existential Exploitation (∃E)

$$n. \quad \exists v \mathscr{A}$$
$$n + p. \quad \mathscr{A}[c/v] \qquad \exists E, n$$

Here c must be a constant new to the proof.

Suppose that we have the information that someone in our department is selling trade secrets to a competitor. We don't know who this person is—or, perhaps, who these people are—but we do want to reason from what we know to find out. We know that at least one person has been selling secrets; our reasoning and our communication will proceed much more readily if we give this person some name—*John Doe,* say, or just *the mole*—so that we can refer to him or her in various contexts. We can't simply say, "Someone has been selling our trade secrets. Someone must have joined the department around the middle of 1981, because that's when secrets began to leak." Nothing here indicates that the two 'someones' are the same. To relate these assertions to the same individual, we must have a way of referring to that person. We can do so by introducing a name. It's critical that the name we choose be new. If Sarah Freeland is the head of the department, and we decide to call the seller of trade secrets *Sarah Freeland,* then utter confusion will result.

The system of rules in this chapter is sound in the sense that the rules never lead us astray; they never allow us to prove a formula that isn't valid or permit us to establish the validity of an invalid argument form. Nevertheless, the existential exploitation rule *seems* unsound; it justifies the inference from $\exists v \mathscr{A}$ to $\mathscr{A}[c/v]$, so long as c is new to the proof. Our demand for a new constant prevents the rule from doing any harm.

Most of our rules have been *truth-preserving:* The truth of the premises of the rules guarantees the truth of their conclusions. Existential exploitation, however, is not truth-preserving. We should hardly be able to argue, 'Some philosophers have been Nazis. Therefore, Aristotle was a Nazi.' This form of inference is fallacious.

By introducing a new name, we avoid such problems because we contain our use of the name within a portion of the proof. We could not use existential exploitation to establish the validity of the above argument. On the left is the attempted proof within quantification theory; on the right is the English equivalent.

1. $\exists x(Px \ \& \ Nx)$	A	Some philosophers have been Nazis.
2. Pa	A	Aristotle was a philosopher.
3. Show Na		Show that Aristotle was a Nazi.
4. $Pb \ \& \ Nb$	$\exists$E, 1	b was a philosopher and a Nazi.

It seems clear that we will never be able to reach Na. Because the constant a appears in both lines 2 and 3, we can't use existential exploitation to obtain the instance, $Pa \ \& \ Na$, necessary for the conclusion. Although $\exists$E is not truth-preserving, therefore, it never gets us into trouble. The rule is *conservative* in that any formula that does not contain the new constant and that follows from the conclusion of the rule also follows from the premise of the rule.

Mathematicians often introduce, in one breath, existential assertions and names for the objects asserted to exist. Consider these examples from a calculus text:

> Let f be continuous at a. For every number $\epsilon > 0$ we can choose a number $\delta > 0$ so that $|f(x) - f(a)| < \epsilon$ for all $x \in A$ with $|x - a| < \delta$. . . .

> Since U is open, there is an open rectangle B with $f(a) \in B \subset U$.[3]

Or consider this, from a text on Lebesgue integration:

> Let $\{s_n\}$ converge to s. Take $\epsilon = 1$ in the definition of convergence; then there exists an integer N such that $|s_n - s| < 1$ for all $n \geq N$, i.e. $s - 1 < s_n < s + 1$ for all $n \geq N$.[4]

These passages combine the introduction of an existential assertion with its exploitation in one step. They name an object asserted to exist in the very act of making that assertion: "there exists an integer N," "there is an open rectangle B" and "we can choose a number δ." Note that, in every instance, the names introduced in this way haven't appeared before; we have no independent information about N, B, or δ. It would be very different—and outrageous—if these passages were to say instead, "there exists an integer 43" or "we can choose a number π."

The third rule for quantifiers is universal exploitation. If we know that something is true about every object, then we can conclude that it is true for each particular object that we consider. If God loves everyone, then God loves me, you, and the Earl of Roxburgh. If Jane likes everyone whom she meets, then she likes you, if she's met you; she likes me, if she's met me; and so on. The rule of universal exploitation states that, from a universally quantified formula, we may infer any of its instances.

	Universal Exploitation (∀E)

n. $\dfrac{\forall v \mathscr{A}}{}$

n + p. $\mathscr{A}[\ell / v]$ ∀E, n

Here ℓ may be any closed term.

This rule does not require us to use a new constant. In fact, it is generally silly to use a new constant when we apply ∀E. There is no point to introducing a new name unless no constants appear in the proof at all up to this line. If constants a and b appear earlier, from a formula $\forall x F x$ we can infer Fa, or Fb, or both. From a formula $\forall x G x x$, we can infer Gaa or Gbb. And, from $\forall x \forall y H x y$, we can obtain $\forall y H a y$ or $\forall y H b y$ and, in another step, any of Haa, Hab, Hba, and Hbb. We could also, of course, infer similar formulas with other constants. Unless we are forced to introduce those constants in other ways, however, using them to exploit a universal formula serves no purpose.

To see how these rules work, let's demonstrate the validity of a simple argument: 'Something's upsetting John; whatever upsets John upsets Edna; so, something's upsetting Edna.'

1.	$\exists x F x j$	A
2.	$\forall y (F y j \rightarrow F y e)$	A
3.	Show $\exists z F z e$	
4.	Faj	∃E, 1
5.	$Faj \rightarrow Fae$	∀E, 2
6.	Fae	→E, 4, 5
7.	$\exists z F z e$	∃I, 6

Problems

Use deduction to show that these arguments are valid.

1. God created everything. So, God created Pittsburgh.

2. God created everything. So, God created Himself.

3. God created everything. So, God created something.

4. God created everything. So, something created God.

▶ **5.** Nothing coherent ever baffles me. This course baffles me. So, this course is incoherent.

6. All writers who express nationalism in their writing are trying to achieve political aims. Some American writers express nationalism in what they write. Thus, some writers trying to achieve political aims are Americans.

7. Everyone who understands the nature of the radical Islamic movement recognizes that it threatens the existing Arab regimes of the Middle East. Some people in the State Department understand the nature of radical Islam, so some State Department personnel realize that the movement threatens existing Arab regimes.

8. Some plants that are widely cultivated in this area cannot survive on rainfall alone. All plants native to this area can survive on rainfall alone. It follows that some plants that are not native to this area are widely cultivated here.

9. Some utility companies are predicting brownouts in their service regions this summer. No utilities that can easily and affordably purchase power from other utilities are predicting brownouts this summer. Hence, some utility companies cannot easily and affordably purchase power from other utilities.

▶ 10. Some computer programs used for processing natural language are written in PROLOG. Nothing written in PROLOG relies heavily on the notion of a list. Consequently, not all computer programs used for processing natural language rely heavily on lists.

11. Each person who came to the company party was carefully observed by the company president. Some people who came to the company party became obviously drunk. Thus, the company president observed some obviously drunk people.

12. Nothing written by committee is easy to write or easy to read. Some documents written by committee are nevertheless extremely insightful. So, some extremely insightful documents are not easy to read.

13. Some of the cleverest people I know are clearly insane. Any of the cleverest people I know could prove that this argument is valid. Hence, some people who could prove this argument valid are clearly insane.

14. There are cities in the Sun Belt that are experiencing rapid growth but that are not ranked as very desirable places to live. Every city that experiences rapid growth has to raise taxes. Therefore, some cities that will have to raise taxes are not ranked as very desirable places to live.

▶ 15. Some analysts insist that we are in the midst of a historic bull market, but others say that the market will soon collapse. Nobody who is expecting the market to collapse is recommending anything but utility stocks. None who believe that M1 controls the direction of the economy contend that we are in the midst of a historic bull market. Thus, some analysts are recommending only utility stocks, but some don't believe that M1 controls the economy's direction.

16. If nobody comes forward and confesses, then someone will be punished. Therefore, someone will be punished if he doesn't confess.

17. One prosecutor can convict another only if he or she can convict everyone that prosecutor can. There are prosecutors who can convict each other. So, there are prosecutors who can convict themselves.

18. The unrestricted axiom of abstraction states that there is a set of all and only those objects satisfying any open sentence. Bertrand Russell proved the axiom inconsistent by using the open sentence '. . . is not a member of itself.' Show, following Russell, that $\exists x \forall y (y \in x \leftrightarrow y \notin y)$ is contradictory by proving its negation.

Use deduction to establish the validity of these argument forms.

19. $\exists x Fx$
    $\therefore \forall x Gx \rightarrow \exists x(Fx \ \& \ Gx)$

20. $\exists x(Fx \ \& \ Gx)$
    $\forall x(Gx \rightarrow \neg Hx)$
    $\therefore \exists x(\neg Hx \ \& \ Fx)$

21. $\exists y Fyy$
    $\exists x \forall z Gxz$
    $\therefore \exists x \exists y(Gyx \ \& \ Fxx)$

22. $\exists x \exists y Fxy$
    $\forall x \forall y(Fxy \leftrightarrow (Gx \ \& \ \neg Gy))$
    $\therefore \exists x Gx \ \& \ \exists x \neg Gx$

23. $\exists x Gx \ \& \ \exists x \neg Gx$
    $\forall x \forall y(Fxy \leftrightarrow (Gx \ \& \ \neg Gy))$
    $\therefore \exists x \exists y Fxy$

24. $\forall x(Fx \rightarrow Gx)$
    $\therefore \exists x \neg Fx \lor \exists x Gx$

25. $\forall x \forall y(Fxy \leftrightarrow (Gx \ \& \ \neg Gy))$
    $\exists x \exists y(Fxy \ \& \ Fyx)$
    $\therefore \exists x(Gx \ \& \ \neg Gx)$

26. $\exists x \exists y(Fx \ \& \ Gyx)$
    $\forall x \forall y(Gxy \rightarrow (Hx \ \& \ Jyx))$
    $\therefore \exists x \exists y((Fx \ \& \ Hy) \ \& \ Jxy)$

27. $\forall x \forall y \forall z((Fxy \ \& \ Fxz) \rightarrow Fyz)$
    $\exists x \exists y(Fxy \ \& \ \neg Fyx)$
    $\therefore \neg \forall x Fxx$

7.2 Universal Proof

Introducing a universal formula, in this system, requires a new method of proof. We already have three proof techniques: direct proof, indirect proof, and conditional proof. Quantificational logic adds *universal proof*.

Universal Proof

n.   Show $\forall v \mathscr{A}$
n + 1. ⎡ Show $\mathscr{A}[c/v]$
        ⎣ ⎡ ⋮

Here c must be a constant new to the proof.

To prove a universal conclusion, in other words, we prove an instance of it. The instance must result from substituting a constant new to the proof for the quantified variable. Since no information regarding the new constant will appear anywhere earlier in the proof, it seems to stand for no object in particular. It represents an arbitrarily chosen object. Because the proof puts no constraints on it, absolutely any object could play this role. Consequently, though we prove something about c, we have shown how to prove it about anything. And this justifies our drawing a universal conclusion.

This method, too, corresponds closely to mathematical practice. Consider this example of a theorem and proof from a standard calculus text:

> If a function has a derivative which is zero at each point of an interval, the function is constant on that interval.... [Proof:] Suppose $f'(x) = 0$ for each x on an interval....[5]

Or this, from a well-known high school geometry textbook:

> *Theorem 9–2.* In a plane, two lines are parallel if they are both perpendicular to the same line.
> *Proof.* Given that $L_1 \perp L$ at P and $L_2 \perp L$ at Q. It is given that L_1 and L_2 are coplanar. We need to show that they do not intersect.
> Suppose that L_1 intersects L_2 at a point R. Then there are two perpendiculars from R to L. By Theorem 6–4, this is impossible. Therefore, $L_1 \parallel L_2$.[6]

These are universal proofs. The former establishes a result about all functions by proving something about an arbitrary function, f. The latter derives a conclusion about any two lines from reasoning about two arbitrarily selected lines, L_1 and L_2. (Within this universal proof there is an indirect proof. The proof also combines an existential exploitation with another step: ". . . at a point R.")

It might seem that this form of proof allows us to prove very silly arguments valid. It lets us derive a universal formula from one of its instances. So, can't we show that, if Fred loves the Go-Gos, everybody loves the Go-Gos? Fortunately, no, because of the new-constant requirement.

1. *Lfg*	A	Fred loves the Go-Gos.
2. Show $\forall x Lxg$		Show everybody loves the Go-Gos.
3. Show *Lag*		Show *a* loves the Go-Gos.

We can't go from the information that Fred loves the Go-Gos to the conclusion that some arbitrarily selected *a* does.

We'll consider this inference pattern as an example of a universal proof: 'All F are G; everything is F; so, everything is G.' To establish its validity, we construct a universal proof:

$$
\begin{array}{lll}
1. & \forall x(Fx \rightarrow Gx) & \text{A} \\
2. & \forall xFx & \text{A} \\
3. & \text{Show } \forall xGx & \\
4. & \quad \text{Show } Ga & \\
5. & \qquad Fa & \forall \text{E, 2} \\
6. & \qquad Fa \rightarrow Ga & \forall \text{E, 1} \\
7. & \qquad Ga & \rightarrow \text{E, 5, 6}
\end{array}
$$

To show that everything is G, we show that some arbitrarily chosen object, a, is G.

Problems

Using deduction, show that these arguments are valid.

1. Anything you think you can achieve, you can achieve. You should try to achieve everything you can achieve. So, anything you think you can achieve, you should try to achieve.

2. Every team that finishes in last place declares its next year a rebuilding year. So, no teams from Chicago will declare next year a rebuilding year only if no Chicago team finishes in last place.

3. No Ivy League colleges have tuitions of under $8,000 a year. Every state-affiliated college has a tuition under $8,000 per year. A college is private if and only if it is not state-affiliated. It follows that every Ivy League college is private.

4. No mammals but bats can fly. Every commonly kept house pet is a mammal, but none are bats. So, nothing that can fly is a commonly kept house pet.

▶ 5. Nothing stupid is difficult. Everything you can do is stupid; anything that isn't difficult, I can do better than you. So, anything you can do, I can do better.

6. Anybody who is reflective despises every demagogue. Anybody who is a demagogue despises everyone. Thus, since there are demagogues, every reflective person despises someone who, in turn, despises him or her.

7. There are no good books that do not require their readers to think. Every book that has inspired acts of terror has been inflammatory. No inflammatory books require their readers to think. Therefore, all books that have inspired acts of terror are no good.

8. Anyone with some brains can do logic. Nobody who has no brains is fit to program computers. No one who reads this book can do logic. So, no one who reads this book is fit to program computers.

9. A person is humble if and only if he or she doesn't admire himself or herself. It follows that nobody who admires all humble people is humble.

▶ 10. All Frenchmen are afraid of Socialists, and Socialists fear only Communists. Thus, every French Socialist is a Communist.

11. All horses are animals. So, all heads of horses are heads of animals.

12. A psychiatrist can help all those who cannot help themselves. So, a psychiatrist can help someone who can help himself or herself.

13. All Don Juans love all women. All who have a conscience treat all whom they love well. If some Don Juans have consciences, therefore, all women are treated well.

14. An Olympic athlete could outrun everyone on our team. Since none on our team can outrun themselves, no one on our team is an Olympic athlete.

15. A person is famous if and only if everyone has heard of him or her. So, all famous people have heard of each other.

16. The Longhorns can beat everyone who can beat everyone the Longhorns can. So, the Longhorns can beat themselves.

17. Mary was Jesus's parent. Mary's parents were born with the taint of original sin. But Jesus was not tainted by original sin. Therefore, it's not true that everyone who has a parent born with the taint of original sin is also so tainted.

18. Popeye and Olive Oyl like each other, since Popeye likes everyone who likes Olive Oyl, and Olive Oyl likes everyone.

19. I like everyone who likes everyone I like. So, there are people that I like.

▶ 20. The government chooses to do x rather than y just in case it doesn't choose to do y over x. A person has veto power just in case the government can choose to do x over y only if that person doesn't prefer y to x. A person is a dictator just in case the government chooses to do x rather than y if he prefers x to y. Consequently, everyone with veto power is a dictator.

These are the syllogistic patterns that Aristotle considered valid, together with their medieval names. Show that each is valid in quantification theory. (Some require extra assumptions; they are listed in parentheses.)

21. Barbara: Every M is L; Every S is M; ∴ Every S is L.

22. Celarent: No M is L; Every S is M; ∴ No S is L.

23. Darii: Every M is L; Some S is M; ∴ Some S is L.

24. Ferio: No M is L; Some S is M; ∴ Some S is not L.

25. Cesare: No *L* is *M*; Every *S* is *M*; ∴ No *S* is *L*.

26. Camestres: Every *L* is *M*; No *S* is *M*; ∴ No *S* is *L*.

27. Festino: No *L* is *M*; Some *S* is *M*; ∴ Some *S* is not *L*.

28. Baroco: Every *L* is *M*; Some *S* is not *M*; ∴ Some *S* is not *L*.

29. Darapti: Every *M* is *L*; Every *M* is *S*; (There are *M*'s;) ∴ Some *S* is *L*.

30. Felapton: No *M* is *L*; Every *M* is *S*; (There are *M*'s;) ∴ Some *S* is not *L*.

31. Disamis: Some *M* is *L*; Every *M* is *S*; ∴ Some *S* is *L*.

32. Datisi: Every *M* is *L*; Some *M* is *S*; ∴ Some *S* is *L*.

33. Bocardo: Some *M* is not *L*; Every *M* is *S*; ∴ Some *S* is not *L*.

34. Ferison: No *M* is *L*; Some *M* is *S*; ∴ Some *S* is not *L*.

Medieval logicians added other syllogistic patterns to those Aristotle explicitly held valid. Show that these "subaltern moods" are valid, at least with the added assumptions in parentheses.

▶ **35.** Barbari: Every *M* is *L*; Every *S* is *M*; (There are *S*'s;) ∴ Some *S* is *L*.

36. Celaront: No *M* is *L*; Every *S* is *M*; (There are *S*'s;) ∴ Some *S* is not *L*.

37. Cesaro: No *L* is *M*; Every *S* is *M*; (There are *S*'s;) ∴ Some *S* is not *L*.

38. Camestros: Every *L* is *M*; No *S* is *M*; (There are *S*'s;) ∴ Some *S* is not *L*.

Theophrastus, who succeeded Aristotle as head of the Lyceum, also added additional syllogistic principles. Show that these, too, are valid, with the added assumptions in parentheses.

39. Baralipton: Every *M* is *L*; Every *S* is *M*; (There are *S*'s;) ∴ Some *L* is *S*.

▶ **40.** Celantes: No *M* is *L*; Every *S* is *M*; ∴ No *L* is *S*.

41. Dabitis: Every *M* is *L*; Some *S* is *M*; ∴ Some *L* is *S*.

42. Fapesmo: Every *M* is *L*; No *S* is *M*; (There are *M*'s;) ∴ Some *L* is not *S*.

43. Frisesomorum: Some *M* is *L*; No *S* is *M*; ∴ Some *L* is not *S*.

These arguments both have the form of *argumenta a recto ad obliquum,* in the terminology of Joachim Junge, who discussed such arguments in his 1638 textbook. They played an important role in leading logicians beyond Aristotelian logic to a comprehensive theory of relations. Show that each is valid.

44. Knowledge is a conceiving; ∴ The object of knowledge is an object of conception (Aristotle, *Topics*).

▶ **45.** A circle is a figure; ∴ Whoever draws a circle draws a figure (Junge).

These principles concern the distribution of quantifiers over sentential connectives. Quine has referred to them as *rules of passage*. Show that each is valid. (Throughout, $\mathscr{A}$ is any formula not containing x.)

46. $\forall x(Fx \ \& \ \mathscr{A}) \leftrightarrow (\forall xFx \ \& \ \mathscr{A})$ **47.** $\exists x(Fx \ \& \ \mathscr{A}) \leftrightarrow (\exists xFx \ \& \ \mathscr{A})$

48. $\forall x(Fx \lor \mathscr{A}) \leftrightarrow (\forall xFx \lor \mathscr{A})$ **49.** $\exists x(Fx \lor \mathscr{A}) \leftrightarrow (\exists xFx \lor \mathscr{A})$

50. $\forall x(Fx \rightarrow \mathscr{A}) \leftrightarrow (\exists xFx \rightarrow \mathscr{A})$ **51.** $\forall x(\mathscr{A} \rightarrow Fx) \leftrightarrow (\mathscr{A} \rightarrow \forall xFx)$

52. $\exists x(Fx \rightarrow \mathscr{A}) \leftrightarrow (\forall xFx \rightarrow \mathscr{A})$ **53.** $\exists x(\mathscr{A} \rightarrow Fx) \leftrightarrow (\mathscr{A} \rightarrow \exists xFx)$

In Chapter 9, we asserted that the order of existential quantifiers within a string of such quantifiers makes no difference; the same is true of universal quantifiers. Illustrate this by showing valid both these principles:

54. $\exists x\exists yFxy \leftrightarrow \exists y\exists xFxy$ ▶ **55.** $\forall x\forall yFxy \leftrightarrow \forall y\forall xFxy$

Although, in general, we can't switch existential and universal quantifiers to reach an equivalent formula, we can do so in special circumstances. Without using any rules of passage above, show that each of these switches is legitimate.

56. $\exists x\forall yFxy; \ \therefore \ \forall y\exists xFxy$

57. $\forall x\exists y(Fx \ \& \ Gy); \ \therefore \ \exists y\forall x(Fx \ \& \ Gy)$*

58. $\forall x\exists y(Fx \lor Gy); \ \therefore \ \exists y\forall x(Fx \lor Gy)$*

59. $\forall x\exists y(Fx \rightarrow Gy); \ \therefore \ \exists y\forall x(Fx \rightarrow Gy)$*

▶ **60.** $\forall x\exists y(Gy \rightarrow Fx); \ \therefore \ \exists y\forall x(Gy \rightarrow Fx)$*

Use deduction to establish the validity of these argument forms.

61. $\exists xFx \rightarrow \exists xGx; \ \therefore \ \exists x(Fx \rightarrow Gx)$

62. $\exists xFx \lor \exists xGx; \ \therefore \ \exists x(Fx \lor Gx)$

63. $\exists x(Fx \lor Gx); \ \therefore \ \exists xFx \lor \exists xGx$

64. $\exists xFx \rightarrow \forall y(Gy \rightarrow Hy); \exists xJx \rightarrow \exists xGx; \ \therefore \ \exists x(Fx \ \& \ Jx) \rightarrow \exists zHz$

▶ **65.** $\exists xFx \lor \exists xGx; \forall x(Fx \rightarrow Gx); \ \therefore \ \exists xGx$

66. $\forall x((Fx \ \& \ Gx) \rightarrow Hx); Ga \ \& \ \forall xFx; \ \therefore \ Fa \ \& \ Ha$

67. $\forall x(\exists yFyx \rightarrow \forall zFxz); \ \therefore \ \forall y\forall x(Fyx \rightarrow Fxy)$

68. $\forall x(Fx \rightarrow \forall y(Gy \rightarrow Hxy)); \exists x(Fx \ \& \ \exists y\neg Hxy); \ \therefore \ \exists x\neg Gx$

69. $\neg\exists x(Fx \ \& \ Gx); \ \therefore \ \forall x(Fx \rightarrow \neg Gx)$

70. $\forall x(Fx \rightarrow \neg Gx); \ \therefore \ \neg\exists x(Fx \ \& \ Gx)$

7.3 Derivable Rules for Quantifiers

By adding these three rules and universal proof to our natural deduction system for sentential logic, we get a system with all the power we need. With it, we can demonstrate the validity of any valid argument form in quantification theory. Nevertheless, we can increase its efficiency and naturalness considerably by adding some further rules. This section presents some derivable rules that, while theoretically superfluous, make the proof system more pleasant to work with.

Combining Rules

We can apply quantifier rules several times in a single step. Suppose, for example, that we want to take instances of $\forall x \forall y F xy$. The universal exploitation rule requires us to move first to the instance $\forall y F ay$, say, and then to its instance $F aa$. But we can substitute any constant for both x and y here, since both are universally quantified. So, it's easy to perform the operation in one step. We can move directly from $\forall x \forall y F xy$ to an instance of an instance of it, $F aa$. Similarly, we can move to $F ab$, $F ba$, $F bb$, $F ac$, and so on. We can record that we've exploited a series of n universal quantifiers at once by writing not $\forall E$ but $\forall E^n$. In this case, then, the application would look like:

$$\begin{array}{ll} \text{m.} & \forall x \forall y F xy \\ \text{m + n.} \;\; F aa & \qquad \forall E^2, \text{m} \end{array}$$

We can readily do the same for series of existential quantifiers, provided that we replace each quantified variable with a new constant. So, we may move immediately from $\exists x \exists y \exists z (F xy \,\&\, F yz \,\&\, F zx)$ to $F ab \,\&\, F bc \,\&\, F ca$, citing the rule we've applied as $\exists E^3$.

We can also compress a sequence of universal proofs into a single proof in a similar way. We can prove a formula with an initial string of n universal quantifiers by proving an instance of an instance . . . of an instance of the formula, using n different new constants to replace the n quantified variables. For example, we can prove $\forall x \forall y (F xy \rightarrow \neg F yx)$ by proving $F ab \rightarrow \neg F ba$, where neither a nor b have appeared earlier in the proof.

Although these combinations of rules or proofs are convenient, it's not a good idea to combine exploitations of universal and existential quantifiers. While $\forall E^2$ and $\exists E^4$ are fairly easy to follow, something like $\forall E^2 \exists E^3 \forall E$ would be extremely difficult to apply or understand.

Rewriting Variables

Variables have no independent meanings. The formulas $\forall x F x$ and $\forall z F z$ function in logically similar ways; so do $\exists y G y$ and $\exists w G w$. Indeed, it's easy to show that these pairs are equivalent. Here are the proofs in one direction:

1. $\forall x F x$	A		1. $\exists y G y$	A	
2. Show $\forall z F z$			2. Show $\exists w G w$		
3. ⌈ Show Fa			3. ⌈ Ga	$\exists$E, 1	
4. ⌊ ⌊ Fa	$\forall$E, 1		4. ⌊ $\exists w G w$	$\exists$I, 3	

The proofs of the other directions follow exactly the same pattern.

Thus, we can substitute one variable for another throughout a formula. The only restriction we must observe is that we should not introduce into a formula a variable that is already there; otherwise, we could go from the legitimate formula $\forall x \forall y F x y$ to the very different nonformula $\forall x \forall x F x x$. The derivable rule, then, is this:

Variable Rewrite (VR)

n. $\mathscr{A}$
m. $\mathscr{A}[v/u]$ VR, n

Here v is foreign to $\mathscr{A}$.

Exploiting Negations

It's extremely useful to have a direct way of dealing with negations of quantified formulas. Our rules allow us to attack formulas with quantifiers as main connectives in one step. But, if a quantifier is preceded by a negation sign, the proof strategy becomes much more complicated. Luckily, negations of quantified formulas are equivalent to formulas with quantifiers as main connectives. Two rules, called *quantifier negation* rules, relate quantified formulas to their negations. In the process, these rules relate the universal and existential quantifiers. In fact, they show how to define each such quantifier in terms of the other.

Quantifier Negation (QN)

n. $\neg\exists v \mathscr{A}$
m. $\forall v \neg \mathscr{A}$ QN, n

n. $\neg\forall v \mathscr{A}$
m. $\exists v \neg \mathscr{A}$ QN, n

Both versions of quantifier negation are invertible. That is, the premise and conclusion are equivalent, and so we can use them in either direction. We can infer $\exists x \neg Fx$ from $\neg \forall x Fx$, and vice versa. Similarly, just by adding a negation sign, we can see that $\exists x Fx$ is equivalent to $\neg \forall x \neg Fx$ and that $\forall x Fx$ is equivalent to $\neg \exists x \neg Fx$. So, we can define the quantifiers in terms of each other.

By deriving even simple applications of these rules from our basic quantifier rules, we can see how much work they can save. These two proofs are necessary to show, for example, that $\exists x \neg Fx$ and $\neg \forall x Fx$ are equivalent.

$$
\begin{array}{lll}
1. & \exists x \neg Fx & A \\
2. & \text{Show } \neg \forall x Fx & \\
3. & \lceil \forall x Fx & \text{AIP} \\
4. & \mid \neg Fa & \exists E, 1 \\
5. & \lfloor Fa & \forall E, 3 \\
\end{array}
$$

$$
\begin{array}{lll}
1. & \neg \forall x Fx & A \\
2. & \text{Show } \exists x \neg Fx & \\
3. & \lceil \neg \exists x \neg Fx & \text{AIP} \\
4. & \mid \text{Show } \forall x Fx & \\
5. & \mid \lceil \text{Show } Fa & \\
6. & \mid \mid \lceil \neg Fa & \text{AIP} \\
7. & \mid \mid \mid \exists x \neg Fx & \exists I, 6 \\
8. & \mid \mid \lfloor \neg \exists x \neg Fx & R, 3 \\
9. & \lfloor \quad \neg \forall x Fx & R, 1 \\
\end{array}
$$

Deriving the other equivalence is similar.

Because QN takes the form of an equivalence, the replacement principle allows us to apply it to portions of formulas as well as entire formulas. Each of the following is thus a legitimate application of QN.

$$
\begin{array}{ll}
\neg \exists x Fxx & \therefore \forall x \neg Fxx \\
\neg \forall x \forall y Gxy & \therefore \exists x \neg \forall y Gxy \\
\forall x \forall y \neg \forall z (Fxz \ \& \ Fyz) & \therefore \forall x \forall y \exists z \neg (Fxz \ \& \ Fyz) \\
\exists x \neg \exists y Gyx & \therefore \exists x \forall y \neg Gyx \\
\end{array}
$$

This rule, too, can be applied several times in a single step. We can abbreviate the first four lines below as the two lines immediately after:

$$
\begin{array}{ll}
\neg \forall x \forall y \forall z ((Rxy \ \& \ Ryz) \rightarrow Rxz) & \\
\exists x \neg \forall y \forall z ((Rxy \ \& \ Ryz) \rightarrow Rxz) & \text{QN} \\
\exists x \exists y \neg \forall z ((Rxy \ \& \ Ryz) \rightarrow Rxz) & \text{QN} \\
\exists x \exists y \exists z \neg ((Rxy \ \& \ Ryz) \rightarrow Rxz) & \text{QN} \\
\end{array}
$$

$$
\begin{array}{ll}
\neg \forall x \forall y \forall z ((Rxy \ \& \ Ryz) \rightarrow Rxz) & \\
\exists x \exists y \exists z \neg ((Rxy \ \& \ Ryz) \rightarrow Rxz) & \text{QN}^3 \\
\end{array}
$$

Problems

Establish the validity of these argument forms by means of deduction.

1. $\forall x(Fx \rightarrow \forall y(Gy \rightarrow Hxy))$
$\forall x(Dx \rightarrow \forall y(Hxy \rightarrow Cy))$
$\therefore \exists x(Fx \& Dx) \rightarrow \forall y(Gy \rightarrow Cy)$

2. $\forall x((Fx \lor Hx) \rightarrow (Gx \& Kx))$
$\neg \forall x(Kx \& Gx)$
$\therefore \exists x \neg Hx$

3. $\exists x(Fx \& \forall y(Gy \rightarrow Hxy))$
$\therefore \exists x(Fx \& (Ga \rightarrow Hxa))$

4. $\forall x \forall y(Gxy \leftrightarrow (Fy \rightarrow Hx))$
$\forall z Gaz$
$\therefore \exists x Fx \rightarrow \exists x Hx$

▶ **5.** $\forall x(\exists y Fxy \rightarrow \exists y \neg Gy)$
$\exists x \exists y Fxy$
$\forall x(Gx \leftrightarrow \neg Hx)$
$\therefore \exists x Hx$

6. $\forall x(Mx \rightarrow Hx)$
$\exists x \exists y((Fx \& Mx) \& (Gy \& Jyx))$
$\exists x Hx \rightarrow \forall y \forall z(\neg Hy \rightarrow \neg Jyz)$
$\therefore \exists x(Gx \& Hx)$

7. $\forall x(\exists y Fxy \rightarrow \forall y Fyx)$
$\exists x \exists y Fxy$
$\therefore \forall x \forall y Fxy$

8. $\forall x(Fx \leftrightarrow Gx)$
$\therefore \forall x Fx \leftrightarrow \forall x Gx$

9. $\forall x(Fx \leftrightarrow Gx)$
$\therefore \exists x Fx \leftrightarrow \exists x Gx$

10. $\exists x(Fx \rightarrow Gx)$
$\therefore \forall x Fx \rightarrow \exists x Gx$

11. $\exists x(Fx \& \forall y(Gy \rightarrow Hy))$
$\forall x(Fx \rightarrow (\neg Lx \rightarrow \neg \exists z(Kz \& Hz)))$
$\therefore \exists x(Kx \& Gx) \rightarrow \exists x Lx$

12. $\exists x \forall y(\exists z Fyz \rightarrow Fyx)$
$\forall x \exists y Fxy$
$\therefore \exists x \forall y Fyx$

13. $\forall x(Kx \rightarrow (\exists y Lxy \rightarrow \exists z Lzx))$
$\forall x(\exists z Lzx \rightarrow Lxx)$
$\neg \exists x Lxx$
$\therefore \forall x(Kx \rightarrow \forall y \neg Lxy)$

14. $\neg \forall x(Hx \lor Kx)$
$\forall x((Fx \lor \neg Kx) \rightarrow Gxx)$
$\therefore \exists x Gxx$

▶ **15.** $\forall x(Fxx \rightarrow Hx)$
$\exists x Hx \rightarrow \neg \exists y Gy$
$\therefore \forall x(Gx \rightarrow \neg \exists z Fzz)$

16. $\forall x(Fx \rightarrow (Gx \lor Hx))$
$\forall x((Jx \& Fx) \rightarrow \neg Gx)$
$\forall x(\neg Fx \rightarrow \neg Jx)$
$\therefore \forall x(Jx \rightarrow Hx)$

17. $\neg \exists x(Hxa \& \neg Gxb)$
$\forall x \neg (Fxc \& Fbx)$
$\forall x(Gex \rightarrow Fxe)$
$\therefore \neg (Hea \& Fec)$

18. $\forall x(\exists y(Ay \& Bxy) \rightarrow Cx)$
$\exists y(Dy \& \exists x((Fx \& Gx) \& Byx))$
$\forall x(Gx \rightarrow Ax)$
$\therefore \exists x(Cx \& Dx)$

19. $\forall x \forall y \forall z((Fxy \& Fyz) \rightarrow Fxz)$
$\neg \exists x Fxx$
$\therefore \forall x \forall y(Fxy \rightarrow \neg Fyx)$

20. $\forall x \forall y \forall z ((Fxy \ \& \ Fyz) \rightarrow \neg Fxz)$
    $\therefore \ \forall x \neg Fxx$

21. $\forall x \forall y \forall z ((Fxy \ \& \ Fyz) \rightarrow Fxz)$
    $\forall x \forall y (Fxy \rightarrow Fyx)$
    $\forall x \exists y Fxy$
    $\therefore \ \forall x Fxx$

22. $\forall x \forall y \forall z ((Fxy \ \& \ Fxz) \rightarrow Fyz)$
    $\forall x \forall y (Fxy \rightarrow Fyx)$
    $\forall x Fxx$
    $\therefore \ \forall x \forall y \forall z ((Fxy \ \& \ Fyz) \rightarrow Fxz)$

23. $\forall x \forall y \forall z ((Fxy \ \& \ Fyz) \rightarrow Fxz)$
    $\forall x \forall y (Fxy \rightarrow Fyx)$
    $\forall x Fxx$
    $\therefore \ \forall x \forall y \forall z ((Fxy \ \& \ Fxz) \rightarrow Fyz)$

24. $\forall x \forall y \forall z ((Fxy \ \& \ Fxz) \rightarrow Fyz)$
    $\therefore \ \forall x \forall y \forall z ((Fxy \ \& \ Fxz) \rightarrow Fzy)$

▶ 25. $\exists x \forall y \neg Fxy$
    $\therefore \ \exists x \forall y \forall z (Fxz \rightarrow Fzy)$

26. $\forall x (Fx \leftrightarrow \forall y Gy)$
    $\therefore \ \forall x Fx \lor \forall x \neg Fx$

27. $Fa \rightarrow (\exists x Gx \rightarrow Gb)$
    $\forall x (Gx \rightarrow Hx)$
    $\forall x (\neg Jx \rightarrow \neg Hx)$
    $\therefore \ \neg Jb \rightarrow (\neg Fa \lor \forall x \neg Gx)$

28. $\forall x (Dx \rightarrow Fx)$
    $\therefore \ Da \rightarrow (\forall y (Fy \rightarrow Gy) \rightarrow Ga)$

29. $\exists x Fx \rightarrow \forall y ((Fy \lor Gy) \rightarrow Hy)$
    $\exists x Hx$
    $\neg \forall z \neg Fz$
    $\therefore \ \exists x (Fx \ \& \ Hx)$

▶ 30. $\exists x (Fx \ \& \ \forall y (Ty \rightarrow Gy))$
    $\forall x (Fx \rightarrow (\exists y (Ay \ \& \ Gy) \rightarrow Bxx))$
    $\exists z (Az \ \& \ Tz)$
    $\therefore \ \exists x Bxx$

31. $\forall x \neg Fxc \rightarrow \exists x Gxb$
    $\therefore \ \exists x (\neg Fxc \rightarrow Gxb)$

32. $\forall x Fx$
    $\therefore \ \neg \exists x Gx \leftrightarrow \neg (\exists x (Fx \ \& \ Gx) \ \& \ \forall y (Gy \rightarrow Fy))$

33. $\exists x (Px \ \& \ \neg Mx) \rightarrow \forall y (Py \rightarrow Ly)$
    $\exists x (Px \ \& \ Nx)$
    $\forall x (Px \rightarrow \neg Lx)$
    $\therefore \ \exists x (Nx \ \& \ Mx)$

34. $\forall x (Fx \rightarrow \neg \exists y (Gy \ \& \ Hxy))$
    $\forall x (Fx \rightarrow \exists y (Fy \ \& \ Hxy))$
    $\therefore \ \forall x \forall y (Hxy \rightarrow Hyx) \rightarrow \forall x \neg (Fx \ \& \ Gx)$

35. $\forall x \forall y (Fxy \rightarrow Fyx)$
    $\therefore \ \forall x \forall y (Fxy \leftrightarrow Fyx)$

36. $\forall x \neg Fxx$
    $\therefore \ \neg \exists x \forall y (Fyx \leftrightarrow \exists z \forall w ((Fwz \rightarrow Fwy) \ \& \ \neg Fzy))$

37. $\forall x\forall y((Ax \ \& \ By) \rightarrow Cxy)$
$\exists y(Fy \ \& \ \forall z(Hz \rightarrow Cyz))$
$\forall x\forall y\forall z((Cxy \ \& \ Cyz) \rightarrow Cxz)$
$\forall x(Fx \rightarrow Bx)$
$\therefore \ \forall z\forall y((Az \ \& \ Hy) \rightarrow Czy)$

38. $\forall z(\neg Hz \leftrightarrow \forall x(Fx \ \& \ Gz))$
$\forall x\exists y(Gy \ \& \ Fx)$
$\therefore \ \neg \forall xHx$

39. Say that an archetypal pig is something such that, if anything at all is a pig, *it* is. Show that there is an archetypal pig.

40. Say that something is truly ugly just in case, if it is beautiful, then anything is. Show that some things are truly ugly.

Notes

[1] This rule was first formulated by the English philosopher William of Ockham in the fourteenth century.

[2] The American philosopher W. V. Quine first formulated existential exploitation in this way in 1950, in his *Methods of Logic* (Cambridge: Harvard University Press, 1950, 1982).

[3] Michael Spivak, *Calculus on Manifolds* (Menlo Park, Calif.: W. A. Benjamin, 1965): 13 and 12.

[4] Alan J. Weir, *Lebesgue Integration and Measure* (Cambridge: Cambridge University Press, 1973): 108.

[5] Angus E. Taylor, *Calculus* (Englewood Cliffs, N.J.: Prentice-Hall, 1959): 71.

[6] Edwin E. Moise and Floyd L. Downs, Jr., *Geometry* (Menlo Park, Calif.: Addison-Wesley, 1967): 230.

8

IDENTITY, FUNCTION SYMBOLS, AND DESCRIPTIONS

his chapter extends our system of quantificational logic to include a new binary predicate and two new ways of forming singular terms. The new predicate, identity, represents the familiar notion of equality in mathematics and an equally familiar concept in natural language. In English, we most often express this concept by using the word *is*. One set of new singular terms will be formed with the help of *function symbols*. Function symbols occur frequently in mathematics; $+$, $-$, $\div$, $\int$, and ∂ all represent functions. Such English expressions as 'mother of,' 'diameter of,' 'length of,' 'kinetic energy of,' 'home town of,' and 'grade of' also represent functions. We've discussed functions at several points in earlier chapters; we will now develop ways of representing them explicitly in our formal language. The other set of new singular terms will be *definite descriptions*. Such phrases as 'the winner of the Nobel Prize in physics in 1966,' 'the year the Saint Louis Browns finished only a game behind the Yankees,' and 'the first girl I ever loved' are definite descriptions in English.

In our quantification theory as it stands, we can already represent identity by using a dyadic predicate and adopting some special axioms concerning it. The theory, furthermore, allows us to represent functions and definite descriptions by using polyadic predicates and some accompanying assumptions. By adding identity, function symbols, and descriptions, however, we can use quantification theory to formalize very naturally a wide variety of

English arguments and mathematical theories. In certain subtle ways, these extensions also add to the power of quantificational logic.

8.1 Identity

To symbolize identity, we'll use the symbol $=$, which mathematicians generally use for the same concept. (For negations of identity formulas, we'll use the symbol $\neq$ to abbreviate $\neg \ldots = \ldots$. Thus, we'll generally write $\neg a = b$ as $a \neq b$.) Obviously, having such a predicate in our logic allows us to work with mathematical formulas with a minimum of translation or alteration. It also allows us to render many English sentences very simply and naturally:

(1) a. Tully is Cicero.
   b. Austin is the capital of Texas.
   c. Ronald Reagan is the President of the United States.
   d. The wealthiest town in the United States is West Hartford, Connecticut.
   e. Baltimore was Babe Ruth's birthplace.

In each of these sentences, the verb *to be* expresses identity. This, of course, isn't that verb's only job. In 'Dan's car is red,' 'Phyllis is angry,' and 'Peter is coming to town tomorrow,' *is* expresses predication. Aristotle was perhaps the first person to recognize the distinction between these two roles of *to be*. Philosophers today sometimes refer to this difference as the contrast between the *is* of identity and the *is* of predication. A primary use of the identity symbol in logic is to formalize the *is* of identity.

Adding the identity symbol to our language requires only one small revision in our formation rules. We must add the rule that, where c and d are constants, $c = d$ is a formula. (In fact, it's an atomic formula.) The required addition to our semantic principles is similarly trivial. A formula of the form $c = d$ is true on an interpretation just in case the interpretation function assigns the same object to both c and d.

Numerical Expressions

Identity allows us to translate a wide and surprising variety of English constructions. We can easily express numerical determiners. In quantificational logic, the existential quantifier means, in essence, 'at least one.' But how can we express, for instance, 'at least two'? We can use two existential quantifiers, but this, in itself, doesn't suffice; $\exists x \exists y (Fx \,\&\, Fy)$ doesn't rule out the possibility that x and y stand for the same object. To say that there are at least two F's, we need to say that there is an object x and there is another object y, both of which are F's. We can say that x and y are different by writing $x \neq y$. So, the appropriate symbolization is $\exists x \exists y (Fx \,\&\, Fy \,\&\, x \neq y)$.

To say that there are at least three F's, we need to write three existential quantifiers and three negated identities to guarantee that all the objects are distinct. The correct translation is correspondingly $\exists x\exists y\exists z(Fx \& Fy \& Fz \& x \neq y \& y \neq z \& x \neq z)$. We can apply the same strategy for any number of objects.

There are at least ... F's	Translation
One	$\exists xFx$
Two	$\exists x\exists y(Fx \& Fy \& x \neq y)$
Three	$\exists x\exists y\exists z(Fx \& Fy \& Fz \& x \neq y \& y \neq z \& x \neq z)$
Four	$\exists x\exists y\exists z\exists w(Fx \& Fy \& Fz \& Fw \& x \neq y$ $\& x \neq z \& x \neq w \& y \neq z \& y \neq w \& z \neq w)$
n	$\exists x_1 \ldots \exists x_n(Fx_1 \& \ldots \& Fx_n \& x_1 \neq x_2$ $\& x_1 \neq x_3 \& \ldots \& x_1 \neq x_n \& x_2 \neq x_3 \& \ldots$ $\& x_2 \neq x_n \& \ldots \& x_{n-1} \neq x_n)$

We can thus symbolize numerical determiners of the form 'at least n.' Notice, however, that the complexity of the resulting formulas increases very quickly.

We can also translate sentences containing determiners of the form 'at most n.' Suppose, for example, that we want to say that there is at most one omnipotent being. This assertion does not imply that there is an omnipotent being; it merely states that there is no more than one. The assertion is the negation of 'There are at least two omnipotent beings.' So, we might try to negate a formula of the kind we've just seen, obtaining $\neg\exists x\exists y(Ox \& Oy \& x \neq y)$. This formula is equivalent to $\forall x\forall y((Ox \& Oy) \rightarrow x = y)$, which says that, for any x and y, if x and y are both omnipotent, then x and y are identical. It says, that is, that all omnipotent beings are the same. But that means that there is at most one omnipotent being.

Similarly, if we want to translate 'There are at most two superpowers,' we can treat it as the negation of 'There are at least three superpowers,' obtaining $\neg\exists x\exists y\exists z(Sx \& Sy \& Sz \& x \neq y \& x \neq z \& y \neq z)$. But this is equivalent to $\forall x\forall y\forall z((Sx \& Sy \& Sz) \rightarrow (x = y \vee x = z \vee y = z))$, which says that, if x, y, and z are all superpowers, two of them must be identical.

This strategy works for any value of n.

There are at most ... F's	Translation
One	$\forall x\forall y((Fx \& Fy) \rightarrow x = y)$
Two	$\forall x\forall y\forall z((Fx \& Fy \& Fz) \rightarrow (x = y \vee x = z \vee y = z))$
Three	$\forall x\forall y\forall z\forall w((Fx \& Fy \& Fz \& Fw)$ $\rightarrow (x = y \vee x = z \vee x = w \vee y = z \vee y = w \vee z = w))$
$n - 1$	$\forall x_1 \ldots \forall x_n((Fx_1 \& \ldots \& Fx_n) \rightarrow (x_1 = x_2$ $\vee x_1 = x_3 \vee \ldots \vee x_1 = x_n \vee x_2 = x_3 \vee \ldots \vee x_2$ $= x_n \vee \ldots \vee x_{n-1} = x_n))$

Finally, we can translate numerical determiners of the form 'exactly n.' Since 'There are exactly n F's' is equivalent to a conjunction of 'There are at least n F's' and 'There are at most n F's,' we can devise such translations simply by conjoining the translations given in the above tables. But there is an easier way. If we want to say that there is exactly one God, we can say that there is a God, and anything that is a God is identical with it. If we want to say that there are exactly two great American writers, we can say that there are at least two and, moreover, that any great American writer must be identical with one or the other. This suggests the strategy:

There are exactly ... F's	Translation
Zero	$\forall x \neg Fx$
One	$\exists x \forall y(Fx \ \& \ (Fy \rightarrow y = x))$
Two	$\exists x \exists y \forall z(Fx \ \& \ Fy \ \& \ x \neq y \ \& \ (Fz \rightarrow (z = x \lor z = y)))$
Three	$\exists x \exists y \exists z \forall w(Fx \ \& \ Fy \ \& \ Fz \ \& \ x \neq y \ \& \ x \neq z \ \& \ y \neq z$ $\& \ (Fw \rightarrow (w = x \lor w = y \lor w = z)))$
n	$\exists x_1 \ldots \exists x_n \forall y(Fx_1 \ \& \ldots \& \ Fx_n$ $\& \ x_1 \neq x_2 \ \& \ x_1 \neq x_3 \ \& \ldots \& \ x_1 \neq x_n \ \& \ x_2 \neq x_3$ $\& \ldots \& \ x_2 \neq x_n \ \& \ldots \& \ x_{n-1} \neq x_n$ $\& \ (Fy \rightarrow (y = x_1 \lor \ldots \lor y = x_n)))$

Anaphora

Identity also allows us to symbolize sentences containing the words *other, another,* and *else.* Typically, these words are anaphoric in much the way pronouns are. That is, they refer implicitly to something introduced in prior discourse. If we say 'John admires himself, but he hates everybody else,' we mean that John hates everybody other than John. A good symbolization of the latter sentence, then, is $\forall x(x \neq j \rightarrow Hjx)$, which says that everyone who isn't John is hated by John. Of course, the antecedent discourse may have introduced some other referent. If the preceding sentence were 'John loves Susan,' we could take 'everybody else' to mean "everybody other than Susan" or "everybody other than John and Susan." An appropriate symbolization is $\forall x(x \neq s \rightarrow Hjx)$ or $\forall x((x \neq s \ \& \ x \neq j) \rightarrow Hjx)$. Unfortunately for the trans-lator, natural languages contain referential ambiguities and determine the range of possible interpretations in very complex and still poorly understood ways. In many cases, only one interpretation is possible. If we say 'A passenger shouted, "This is a hijacking!" and another waved a pistol overhead,' we clearly mean that a passenger other than the one who shouted did the pistol-waving. Introducing the obvious predicates, then, we can translate this sentence as $\exists x \exists y \exists z(Px \ \& \ Sx \ \& \ Py \ \& \ y \neq x \ \& \ Gz \ \& \ Wyz)$ with all quantifiers in front, or, more naturally, as $\exists x((Px \ \& \ Sx) \ \& \ \exists y(Py \ \& \ y \neq x \ \& \ \exists z(Gz \ \& \ Wyz)))$.

Superlatives

Identity also allows us to translate superlatives such as *fastest* and *most interesting* into formulas containing predicates that represent comparatives such as *faster* and *more interesting*. Suppose that *F* symbolizes 'is faster than.' To express 'This is the fastest car,' we can say that it's faster than every other car. That is, we can say that this car is faster than every car not identical with it. So, an appropriate translation is $\forall x((Cx \ \& \ x \neq a) \rightarrow Fax)$, where *a* stands for 'this car.' (We might also want to add a subformula *Ca*, to represent 'This is a car.') If we want to say 'this is the slowest car,' we can say that every other car is faster than it: $\forall x((Cx \ \& \ x \neq a) \rightarrow Fxa)$.

Notice that, if this is the fastest car, then it follows that no car is faster than it. But these two statements aren't equivalent. Suppose that several cars tie for the title of fastest car in the universe. No car is faster than any of them, but none is the fastest car; each will be one of the fastest, but not the fastest.

Superlatives thus translate readily into formulas containing dyadic predicates representing correlated comparatives. We must be very cautious in translating, however, to avoid confusing the direction of the comparison and mixing up *more* and *less:*

Statement	Translation
a is more *F* than *b*	Fab
a is the most *F*	$\forall x(x \neq a \rightarrow Fax)$
a is the least *F*	$\forall x(x \neq a \rightarrow Fxa)$
a is less *G* than *b*	Gab
a is the most *G*	$\forall x(x \neq a \rightarrow Gxa)$
a is the least *G*	$\forall x(x \neq a \rightarrow Gax)$

The order of the variable and constant in this formula makes all the difference between *more* and *less,* and *most* and *least.* In our examples above, we had to add a predicate to these schemas; we wanted to say, not that this was the fastest *thing,* but that this was the fastest *car.*

Only

Finally, identity permits us to symbolize many sentences containing the word *only.* Recall that we can translate sentences of the form 'Only *F G*' as $\forall x(\neg Fx \rightarrow \neg Gx)$ or, equivalently, $\forall x(Gx \rightarrow Fx)$. These are equivalent, that is, to 'No non-*F G*' or 'All *G F*.' Identity allows us to translate as well sentences involving *only* in combination with singular terms. Consider, for example, 'Only Elmo got fired.' We could paraphrase this as 'Only those identical with Elmo got fired.' So, applying the strategy we devised for combinations of *only* and a general term, this sentence should be equivalent to 'All who got fired were identical with Elmo,' or, in symbolic terms, $\forall x(Fx \rightarrow x = e)$.

Equivalently, we could say that nobody but Elmo got fired: $\forall x(x \neq e \rightarrow \neg Fx)$ or $\neg \exists x(Fx \lor x \neq e)$ or $\forall x(\neg Fx \lor x = e)$.

These formulas seem surprising, for they do not imply that Elmo got fired. 'Only Elmo got fired' is at least extremely misleading, if not false, if Elmo didn't lose his job. Translating 'Only Elmo got fired' as $Fe \;\&\; \forall x(Fx \rightarrow x = e)$ — "Elmo, and only Elmo got fired"—thus better harmonizes with certain linguistic intuitions.

Problems

Use identity to symbolize these sentences in QL. If a sentence is ambiguous, explain why.

1. April is the cruelest month (T. S. Eliot).

2. Twice no one dies (Thomas Hardy).

3. Philosophy is the highest music (Plato).

4. To work for the common good is the greatest creed (Albert Schweitzer).

▶ 5. This poem is the reader and the reader this poem (Ishmael Reed).

6. Liberty is always dangerous, but it is the safest thing we have (Harry Emerson Fosdick).

7. Action is the last resource of those who know not how to dream (Oscar Wilde).

8. The seed ye sow, another reaps; the wealth ye find, another keeps; the robes ye weave, another wears; the arms ye forge, another bears (Percy Bysshe Shelley).

9. The most precious thing a parent can give a child is a lifetime of happy memories (Frank Tyger).

▶ 10. That action is best, which procures the greatest happiness for the greatest number (Francis Hutcheson).

11. The worst-tempered people I've ever met were people who knew they were wrong (Wilson Mizner).

12. The most valuable executive is one who is training somebody to be a better man than he is (Robert Ingersoll).

13. I hold that man in the right who is most closely in league with the future (Henrik Ibsen).

14. An executive organization, like a chain, is no stronger than its weakest link (Robert Patterson).

▶ 15. In cases of difficulty and when hopes are small, the boldest counsels are the safest (Livy).

16. Behold, the people is one, and they have all one language; . . . and now nothing will be restrained from them, that they have imagined to do (Genesis 11:6).

17. Once upon a time there were Three Bears, who lived together in a house of their own, in a wood (Robert Southey).

18. He prayeth best, who loveth best all things both great and small (Samuel Taylor Coleridge).*

Sentences 19, 20, 23, and 24 are cited by James McCawley[1]; each involves *only* in combination with singular terms.

19. Only Lyndon pities himself.

▶ 20. Only Lyndon pities Lyndon.

21. Lyndon pities only Lyndon.

22. Lyndon pities only himself.

23. Only Lyndon pities only Lyndon.*

24. Only Lyndon pities only himself.*

▶ 25. Only Lyndon pities only those who pity only Lyndon.*

8.2 Tableau Rules for Identity

Semantic tableaux extend to cover identity with the addition of two rules: identity left and identity right. Though modern thinking about the logic of identity goes back to Leibniz, Aristotle, in some of his earliest logical writings, formulated principles about identity that underlie these tableau rules. The first principle is very simple: Everything is identical with itself. Socrates is Socrates; France is France; and $2 = 2$. The second principle, sometimes called the principle of the indiscernibility of identicals, says that, if $a = b$, then whatever is true of a is true of b, and conversely. This principle is also called the substitutivity of identicals, because it implies that, if $a = b$, then we can substitute a for b (or vice versa) in any sentence or formula without changing its truth value. For example, Cassius Clay is Muhammad Ali. It follows that, if Muhammad Ali is a great boxer, then so is Cassius Clay. Similarly, Lew Alcindor and Kareem Abdul-Jabbar are the same person. So, if Kareem Abdul-Jabbar plays for the Lakers, we can infer that Lew Alcindor plays for the Lakers. We can take a more interesting case: if Lew Alcindor is his own agent, then so is Kareem Abdul-Jabbar. But we can also conclude that Lew Alcindor is Kareem Abdul-Jabbar's agent and that Kareem Abdul-

Jabbar is Lew Alcindor's agent. Thus, we don't need to make the substitution every time it would be legitimate.

The second principle inspires the rule for identity on the left. Suppose that, where t and t' are two terms, we find $t = t'$ on the left side of a tableau branch. We can substitute t and t', or vice versa, in any formula on that branch. The substitution need not occur every time it is possible. In this way, the rule is like the deduction rule of existential introduction and unlike the other tableau and deduction rules. $\mathscr{A}[t//t']$ will be any result of substituting t for some or all occurrences of t' in $\mathscr{A}$.

Identity Left (=L)

$$\begin{array}{r|l} *\,t = t' & \\ \mathscr{A} & \\ \mathscr{A}[t//t'] & \\ (\text{or } \mathscr{A}[t'//t]) & \end{array} \qquad\qquad \begin{array}{r|l} *\,t = t' & \\ & \mathscr{A} \\ & \mathscr{A}[t//t'] \\ & (\text{or } \mathscr{A}[t'//t]) \end{array}$$

The first principle, that everything is identical with itself, inspires the rule for identity on the right. This rule is extremely simple. Any statement of self-identity ($a = a$) is always true. So, if such a formula appears on the right side of a branch, we've reached a contradiction; the branch closes.

Identity Right (=R)

$$\begin{array}{l|l} & t = t \\ \text{Cl} & \end{array}$$

As examples showing how these rules work, let's consider three standard properties of identity. The first is reflexivity: Everything is identical with itself. This is simply our first principle and so should be easy to show valid with the rule for identity on the right:

$$\begin{array}{l|l} & \forall x(x = x) \;\checkmark \\ & a = a \\ \text{Cl} & \end{array}$$

A second property is symmetry: If $x = y$, then $y = x$. The order in which terms appear in the identity relation makes no difference. The rule for identity on the left makes this property easy to establish:

$$\begin{array}{c|l}
 & \forall x \forall y(x = y \rightarrow y = x) \quad \checkmark \\
 & a = b \rightarrow b = a \quad \checkmark \\
* \; a = b & \\
 & b = a \\
 & a = a \\
\hline
 & \text{Cl}
\end{array}$$

We reach $a = a$ on the right by substituting a for the b in $b = a$.

A third property is transitivity. If $x = y$ and $y = z$, then $x = z$. This, too, is easy to establish. We may begin by applying $\forall$R three times, using some sentential rules, and then applying $=$L, substituting a for b in $b = c$ to obtain $a = c$:

$$\begin{array}{r|l}
 & \forall x \forall y \forall z((x = y \; \& \; y = z) \rightarrow x = z) \quad \checkmark \\
 & (a = b \; \& \; b = c) \rightarrow a = c \quad \checkmark \\
\checkmark \;\; a = b \; \& \; b = c & \\
 & a = c \\
* \; a = b & \\
b = c & \\
a = c & \\
\hline
 & \text{Cl}
\end{array}$$

These tableau rules are not difficult to use. They suffice for the logic of identity; we can close a tableau for any valid formula or argument form in our expanded quantificational language using just the rules provided so far. Obviously, however, the $=$L rule offers a large number of choices. It licenses substitutions, but it doesn't tell us which substitutions are worth making. In general, we'll follow the strategy we used to apply $\forall$L and $\exists$R, which also offer choices among various possible instances. We'll use $=$L to obtain formulas that close the tableau or have some subformulas in common with other formulas already present or obtainable on the branch.

Problems

Consider these three sentences: (a) There is at most one object. (b) There is exactly one object. (c) Everything is the same as everything else.

1. Are (a) and (b) equivalent?

2. Are (b) and (c) equivalent?

3. Are (c) and (a) equivalent?

Evaluate these arguments for validity.

4. Joan was born in 1952, but Fran was born in 1955. So, Joan isn't Fran.

▶ **5.** Frank played in the tournament and won; Joe played, but didn't win. So, at least two people played in the tournament.

6. Everybody other than Admiral Inman has a boss. Thus, given any two people, at least one of them has a boss.

7. Everyone suspects someone, but no one suspects himself or herself. So everyone suspects someone else.

8. Everything is the same as everything else. So, either everything is good, or nothing is.

9. There are at most two ultimate forces at work in the universe. So, one of these three contentions must be true: (a) Every ultimate force at work in the universe is evil; (b) Satan controls every evil ultimate force in the universe; or (c) Satan controls no evil ultimate force in the universe.

10. Everyone is afraid of Mr. Hyde, but Mr. Hyde is afraid only of Dr. Jekyll. Therefore Mr. Hyde is Dr. Jekyll.

Using tableaux, evaluate these argument forms for validity.

11. $\forall x \forall y((Fy \,\&\, y = x) \rightarrow Fx)$
∴ $\forall y(Fy \leftrightarrow \exists x(x = y \,\&\, Fx))$

12. $\forall x \forall y \; x = y$
∴ $\forall x \forall y \forall z(z = x \vee z = y)$

13. $\exists x \forall y \; y = x$
∴ $\forall x \exists y \forall z(y \neq x \rightarrow (z \neq x \rightarrow z = y))$

14. $\exists x \exists y \forall z(z = x \vee z = y)$
∴ $\forall x \forall y \forall z(x \neq y \rightarrow (x = z \vee y = z))$

▶ **15.** $\exists x \forall y(y \neq x \rightarrow Fy)$
∴ $\forall x \forall y(x \neq y \rightarrow (Fx \vee Fy))$

16. $\forall x \forall y(x \neq y \rightarrow (Fx \vee Fy))$
∴ $\exists x \forall y(y \neq x \rightarrow Fy)$

17. $\exists x(Fax \,\&\, \neg Fbx)$
∴ $a \neq b$

18. $\forall x \forall y \forall z(x \neq y \rightarrow (x = z \vee y = z))$
∴ $\exists x \exists y \forall z(z = x \vee z = y)$

19. $\exists x \exists y(Fx \,\&\, Fy \,\&\, x \neq y)$
∴ $\exists x \exists y \exists z(Fx \,\&\, Fy \,\&\, Fz \,\&\, x \neq y \,\&\, y \neq z \,\&\, x \neq z)$

20. $\exists x \exists y \exists z(Fx \,\&\, Fy \,\&\, Fz \,\&\, x \neq y \,\&\, y \neq z \,\&\, x \neq z)$
∴ $\exists x \exists y(Fx \,\&\, Fy \,\&\, x \neq y)$

Evaluate these arguments for validity by using tableaux.

21. God created all other existing things. But everything was created by something. Therefore, God was created by one of his own creations.

22. Nobody can fully love more than one person. Each person is fully loved by somebody or other. Thus, everybody fully loves somebody or other.

23. Any administrator can talk less meaningfully than any less experienced one. So, if there is a most experienced administrator, there is an administrator who can talk less meaningfully than any other.

24. No number is greater than itself; given any number, there is another that is greater. Therefore, there is no greatest number.

▶ 25. No number is greater than itself; given any number, there is another that is greater. Therefore, there is no smallest number.

The playwright Anton Chekhov wrote, "The more refined one is, the more unhappy." Does it follow that:

26. The most refined are the most unhappy?

27. The least refined are the happiest?

28. The happiest are the least refined?

29. The unhappiest are the most refined?

30. The happier one is, the less refined one is?

Novelist John Galsworthy wrote, "Idealism increases in direct proportion to one's distance from the problem." It seems to follow that one person is more idealistic than another if and only if he or she is farther from the problem. Does this assertion follow from:

31. The most idealistic are the farthest from the problem?

32. The least idealistic are the closest to the problem?

Does Galsworthy's assertion imply that:

33. The least idealistic are closest to the problem?

34. Those closest to the problem are the least idealistic?

▶ 35. The most idealistic are those farthest from the problem?

8.3 Deduction Rules for Identity

The two principles of self-identity and the indiscernibility of identicals underlie natural deduction rules for identity. The rule for identity introduction is extremely simple: Whenever you wish, in the course of a proof, you may record $t = t$, for any term t.

Identity Introduction ($=$ I)

$$\text{n. } \ell = \ell \qquad = \text{I}$$

This rule allows a very quick proof of the principle that everything is self-identical:

1. Show $\forall x\ x = x$
2. ⌐Show $a = a$
3. └$[a = a]$ $=$ I

The rule of identity exploitation resembles closely the tableau rule for identity on the left. It states that, for any terms ℓ and ℓ', $\ell = \ell'$ justifies substituting ℓ for ℓ' or ℓ' for ℓ, in any formula free at that point in the proof.

Identity Exploitation ($=$ E)

$$\text{n. } \ell = \ell'$$
$$\text{m. } \mathscr{A}$$
$$\text{p. } \mathscr{A}[\ell//\ell'] \text{ (or } \mathscr{A}[\ell'//\ell]) \qquad = \text{E, n, m}$$

Here $\mathscr{A}[\ell//\ell']$ is any result of substituting ℓ for some or all occurrences of ℓ' throughout $\mathscr{A}$. If $\mathscr{A}$ is Fdd, for instance, $\mathscr{A}[c//d]$ could be Fcd, Fdc, or Fcc.

This rule allows us to prove that identity is symmetric: that the order of the terms in an identity statement, in other words, makes no difference.

1. Show $\forall x \forall y(x = y \rightarrow y = x)$
2. ⌐Show $a = b \rightarrow b = a$
3. │⌐$a = b$ ACP
4. ││$a = a$ $=$ I (or, $=$ E, 3, 3)
5. └└$b = a$ $=$ E, 3, 4

We can reach line 4 here either by introducing $a = a$ by identity introduction or by substituting a for b in $a = b$.

Identity exploitation also allows us to prove that identity is transitive.

1. Show $\forall x \forall y \forall z((x = y\ \&\ y = z) \rightarrow x = z)$
2. ⌐Show $(a = b\ \&\ b = c) \rightarrow a = c$
3. │⌐$a = b\ \&\ b = c$ ACP
4. ││$a = b$ &E, 3
5. ││$b = c$ &E, 3
6. └└$a = c$ $=$ E, 4, 5

We begin by using a triple universal proof. We reach the crucial line, 6, by using identity exploitation. We can think of ourselves as substituting a for b in $b = c$ or as substituting c for b in $a = b$. Either way, we end up with $a = c$.

As a final example, suppose we are faced with the argument, 'John is selfish. But everybody else is selfish, too. So, all people are selfish.' Assuming that John is a person, we can let our universe of discourse consist just of people, and so translate this as follows:

Sj
$\forall x(x \neq j \rightarrow Sx)$
$\therefore \forall x Sx$

To show that this argument form is valid, we can construct a universal proof, which contains an indirect proof.

$$
\begin{array}{lll}
1. & Sj & \text{A} \\
2. & \forall x(x \neq j \rightarrow Sx) & \text{A} \\
3. & \text{Show } \forall x Sx & \\
4. & \quad \lceil \text{Show } Sa & \\
5. & \quad \mid \lceil \neg Sa & \text{AIP} \\
6. & \quad \mid \mid a \neq j \rightarrow Sa & \forall \text{E, 2} \\
7. & \quad \mid \mid a = j & \rightarrow \text{E*, 6, 5} \\
8. & \quad \mid \lfloor Sa & = \text{E, 7, 1}
\end{array}
$$

We want to show that some arbitrary person—say Ann—is selfish. So, we assume, for purposes of indirect proof, that she isn't selfish. Since everybody other than John is selfish, but Ann isn't, Ann must be John. But John is selfish as well; so if Ann is John, then, by identity exploitation, she's selfish. But this contradicts our assumption that she isn't selfish.

Problems

We can say that there is one and only one God, in symbolic terms, by writing the formula $\exists x \forall y(y = x \leftrightarrow Fy)$. Show that each of the following is a consequence of this formula.

1. $\exists x(Fx \ \& \ Gx) \leftrightarrow \forall x(Fx \rightarrow Gx)$

2. $\forall x Gx \rightarrow \exists x(Fx \ \& \ Gx)$

3. $\exists x(Fx \ \& \ Gxx) \leftrightarrow \exists x \exists y(Fx \ \& \ Fy \ \& \ Gxy)$

4. $\exists x(Fx \ \& \ (Ga \rightarrow Hx)) \leftrightarrow (Ga \rightarrow \exists x(Fx \ \& \ Hx))$

▶ **5.** $\exists x(Fx \ \& \ (Gx \rightarrow Ha)) \leftrightarrow (\exists x(Fx \ \& \ Gx) \rightarrow Ha)$

6. $\exists x(Fx \ \& \ \neg Gx) \leftrightarrow \forall x(Fx \rightarrow \neg Gx)$

7. $\exists x(Fx \ \& \ \forall y Gyx) \leftrightarrow \forall x \exists y(Fy \ \& \ Gxy)$

8. $\forall x \forall y((Fx \ \& \ Fy) \rightarrow x = y)$

9. $\forall x Fx \rightarrow \forall x \forall y\ x = y$

10. $\exists x \exists y \forall z(x \neq y\ \&\ (z = x \vee z = y)) \leftrightarrow \exists x \forall y(y = x \leftrightarrow \neg Fy)^*$

8.4 Function Symbols

Function or operation symbols combine with constants, variables, and other function symbols to form singular terms. They are extremely common .in mathematics. The basic arithmetic operations of addition, subtraction, multiplication, and division, for example, are all functions. Function symbols also translate many English possessives and *of* constructions:

John's birthday	the length of this line
Barbara's father	the truth value of the sentence
Jill's house	the mass of an electron
somebody's BMW	the trunk of my Camaro
nobody's honor	the President of the United States

Using function symbols, we can translate each of these expressions into quantification theory.

All functions take a certain number of arguments, that is, inputs. A function taking just one input is a singulary function. Binary functions take two inputs; in general, *n*-ary functions take *n* inputs. Addition, multiplication, subtraction, and division are all binary functions. Taking a square and taking a square root are both singulary functions.

Let's agree to call constants, variables, function terms, and definite descriptions *singular terms,* or, more simply, *terms.* To form a *function term,* we must concatenate an *n*-ary function symbol with *n* constants, variables, or other terms. The string of *n* terms must be enclosed in parentheses and separated by commas. Where f, g, and h are singulary, binary, and ternary function symbols, respectively, we can form the function terms

f(a)	g(a, a)	h(a, b, c)
f(x)	g(a, x)	h(x, y, z)
f(f(b))	g(y, b)	h(a, z, b)
f(f(y))	g(x, y)	h(f(a), f(b), f(c))
f(g(a, a))	g(a, g(a, a))	h(g(x, x), x, x)
f(h(x, y, z))	g(f(x), h(b, f(y), z))	h(h(x, y, z), f(z), g(z, x))

Notice that function terms may appear inside other function terms.

Function symbols of three or more places are usually written in front of their argument terms, with parentheses marking the boundaries. But singulary and binary function terms may be written in other ways. We may write binary function symbols between their argument terms, supplying parentheses only when necessary to avoid ambiguity. Thus, mathematicians generally write $2 + 3$ rather than $+(2, 3)$ and $a \cup b$ rather than $\cup(a, b)$.

Indeed, we do the same in sentential logic. Sentential connectives express truth functions, but we write the binary connectives between the formulas they link. Thus, we write the formulas on the left rather than the versions on the right:

$$p \vee q \qquad \vee(p, q)$$
$$(p \rightarrow q) \mathbin{\&} r \qquad \mathbin{\&}(\rightarrow(p, q), r)$$
$$r \leftrightarrow (p \mathbin{\&} \neg r) \qquad \leftrightarrow(r, \mathbin{\&}(p, \neg q))$$

In fact, since we know which connectives are singulary and which are binary, we don't need the parentheses in the expressions on the right. So we could express the formulas more simply:

$$p \vee q \qquad \vee(p, q) \qquad \vee pq$$
$$(p \rightarrow q) \mathbin{\&} r \qquad \mathbin{\&}(\rightarrow(p, q), r) \qquad \mathbin{\&} \rightarrow pqr$$
$$r \leftrightarrow (p \mathbin{\&} \neg r) \qquad \leftrightarrow(r, \mathbin{\&}(p, \neg q)) \qquad \leftrightarrow r \mathbin{\&} p \neg q$$

The third column shows the form in which computers generally read logical formulas and other instructions. A group of Polish logicians developed and wrote in this notation earlier in this century; for this reason, it's called *Polish notation*.

To incorporate function terms into our language, we need to change its formation rules slightly. Officially, we'll maintain the standard notation for function terms, though we'll allow ourselves to use forms in more frequent use when that proves convenient. Lowercase letters from the middle of the alphabet—f, g, h, i, etc., with or without subscripts—will serve as our function symbols. We'll say that a function term is *open* if it contains some occurrences of variables, and *closed* if it doesn't. A *closed term* is either a constant or a closed function term. We may amend our formation rules to read:

Formation Rules

An *n*-ary predicate followed by *n* closed terms is a formula.

An *n*-ary function term followed by *n* constants (in parentheses, separated by commas) is a closed term.

If t is a closed function term containing a constant, c, and t' is a closed term, then $t[t'/c]$ is also a closed function term.

If t and t' are closed terms, then $t = t'$ is a formula.

If $\mathcal{A}$ and $\mathcal{B}$ are formulas, then $\neg\mathcal{A}$, $(\mathcal{A} \mathbin{\&} \mathcal{B})$, $(\mathcal{A} \vee \mathcal{B})$, $(\mathcal{A} \rightarrow \mathcal{B})$, and $(\mathcal{A} \leftrightarrow \mathcal{B})$ are formulas.

If $\mathcal{A}$ is a formula with a closed term, t, and v is a variable not in $\mathcal{A}$, then $\exists v \mathcal{A}[v/t]$ and $\forall v \mathcal{A}[v/t]$ are formulas.

Every formula may be constructed by a finite number of applications of these rules.

Semantically, function symbols stand for functions on the domain; closed function terms stand for individual elements of the domain. Function terms thus differ sharply from formulas. Formulas are either true or false; closed function terms stand for objects. Connectives may join formulas but not function terms: $f(a) \rightarrow g(b, c)$, which might correspond to the English 'If Ann's mother, then Bill and Carlotta's anniversary,' makes no sense. Similarly, function terms may flank the identity sign, but formulas may not. A formula such as $f(a) = g(b, c)$ could represent an English sentence such as 'Ann's birthday is Bill and Carlotta's anniversary,' but $\exists x F x = Ga$ is incoherent. The closest English rendering of it would be something like 'There is a fish is Al is crazy.'

The fact that function symbols stand for functions on the domain has a very important consequence. Functions are relations between objects. But two things distinguish functions from other relations. First, given some input to the function, we always obtain an output. Any object stands in the appropriate relation to some object. Second, given an input, we always obtain a unique output. Each object stands in the appropriate relation to only one object. Together these requirements imply that, given any input, we must obtain exactly one output. We may refer to the first requirement as the *existence* requirement; the second, as the *uniqueness* requirement. Because function symbols stand for functions, we may use a function symbol to represent an English expression only when these requirements are satisfied.

Mathematical expressions are easiest to evaluate in light of these requirements. Consider addition. Any two numbers have exactly one sum, so the requirements are satisfied. Or, consider multiplication. Any two numbers have a unique product, so multiplication is a function, and we may represent it with a function symbol.

Subtraction and division are slightly more complex. If we include negative numbers, then subtraction is a function: Any two numbers, taken in a given order, have a unique difference. Without negative numbers, however, the existence requirement fails; no positive number is $3 - 5$. Division is almost a function, but not quite, even if we include fractions, since, for any n, $n \div 0$ is undefined. Division is a function only on the nonzero numbers.

Satisfying these requirements is important, because, armed with a function symbol, we can prove existence and uniqueness. The use of a function symbol thus presupposes them. To see this, we need to know how to handle function terms within quantification theory. We need no new rules. Recall our definition:

> **DEFINITION.** A formula $\mathscr{A}[t/v]$ is an *instance of* a formula $\exists v \mathscr{A}$ or $\forall v \mathscr{A}$ iff it results from dropping the main connective of that formula and substituting a closed term t for every occurrence of the variable v throughout $\mathscr{A}$.

We can form instances, then, by dropping quantifiers and substituting closed function terms as well as constants for variables. Both Fa and $Ff(a)$ count as instances of $\exists x F x$.

Since ∀L and ∃R apply to terms in general, we may use instances with closed function terms when applying them. But ∀R and ∃L require that we use a constant new to the branch in taking an instance. That requirement still stands. Both rules require that we introduce a name about which we have no prior information. If we were to introduce a function term rather than a constant, we might be importing information. Not everyone, for example, is a mother. So, to go from the information that someone in the company has been selling trade secrets to the conclusion that somebody's mother has been selling trade secrets would be to slip in the completely unwarranted assumption that the scoundrel is a woman with children.

Exactly the same alterations apply to our deduction rules. ∃I and ∀E both allow us to work with any instance of a formula. Universal exploitation permits us to deduce from a universal formula any of its instances, while existential introduction allows us to deduce an existential from any of its instances. We cannot use function terms in applying our existential exploitation rule or our universal proof method, however, for precisely the reasons that we can't do so when applying the ∀R and ∃L tableau rules. Existential exploitation requires that we use a constant completely new to the proof in taking an instance. Universal proof similarly allows us to prove a universal formula by proving an instance with a new constant. Whenever our rules have required a new constant, then, they will continue to do so: ∀R, ∃L, ∃E, and universal proof all require a new constant.

It's easy to see that we can prove the existence and uniqueness of values of functions, given any set of inputs in the domain. The existence requirement states that a function must yield a value for any input. We can symbolize this as $\forall x \exists y \ y = f(x)$. This is valid, as the following tableau and proof show:

1.	Show $\forall x \exists y \ y = f(x)$		$\forall x \exists y \ y = f(x)$ ✓
2.	Show $\exists y \ y = f(a)$		$\exists y \ y = f(a)$*
3.	$f(a) = f(a)$	=I	$f(a) = f(a)$
4.	$\exists y \ y = f(a)$	∃I, 3	CI

Notice that, in applying ∃R in this tableau, we substituted a closed function term for a variable. Similarly, in applying ∃I in the proof, we substituted a variable for a closed function term. In both cases, the instance of the relevant quantified formula resulted from substituting a closed function term rather than a constant for the quantified variable.

Uniqueness is almost as easy to prove valid. The uniqueness requirement states that for any input to the function there is a unique output. We can symbolize this by construing the requirement as stating that there is at most one output to the function, given any input: $\forall x \forall y \forall z ((y = f(x) \ \& \ z = f(x)) \rightarrow y = z)$.

1. S̶h̶o̶w̶ $\forall x \forall y \forall z((y = f(x) \,\&\, z = f(x)) \rightarrow y = z)$
2. ⌈S̶h̶o̶w̶ $(b = f(a) \,\&\, c = f(a)) \rightarrow b = c$
3. ⌈$b = f(a) \,\&\, c = f(a)$ ACP
4. $b = f(a)$ &E, 3
5. $c = f(a)$ &E, 3
6. ⌊⌊$b = c$ = E, 4, 5

                         $\forall x \forall y \forall z((y = f(x) \,\&\, z = f(x)) \rightarrow y = z)$ ✓
                         $(b = f(a) \,\&\, c = f(a)) \rightarrow b = c$ ✓

✓ $b = f(a) \,\&\, c = f(a)$

                       $b = c$

        *$b = f(a)$
         $c = f(a)$
         $b = c$

          Cl

To see why observing these requirements in translation from natural language is so important, consider these arguments.

(2) Your mother is your parent.
    Your father is your parent.
    ∴ Your mother is your father.

Although we can reasonably translate 'mother' and 'father' as function symbols, we can't so translate 'parent'; among humans, parents come in pairs. 'Parent' thus violates the uniqueness requirement. Argument (2) is certainly invalid. But it would appear valid if we were to render 'parent' as a function symbol. All that we need to establish the validity of the argument form is the principle of the indiscernibility of identicals, expressed in the rule =L or =E:

1. $m(a) = g(a)$ A   * $m(a) = g(a)$
2. $f(a) = g(a)$ A $f(a) = g(a)$
3. S̶h̶o̶w̶ $m(a) = f(a)$
4. ⌈$m(a) = f(a)$ =E, 1, 2 $m(a) = f(a)$ $m(a) = f(a)$
                                       Cl

To see the trouble that arises if we ignore the existence requirement, consider this terrible argument, surely belied by the baby boom:

(3) No babies have children.
    ∴ There are no babies.

We shouldn't translate 'child of' as a function symbol, since it fails to meet either requirement. Many people have no children, while others have more than one. If we nevertheless symbolize the argument using a function symbol, the resulting argument form will be valid:

1. $\forall x(Bx \rightarrow \neg\exists y\ y = g(x))$ A
2. Show $\neg\exists x Bx$
3. $\lceil \exists x Bx$ AIP
4. $\vert\ Ba$ $\exists$E, 3
5. $\vert\ Ba \rightarrow \neg\exists y\ y = g(a)$ $\forall$E, 1
6. $\vert\ \neg\exists y\ y = g(a)$ $\rightarrow$E, 5, 4
7. $\vert\ \forall y\ y \neq g(a)$ QN, 6
8. $\vert\ g(a) \neq g(a)$ $\forall$E, 7
9. $\lfloor\ g(a) = g(a)$ =I

Using function symbols in translation is legitimate, strictly speaking, only if the existence and uniqueness requirements are satisfied. For this reason, many mathematicians demand a proof of existence and uniqueness before they admit a function or operation symbol into a mathematical theory.

Nevertheless, we can relax these requirements somewhat when translating natural-language arguments. Many function symbols satisfy existence and uniqueness on only a portion of the domain. If we use f to symbolize 'the Social Security number of,' for example, we must restrict its application to constants denoting people. The domain will contain both people and numbers, and we don't want to commit ourselves to talking about the Social Security numbers of numbers. This is an informal restraint; we could impose it formally by restricting our logical rules when we use function symbols. (In Chapter 12, we will formulate similar but slightly broader restrictions, for very similar reasons.) Alternatively, we might let any intuitively silly function term, such as 'the Social Security number of 17,' name some specially designated "dummy" object.

Problems

Symbolize these sentences in QL with identity and function symbols.

1. A man's judgment and his conscience is the same thing (Thomas Hobbes).

2. Our birth is but a sleep and a forgetting (William Wordsworth).

3. The painter's brush consumes his dreams (William Butler Yeats).

4. If the mind of the teacher is not in love with the mind of the student, he is simply practicing rape, and deserves at best our pity (Adrienne Rich).

▶ 5. Nobody loves me but my mother, and she could be jivin' too (B. B. King).

6. A fool despises his father's instruction (Proverbs 15:5).

7. The friend I can trust is the one who will let me have my death. The rest are actors who want me to stay and further the plot (Adrienne Rich).

8. If you see in any situation only what everybody else can see, you can be said to be so much a representative of your culture that you are a victim of it (S. I. Hayakawa).

9. In a hierarchy every employee tends to rise to his level of incompetence (Laurence J. Peter).

▶ 10. Common sense holds its tongue (Proverbs 10:19).

11. Not all those who know their minds know their hearts as well (La Rouchefoucauld).

12. All that we send into the lives of others comes back into our own (Edwin Markham).

13. The cleverly expressed opposite of any generally accepted idea is worth a fortune to somebody (F. Scott Fitzgerald).

14. A clever man is wise and conceals everything, but the stupid parade their folly (Proverbs 13:16).

▶ 15. The man with no inner life is the slave of his surroundings (Henri Frederic Amiel).

Use tableaux to evaluate these arguments for validity.

16. Nobody's perfect. Consequently, nobody's mother is perfect.

17. Tanya and Tatiana are the same person. So, Tanya's Social Security number is the same as Tatiana's.

18. Everybody remembers his or her first grade teacher. So, everybody's first grade teacher remembers him or her.

19. Everybody is his or her own best friend. Therefore, each person's best friend is his or her best friend's best friend.

▶ 20. Suppose that the name of the name of a thing is always just the thing's name itself. Then it follows that anything that names anything is a name of itself.

21. Suppose that anything that names anything is a name of itself. It follows that the name of the name of a thing is always just the thing's name itself.

22. For every man there is a priest who is his intermediary with God. Every priest is a man. No one is intermediary between God and God. Therefore, God is neither man nor priest.

23. Augustus is Superman's father. Augustus knows everybody. Somebody can beat up Superman if and only if he or she can do anything. All failure results from a lack of self-knowledge. Therefore, Superman's father can beat him up.

24. No matter who you are, the being with the ultimate power over your actions is God. One person has responsibility for another's actions just in case that person has ultimate power over the other's acts. A person can be blamed, from a moral point of view, only if he or she has responsibility for his or her own actions. Thus nobody but God can be morally blamed.

▶ 25. Nothing is demonstrable unless the contrary implies a contradiction. Nothing that is distinctly conceivable implies a contradiction. Whatever we conceive as existent, we can also conceive as nonexistent. There is no being, therefore, whose nonexistence implies a contradiction. Consequently, there is no being whose existence is demonstrable (David Hume). [Note: This is really two arguments: (a) Nothing that is distinctly conceivable implies a contradiction. Whatever we conceive as existent, we can also conceive as nonexistent. Therefore, there is no being whose nonexistence implies a contradiction. (b) There is no being whose nonexistence implies a contradiction. Nothing is demonstrable unless the contrary implies a contradiction. Consequently, there is no being whose existence is demonstrable. To give these a fair hearing, assume that the nonexistence of a thing is the contrary of its existence.]

Using either natural deduction or semantic tableaux, show that these argument forms are valid.

26. $\forall x Fx$; $\therefore$ $\forall x(Gxa \rightarrow Fh(x, a))$

27. $\forall x \exists y \exists z\ x = f(y, z)$; $\therefore$ $\forall x \forall y Gf(x, y) \rightarrow \forall x Gx$

28. $\forall x \forall y\ x = y$; $\therefore$ $\forall x\ x = f(x)$

29. $\forall x\ x = f(a)$; $\therefore$ $\forall x\ a = f(x)$

30. $\forall x\ g(f(x)) = x$; $\therefore$ $\forall x \forall y(f(x) = y \rightarrow g(y) = x)$

31. $\forall x \forall y(f(x) = y \rightarrow g(y) = x)$; $\therefore$ $\forall x\ g(f(x)) = x$

32. $\forall x Ff(x)x$; $\therefore$ $\forall x Ff(f(x))f(x)$

33. $\exists x(x \neq a\ \&\ Gx)$; $\therefore$ $\exists x Gx\ \&\ (Ga \rightarrow \exists x \exists y(Gx\ \&\ Gy\ \&\ x \neq y))$

34. $\forall x\ x = f(x)$; $\therefore$ $\forall x\ f(x) = f(f(f(x)))$

▶ 35. $\exists x \exists y(x \neq y\ \&\ \forall z(z = x \vee z = y))$; $\forall x \forall y(f(x) = f(y) \rightarrow x = y)$; $\therefore$ $\forall x \exists y\ x = f(y)$

36. Suppose there are functions f and g such that (a) $\forall x\ f(x, 0) = x$, (b) $\forall x\ f(x, g(x)) = 0$, and (c) $\forall x \forall y\ f(x, y) = f(y, x)$. Show that $\exists x\ x = g(x)$ and that $\exists x\ f(x, x) = x$.

37. Suppose that there is a binary function $\circ$ satisfying the axioms (a) $\forall x \forall y\ y = x \circ y$ and (b) $\forall x \forall y\ x \circ y = y \circ x$. Show that the domain of such a function must include at most one object: $\forall x \forall y\ x = y$.

Leibniz developed a calculus concerning the combination of concepts using a single function for combination and a single predicate "is in" or "is contained by," defined in terms of the combination function. Using Leibniz's symbol $\oplus$ for combination and $<$ for "is in," we can present Leibniz's theory in one definition and two axioms.

> Definition. $\forall x \forall y (x < y \leftrightarrow \exists z\ x \oplus z = y)$
> Axiom 1. $\forall x \forall y\ x \oplus y = y \oplus x$
> Axiom 2. $\forall x\ x \oplus x = x$

Leibniz proceeds to prove the following propositions. Show that they follow from these axioms, together with the associativity principle:

> Axiom 3. $\forall x \forall y \forall z\ x \oplus (y \oplus z) = (x \oplus y) \oplus z$

38. $\forall x\ x < x*$

39. $\forall x \forall y \forall z (x < y \rightarrow z \oplus x < z \oplus y)*$

40. $\forall x \forall y (x \oplus y = x \rightarrow y < x)*$

41. $\forall x \forall y (x < y \rightarrow y \oplus x = y)*$

42. $\forall x \forall y \forall z ((x < y\ \&\ y < z) \rightarrow x < z)*$

43. $\forall x \forall y \forall z (x \oplus y < z \rightarrow y < z)*$

44. $\forall x \forall y \forall z (y < z \rightarrow y < z \oplus x)*$

▶ **45.** $\forall x \forall y ((x < y\ \&\ y < x) \rightarrow x = y)*$

46. $\forall x \forall y \forall z ((x < y\ \&\ z < y) \rightarrow x \oplus z < y)*$

47. $\forall x \forall y \forall z \forall w ((x < y\ \&\ z < w) \rightarrow x \oplus z < y \oplus w)*$

8.5 DEFINITE DESCRIPTIONS

Definite descriptions, in English, appear in one of several forms: (a) noun phrases beginning with *the,* such as 'the Queen of England' and 'the man who shot Liberty Valance'; (b) noun phrases with possessives or *of* phrases, such as 'my friend,' 'Richard's father,' and 'Prince of Peace'; and (c) noun phrases that contain other noun phrases acting as adjectives, such as 'Tuesday afternoon,' 'Indiana University,' 'Austin City Limits,' and 'City of Pittsburgh Police.' Such expressions seem to function in many different ways in natural language. In this section, we try to isolate four functions and explain how to render them in quantification theory.

First, definite descriptions often act as anaphoric devices. Used in this way, they could often be replaced by pronouns. They refer to entities that previous discourse has introduced. Consider, for example, these small discourses. (The definite descriptions are in italic.)

(4) a. Sarah saw a man run from the store. *The man* was holding a shotgun.
   b. Sarah saw a man run from the store. He was holding a shotgun.
   c. Everyone who owns a Cadillac washes *the Cadillac* frequently.
   d. Everyone who owns a Cadillac washes it frequently.
   e. A medieval philosopher had noticed this several hundred years earlier. *The philosopher* was obscure, however, and so his insights had to be rediscovered.
   f. A medieval philosopher had noticed this several hundred years earlier. He was obscure, however, and so his insights had to be rediscovered.

Sentences a, c, and e all contain definite descriptions; b, d, and f contain pronouns in their places. In every case, substituting a pronoun for a definite description has no effect on the meaning of the sentence. These descriptions don't introduce an object into the discourse; the objects they refer to have been introduced earlier. Nor do they imply that the object being discussed is unique; nothing in these sentences denies that many people own Cadillacs or that several medieval philosophers may have made the same forgotten discovery.

These definite descriptions, then, should be symbolized as if they were pronouns. If a pronoun refers to an object for which we have a name, we can simply use the constant that symbolizes the name to symbolize the pronoun. Thus, if our discourse is 'John was tired. He was also worried,' we can translate it into the formulas Tj and Wj. If we have no name for the object, we can translate the pronoun by using a variable within the scope of a quantifier. So if the discourse is 'A man sat down across from Ellen. He looked sad,' we must translate it into a single formula, such as $\exists x(Mx \ \& \ Sxe \ \& \ Lx)$. Most often, in English, an anaphoric definite description refers to an unnamed object, and so must be translated as a variable. The last symbolization, therefore, serves just as well to symbolize 'A man sat down across from Ellen. The man looked sad.'

Second, definite descriptions may introduce objects into the discourse without making any claims about uniqueness. The function of the descriptions in each of these sentences is to introduce objects into discourse:

(5) a. Bernard was an elder. *His daughter* wanted to become a missionary.
   b. Darlene lost *her pencil*.
   c. Lars wanted to go to Luckenbach, but he missed *the turn*.
   d. We went to *the hamburger place on Van Ness*.

Third, definite descriptions may serve to characterize an entity already introduced:

(6) a. Marsha is *my friend*.

b. Angela Grissom is *the lobbyist,* and Jim Jones is *the congressman.*

c. Joyce met with Ms. Hampton, a manager. *The carefully groomed woman* acted interested but refused to give Joyce a loan.

The definite descriptions in (5) introduce entities into the discussion, while those in (6) describe objects already being discussed. In neither (5) nor (6) do the descriptions convey uniqueness, though some of them imply familiarity with the objects they denote. We tend to assume that the speaker has a unique object in mind. But, for all we know from these sentences, Bernard may have several daughters and Darlene may have more than one pencil. Several turns may take one to Luckenbach, and the Hippo and the Hard Rock Cafe are both on Van Ness. Similarly, Marsha may have lots of friends, and there are many lobbyists, congressmen, and carefully groomed women.

We can treat definite descriptions that introduce objects into discourse without claiming uniqueness as existential quantifications. *The,* in these uses, acts much like *a* or *an.* So we can symbolize 'Bernard's daughter' just as we would 'a daughter of Bernard'; 'her pencil' as we would 'a pencil of hers'; and so on.

Definite descriptions that characterize objects already being discussed, without asserting uniqueness, tend to translate as predications of the already introduced entities. Here, *the* also acts as *a* or *an.* We could therefore use existential quantification for these descriptions as well. But just as we can symbolize 'Socrates is a man' as either *Ms* or $\exists x(Mx \,\&\, x = s)$, we can generally omit the existential quantification in favor of a simple predication. Thus, we can translate 'Martha is my friend' as either $\exists x(Fxi \,\&\, x = m)$ or as *Fmi.* We can analogously treat 'Angela is the lobbyist' as *La.* (6)c is slightly more complex, since it both describes an object already introduced and refers back to that object from another sentence. It is, then, both descriptive and anaphoric. Since we have a name for the referent (Ms. Hampton) we can symbolize the sentences as *Mjh & Nh*; $Gh \,\&\, \neg \exists x(Lx \,\&\, Ohjx)$.

Fourth, definite descriptions may specify an object by giving, or purporting to give, a unique characterization of it:

(7) a. In 1858, James Buchanan was *the President of the United States.*

b. *Julia's favorite month* is November.

c. I felt that way, too, *the first time I fell in love.*

d. Did you see *today's Wall Street Journal?*

This use of descriptions has attracted far more attention from philosophers and logicians than any other. There are two ways of understanding descriptions such as those in (7). They appear to be singular terms, picking out a unique object much as a name does. So, one approach to descriptions is to take them as singular terms. Another approach is to take them as something like quantifiers. *The,* after all, is a determiner; we might, pursuing the analogy with *some* and *every,* think of such phrases as 'the President of the United States' as playing a role similar to that of such phrases as 'every frog in the garden.' In this chapter, we develop two approaches to descriptions: one taking them as singular terms and another taking them as quantifierlike expressions. These approaches are not equivalent. Both have some disadvantages, but both explain some important features of definite descriptions.

Descriptions as Singular Terms

To symbolize descriptions as singular terms, we'll introduce a new symbol: $\imath$, the *inverted iota.* We'll read a term such as $\imath x F x$ as 'the x such that Fx' or, more simply, as 'the F.' This symbol, like a quantifier, governs variables. But it converts formulas, not into other formulas, but into singular terms. These terms act just like others in forming formulas. Thus, $\imath x F x = a$, $F\imath y G y b$, and $\exists x(Fx \ \& \ Gx\imath z(Hz \ \& \ Gzx))$ are all acceptable formulas.

We can build descriptions into our symbolic language by expanding the vocabulary to include $\imath$ and altering a few definitions and our formation rules. A *description functor* is $\imath$ together with a variable; the functor is *on* that variable. If $\mathscr{A}$ is a formula with a constant c in which the variable v doesn't occur, then $\imath v \mathscr{A}[v/c]$ is a *description term* or, more simply, a *description.* The *scope* of the functor $\imath v$ in $\imath v \mathscr{A}[v/c]$ is just $\imath v \mathscr{A}[v/c]$. The scope of a functor, in other words, is the entire description.

A *closed term* is either a constant, a closed function term, or a description term. We may amend our formation rules to read:

Formation Rules

An n-ary predicate followed by n closed terms is a formula.

An n-ary function term followed by n constants (in parentheses, separated by commas) is a closed term.

If t is a closed function term containing a constant, c, and t' is a closed term, then $t[t'/c]$ is also a closed function term.

If t and t' are closed terms, then $t = t'$ is a formula.

If $\mathscr{A}$ and $\mathscr{B}$ are formulas, then $\neg\mathscr{A}$, $(\mathscr{A} \ \& \ \mathscr{B})$, $(\mathscr{A} \vee \mathscr{B})$, $(\mathscr{A} \rightarrow \mathscr{B})$, and $(\mathscr{A} \leftrightarrow \mathscr{B})$ are formulas.

If $\mathscr{A}$ is a formula with a closed term, t, and v is a variable not in $\mathscr{A}$, then $\exists v \mathscr{A}[v/t]$ and $\forall v \mathscr{A}[v/t]$ are formulas.

If $\mathscr{A}$ is a formula with a closed term, t, and v is a variable not in $\mathscr{A}$, then $\imath v \mathscr{A}[v/t]$ is a closed term.

Every formula may be constructed by a finite number of applications of these rules.

To see how to translate sentences containing descriptions of this sort, consider sentence (7)a. 'James Buchanan was the President of the United States' is an identity statement. It asserts that 'James Buchanan' and 'the President of the United States' refer to the same person. We can symbolize 'the President of the United States' (taking 'the United States' as a name rather than a description) as $\imath xPxa$. The sentence as a whole, then, becomes $b = xPxa$. 'The author of *Waverley* was Scottish,' in contrast, is an ordinary predication. 'The author of *Waverley*' translates as $\imath xWxa$ ('the x such that x wrote *Waverley*'). Where Sb symbolizes 'b is Scottish,' we can then translate the entire sentence as $S\imath xWxa$. Descriptions may be embedded in other descriptions. We may symbolize 'the husband of the prime minister' as $\imath xHx\imath yPy$—'the x such that x is husband of the y such that y is prime minister'—and 'the dog that bites the hand that feeds him' as $\imath x(Dx \ \& \ Bx\imath y(Hy \ \& \ Fyx))$.

Other uses of definite descriptions are quite complex and beyond the scope of this chapter. The most important are *generic* descriptions, such as 'the whale' in the sentence 'The whale is a mammal.' As we mentioned in the chapter on quantifiers, generic *a* and *the* approximate, but aren't equivalent to, universal quantifiers.

We need no special techniques to handle sentences containing descriptions of the first three kinds. To evaluate the validity of sentences or arguments containing definite descriptions that specify objects uniquely, however, we must deal with description terms. We'll use a tactic that W. V. Quine invented and called *the method of descriptional premises.*[2] The method is very simple. Suppose an argument form contains some descriptions we wish to evaluate. We follow these steps:

1. We replace each description $\imath v \mathscr{A}$ with a constant not already in the argument form. We can call such constants *descriptional constants*.

2. We add to the premises, for each descriptional constant c, a new premise $\forall z(z = c \leftrightarrow \mathscr{A}[z/v])$, where $\imath v \mathscr{A}$ is the description c replaces. These are *descriptional premises*. They express the existence and uniqueness conveyed by the original description.

3. We evaluate the resulting argument form for validity. It will be valid if and only if the original was valid.

   To see how this method works, consider a simple argument:

   (8) The author of *Waverley* was Scott.
   $\therefore$ Scott wrote *Waverley*.

We can translate this into:

   (9) $\imath xWxa = s$
   $\therefore Wsa$

This argument form contains the description $\imath x Wxa$. We begin by replacing it with a new constant, say, b; the premise thus becomes $b = s$. Second, we add the descriptional premise $\forall z(z = b \leftrightarrow Wza)$. This results in:

(10) $b = s$
$\quad \forall z(z = b \leftrightarrow Wza)$
$\quad \therefore Wsa$

Third, we evaluate this argument form. Using natural deduction or semantic tableaux, we can easily show that it is valid. Here, for example, is a natural deduction of the conclusion from the premises:

$$
\begin{array}{lll}
1. & b = s & \text{A} \\
2. & \forall z(z = b \leftrightarrow Wza) & \text{A} \\
3. & \text{Show } Wsa & \\
4. & \lceil s = b \leftrightarrow Wsa & \forall\text{E, 2} \\
5. & \mid s = b & =\text{E}^2, 1, 1 \\
6. & \lfloor Wsa & \leftrightarrow\text{E, 4, 5}
\end{array}
$$

This argument is a slightly more difficult example, with two description terms:

(11) The author of *Waverley* wrote *Ivanhoe*.
$\quad \therefore$ The author of *Ivanhoe* wrote *Waverley*.

This translates into

(12) $W(\imath x Wxa)i$
$\quad \therefore W(\imath x Wxi)a$

The parentheses around the descriptions here are not needed according to our formation rules, but they help to make the structure of the expressions more readable. The argument form contains the description terms $\imath x Wxa$ and $\imath x Wxi$. Despite their similarity, they are different terms, so they require two distinct descriptional constants, say, b and c. Replacing the terms with their associated constants yields the argument form

(13) Wbi
$\quad \therefore Wca$

This form is obviously invalid. But we must add two descriptional premises to obtain

(14) Wbi
$\quad \forall z(z = b \leftrightarrow Wza)$
$\quad \forall z(z = c \leftrightarrow Wzi)$
$\quad \therefore Wca$

Finally, we evaluate this argument form, finding it valid:

$$
\begin{array}{lll}
1. & Wbi & A \\
2. & \forall z(z = b \leftrightarrow Wza) & A \\
3. & \forall z(z = c \leftrightarrow Wzi) & A \\
4. & \text{Show } Wca & \\
5. & \lceil b = c \leftrightarrow Wbi & \forall E, 3 \\
6. & \mid b = c & \leftrightarrow E, 1, 5 \\
7. & \mid c = b & = E^2, 6, 6 \\
8. & \mid c = b \leftrightarrow Wca & \forall E, 2 \\
9. & \lfloor Wca & \leftrightarrow E, 7, 8
\end{array}
$$

Our method of descriptional premises works just as well with semantic tableaux as with deduction.

Descriptions as Quantifierlike Expressions

So far, we discussed approaches to descriptions that take them as singular terms. We can also think of them as quantifierlike expressions. Linguistically, in many circumstances, they tend to act like other quantified noun phrases. So, we might think that we should use ι or a similar symbol, not as a device for forming singular terms, but as something like a quantifier.

One reason for thinking that descriptions sometimes act like quantifiers is that many sentences containing descriptions are ambiguous. We may take sentences such as 'The present king of France is bald' to be false, because they attach a predicate to a description, such as 'the present king of France,' that denotes nothing. 'The present king of France is bald' seems to say that there is one and only one king of France, and he is bald. If there is no king of France, it is false to assert that there is one, and only one, and he is bald.[3] But what about 'The present king of France is not bald'? This sentence can have either of two senses:

(15) a. There is one and only one king of France, and he is not bald.
      b. It is not the case that there is one and only one king of France and he is bald.

The sentence is true if taken in the latter sense but false if taken in the former. Similar ambiguities occur when descriptions cohabit with other connectives; consider 'The number of planets is necessarily greater than seven' and 'The mayor of Philadelphia should attend that conference the next time it's held.' We may read 'The number of planets is necessarily greater than seven' as asserting that nine, the actual number of planets, is necessarily greater than seven, which is true; or we may read it as asserting that there could not have been fewer than eight planets, which is false. Similarly, we may read 'The mayor of Philadelphia should attend that conference the next time it's held' as asserting that whoever is *now* the mayor should attend the conference next time, or that whoever is *then* the mayor should do so.

If such sentences are indeed ambiguous, then we need to think of at least some descriptions as something more than singular terms. If we take a description as akin to a quantifier, then we can capture the two senses in (15) by means of scope, basically on a par with the difference between

(16) Some people don't like broccoli

which is obviously true, and

(17) It's not the case that some people like broccoli

which is false. This forces us to use a description operator in a different capacity.

To avoid confusion, we'll use ι only as an operator that forms singular terms, and we'll use τ—the Greek letter *tau*—as an operator that acts in a quantifierlike way. If $\mathcal{A}$ is a formula with a constant c but containing no occurrences of the variable v, then $(\tau v\!: \mathcal{A}[v/c])$ is a *description quantifier*. If, in addition, $\mathcal{B}$ is a formula with a constant d but without any occurrences of v, then $(\tau v\!: \mathcal{A}[v/c])\mathcal{B}[v/d]$ is a formula. We can read a description quantifier such as $(\tau x\!: Fx)$ as 'the x such that x is F'; a formula such as $(\tau x\!: Fx)Gx$ we can read as 'the x such that x is F is G' or, more simply, 'the F is G.'

Using description quantifiers, we can translate each of these sentences as follows:

(18) a. The author of *Waverley* is Scott.
    b. $(\tau x\!: Wxa)x = s$
    c. The author of *Waverley* is Scottish.
    d. $(\tau x\!: Wxa)Sx$
    e. The present king of France is bald.
    f. $(\tau y\!: Kyf)By$
    g. The president is the commander-in-chief.
    h. $(\tau x\!: Px)(\tau z\!: Cz)x = z$
    i. The husband of the prime minister loves her.
    j. $(\tau x\!: Px)(\tau y\!: Hyx)Lyx$

In symbolizing these sentences, we first translate the descriptions into description quantifiers, using a different variable for each description. We then translate the remainder of the sentence, using just the variable in place of the description. The variable of the description quantifier, in other words, acts just as descriptional constants do when we treat descriptions as singular terms. When translating a sentence with one description embedded inside another, such as (18)i, we must begin with the description that is most deeply embedded and work our way out.

We can now symbolize the two senses of 'The present king of France is not bald.'

(19) a. The present king of France is such that he is not bald.
    b. $(\tau x\!: Kxf)\neg Bx$

c. It is not the case that the present king of France is bald.
d. $\neg (\tau x\!: Kxf)Bx$

In (19)b, the description quantifier is the main connective; the negation falls within its scope. In (19)d, in contrast, the negation is the main connective.

To explain the meaning of the description quantifier, we can say that a formula of the form $(\tau v\!: \mathcal{A})\mathcal{B}$ is true on an interpretation just in case there is one and only one object in the domain satisfying $\mathcal{A}$, and that object also satisfies $\mathcal{B}$. $(\tau v\!: \mathcal{A})\mathcal{B}$ thus says that there is a unique $\mathcal{A}$, and it is $\mathcal{B}$. This is why we must write description quantifiers in a form different from that of the existential and universal quantifiers. $\mathcal{A}$ and $\mathcal{B}$ make different contributions to the truth conditions of $(\tau v\!: \mathcal{A})\mathcal{B}$, and no sentential connective allows us to express that difference. To say that there is a unique $\mathcal{A}$, and it is $\mathcal{B}$, is not equivalent to saying that there is one and only one thing that is both $\mathcal{A}$ and $\mathcal{B}$. 'The margarita I had last night was good' says that I had one and only one margarita last night, and it was good. It implies, but is not equivalent to, the assertion that I had one and only one good margarita last night. This latter assertion is compatible with my having had many margaritas, only one of which was good.

To formulate tableau rules for descriptions, we must begin by asking under what circumstances something of the form $(\tau v\!: \mathcal{A})\mathcal{B}$ is true. We've seen that, if $(\tau v\!: \mathcal{A})\mathcal{B}$ is true, there must be a unique $\mathcal{A}$, and it must be $\mathcal{B}$. The rule for descriptional quantifiers on the left expresses this:

Description Left (τL)

$$\begin{array}{c|} \checkmark \quad (\tau v\!: \mathcal{A})\mathcal{B} \\ \mathcal{B}[c/v] \\ \forall v(v = c \leftrightarrow \mathcal{A}) \end{array}$$

Here c must be a constant new to the branch.

The universal quantification tells us that there is a unique $\mathcal{A}$, namely c, and $\mathcal{B}[c/v]$ says that it is $\mathcal{B}$.

To see how this rule works, let's show that the simple argument

(20) Scott was the author of *Waverley*.
∴ Scott wrote *Waverley*.

is valid. We can translate this, using descriptional quantifiers, as

(21) $(\tau x\!: Wxa)s = x$
∴ Wsa

To test the validity of this on a tableau, we place the premise formula on the left and the conclusion formula on the right.

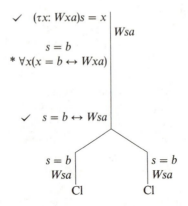

Since every branch closes, the argument form is valid.

To formulate a rule for descriptions on the right, we need to think about what makes something of the form $(\tau v\colon \mathcal{A})\mathcal{B}$ false. There are several possibilities. There may be no $\mathcal{A}$; there may be more than one. Finally, the $\mathcal{A}$ may not be $\mathcal{B}$. We have to take all these into account on the tableau. So the rule is complex:

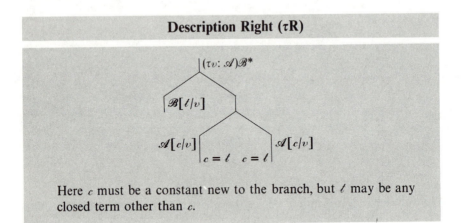

Description Right (τR)

Here c must be a constant new to the branch, but ℓ may be any closed term other than c.

In this rule, ℓ is meant to be the unique $\mathcal{A}$. The rightmost branch corresponds to the possibility that nothing is $\mathcal{A}$; the middle branch, to the possibility that perhaps something else is $\mathcal{A}$. The left branch corresponds to the possibility that perhaps the $\mathcal{A}$ is not $\mathcal{B}$.

To see how this rule works, let's consider the argument we analyzed earlier. Can we infer that the author of *Ivanhoe* wrote *Waverley* from the fact that the author of *Waverley* wrote *Ivanhoe*? Using descriptional quantifiers, we can symbolize this argument as

(22) $(\tau x: Wxa)Wxi$
$\therefore (\tau x: Wxi)Wxa$

As the following tableau shows, this argument is not valid:

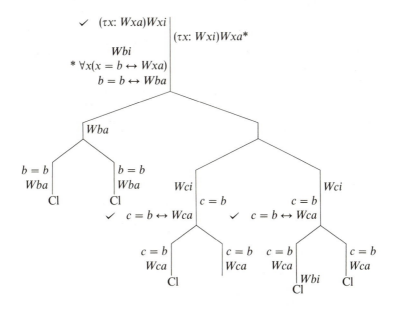

Every branch of this tableau closes but one. On that branch, Wbi and Wci are true, but $c = b$ is false. The interpretation making the premise true and the conclusion false, in other words, is one in which the author of *Waverley* and someone else who didn't write *Waverley* both wrote *Ivanhoe*. So, the possibility that Scott was merely a coauthor of *Ivanhoe* defeats the argument. This shows, incidentally, that our two approaches to descriptions sometimes disagree.

Formulating natural deduction rules for descriptional quantifiers is somewhat easier. If we have established something of the form $(\tau v: \mathscr{A})\mathscr{B}$, we may introduce a new constant to stand for the $\mathscr{A}$, say that it is $\mathscr{B}$, and express the uniqueness of that thing:

Description Exploitation (τE)

n. $(\tau v: \mathscr{A})\mathscr{B}$
n + p. $\mathscr{B}[c/v] \ \& \ \forall v(v = c \leftrightarrow \mathscr{A})$ τE, n

Here c must be a constant new to the proof.

To formulate a rule of description introduction, we may simply invert this:

Description Introduction (τI)

    n. $\mathscr{B}[\ell/v]$
    m. $\forall v(v = \ell \leftrightarrow \mathscr{A})$
    p. $\overline{(\tau v\colon \mathscr{A})\mathscr{B}}$ τI, n, m

Here ℓ may be any closed term.

To demonstrate the use of these rules, let's show that the argument we've just considered is valid if we assume that no more than one person wrote *Ivanhoe*.

1.	$(\tau x\colon Wxa)Wxi$	A
2.	$\forall x \forall y((Wxi \mathbin{\&} Wyi) \to x = y)$	A
3.	Show $(\tau x\colon Wxi)Wxa$	
4.	$\lceil Wbi \mathbin{\&} \forall x(x = b \leftrightarrow Wxa)$	τE, 1
5.	Wbi	&E, 4
6.	$\forall x(x = b \leftrightarrow Wxa)$	&E, 4
7.	Show $\forall x(x = b \leftrightarrow Wxi)$	
8.	$\lceil$Show $c = b \leftrightarrow Wci$	
9.	$\lceil$Show $c = b \to Wci$	
10.	$\lceil c = b$	ACP
11.	$\lfloor Wci$	=E, 5, 10
12.	Show $Wci \to c = b$	
13.	$\lceil Wci$	ACP
14.	$(Wci \mathbin{\&} Wbi) \to c = b$	$\forall$E^2, 2
15.	$Wci \mathbin{\&} Wbi$	&I, 13, 5
16.	$\lfloor c = b$	$\to$E, 14, 15
17.	$\lfloor c = b \leftrightarrow Wci$	$\leftrightarrow$I, 9, 12
18.	$b = b$	=I
19.	$b = b \leftrightarrow Wba$	$\forall$E, 6
20.	Wba	$\leftrightarrow$E, 18, 19
21.	$(\tau x\colon Wxi)Wxa$	τI, 7, 20

We may thus analyze arguments involving descriptions by means of semantic tableaux or natural deduction, whether we take descriptions as singular terms or quantifierlike expressions.

Problems

Symbolize these sentences, taking descriptions as either singular terms or quantifiers. (If any are ambiguous, say so. Some of these cannot be symbolized in one or both approaches; if symbolization is impossible, say why.)

1. The dog saw Mary.

2. The black dog ran.

3. The large black dog lives on Raincreek Parkway.

4. The poodle lives on the next street.

▶ 5. Everest is the tallest mountain.

6. The wolf or coyote escaped.

7. The man in the street listens to E. F. Hutton.

8. The man in the street is my brother.

9. The best man will win.

▶ 10. The person who saw Millie talked to Don.

11. This is the ugliest house on the block.

12. This is the ugliest house on any block in the city.

13. The ugliest house on the block is also the most expensive.

14. The project threatens the Waterman building.

▶ 15. The project threatens the oldest building in the city.

16. The wolf or coyote that knocked the garbage can over has returned.

17. The assistant to the chairman hates me.

18. The assistant to the chairman hates him.

19. The assistant to the chairman hates himself.

▶ 20. James Buchanan has been the only president from Pennsylvania.

21. The man who loves a woman who loves Julio hates Harry.

22. The man who loves a woman who loves him hates Harry.

23. The man who loves a woman who loves him admires her.

24. The pilot who aimed at it hit the MIG that chased him.

▶ 25. He loved the bird who loved the man who shot him with his bow (Samuel Taylor Coleridge).

Determine whether these sentences are valid (a) taking descriptions as singular terms, or (b) taking descriptions as descriptional quantifiers.

26. If the party is over, then some party is over.

27. If the party is over, all parties are over.

28. If the fat lady is singing, every fat lady is singing.

29. If the fat lady is singing, then something that sings is fat.

▶ 30. The author of *Waverley* wrote *Waverley*.

31. The person who hates everybody hates himself.

32. The author of *Waverley* wrote something.

33. The sad clown is a clown.

34. The angry student is angry.

35. The king of France is the king of France.

36. The person who is identical to George Burns is George Burns.

37. The king of France is either bald or not bald.

38. The round square is round.

39. The commander of the platoon commands the platoon.

▶ 40. Either the king of France is bald, or the king of France isn't bald.

Evaluate these arguments as valid or invalid, using one of our two methods.

41. The man you saw yesterday was the thief. The man you saw yesterday bought a ticket for São Paulo. Therefore, the thief bought a ticket for São Paulo.

42. The winged horse is Pegasus. So, some horses have wings.

43. The winged horse is Pegasus. No horses have wings, however. So, Pegasus is mythological.

44. The circus is in town. So, there is at most one circus in town.

45. The circus is in town. So, there is exactly one circus in town.

46. No set has itself as a member. Therefore, the set of all sets is not a set.*

47. The least intelligent student at this university has read this book. Everyone who has read this book can think clearly. Thus, even the least intelligent student at this university can think clearly.*

48. Anyone who can do description problems is more capable than anyone who can't. Only one person here can do description problems. So, the most capable here can do description problems.*

49. The student who is studying philosophy is in love with Brooke. It follows that only one student is studying philosophy.*

50. Any criminal who can break out of Alcatraz is more devious than anyone who can't. Only one criminal who can break out of Alcatraz is on the FBI's ten-most-wanted list. Hence, the criminal on the FBI's ten-most-wanted list who can break out of Alcatraz is identical to the most devious

criminal on that list. (Assume that, if x is more devious than y, y is not more devious than x.)*

51. 'The moon is bright,' we've said, means that there is one and only one moon in our domain of discourse, and it's bright. This suggests that there should be an intimate relationship between $B\imath x\ Mx$ (or $(\tau x\colon Mx)Bx$) and $\exists x \forall y (Bx\ \&\ (y = x \leftrightarrow My))$. Are these equivalent in either of our methods of analyzing descriptions?*

Notes

[1] James McCawley, "A Program for Logic," in Donald Davidson and Gilbert Harman (eds.), *Semantics of Natural Language* (Dordrecht: D. Reidel, 1972): 498–544, p. 511.

[2] See W. V. Quine, *Methods of Logic* (Cambridge: Harvard University Press, 1950), Chapter 26.

[3] This is Bertrand Russell's analysis in "On Denoting," originally published in *Mind* in 1905 and reprinted in R. C. Marsh (ed.), *Logic and Knowledge* (New York: G. P. Putnam's Sons, 1956): 39–56.

PART

III

EXTENSIONS
OF
CLASSICAL
LOGIC

NECESSITY

eople use *modal logic* to evaluate
and justify reasoning about possibility and necessity. Aristotle and medieval
logicians tended to think of possibility, actuality, and necessity as *modes of
truth,* that is, as ways in which sentences could be true or false. The study of
the modes of truth became known as *modal logic.*

9.1 IF

Contemporary interest in modal logic stems partly from unhappiness with
treating conditional sentences in terms of truth functions. Within the framework of truth-functional sentential logic, we can symbolize sentences of the
form 'If $\mathcal{A}$, then $\mathcal{B}$' only as formulas having the structure $\mathcal{A} \rightarrow \mathcal{B}$. As our
definition of the conditional truth function indicates, these formulas are true
whenever $\mathcal{A}$ is false or $\mathcal{B}$ is true. But taking a truth-functional rendering of
if seriously gives rise to a variety of puzzles.

First, a truth-functional analysis leads to the "paradoxes of material
implication." These "paradoxes," though not really contradictions in a logical
sense, show that our definition of the conditional can lead us to count some
bizarre arguments as valid. Both these argument forms are valid in classical
sentential logic:

(1) a. p
$\therefore q \rightarrow p$
b. $\neg p$
$\therefore p \rightarrow q$

However, arguments corresponding to them sound strange:

(2) a. The Colorado River is good for white-water canoeing.
$\therefore$ If a nuclear bomb just exploded over the Rockies, the
Colorado River is good for white-water canoeing.

b. Not many Americans eat Thai food.

∴ If many Americans eat Thai food, sales of antacids will soar.

Neither argument seems valid.

Second, a truth-functional rendering of *if* smiles on a variety of other dubious argument forms. Consider this "proof" that God is dead:[1]

(3) a. It isn't true that, if God is dead, then everything is permitted.

∴ God is dead.

b. $\neg(p \to q)$

∴ p

(3)a sounds outrageous, but (3)b is valid. Also, consider:

(4) a. This gun will fire if you load it and pull the trigger.

∴ Either this gun will fire if you load it, or it will fire if you pull the trigger.

b. $(p \mathbin{\&} q) \to r$

∴ $(p \to r) \lor (q \to r)$

Again, taking an English conditional as truth-functional leads us to count a seemingly silly argument valid.

Third, such a strategy forces us to count some peculiar sentences as logical truths:

(5) a. Either you'll die if you eat tomatoes, or you'll die only if you eat tomatoes.

b. $(p \to q) \lor (q \to p)$

c. Either the patient will die if we operate, or he will die if we don't operate.

d. $(p \to q) \lor (\neg p \to q)$

(5)a and c seem to have forms (5)b and d, respectively, and both forms are classically valid. Yet the sentences in (5) hardly sound like truths of logic. (5)a seems to suggest that one ought to avoid eating tomatoes; (5)c seems to say that the patient is in serious trouble.

The English connectives *if* and *only if* aren't truth-functional, therefore, although the truth-functional conditional approximates them closely within a wide range of cases. Finding a better approximation provides one important motivation for studying modal logic.

A second motivation concerns the concepts of logic itself. The most important idea in logic, perhaps, is that of validity or implication. And implication plainly fails to be truth-functional; knowing whether $\mathscr{A}$ and $\mathscr{B}$ are true or false doesn't suffice to let us determine whether $\mathscr{A}$ implies $\mathscr{B}$. We need information about their truth values, not just in the real world, but in every actual or possible circumstance. To say that $\mathscr{A}$ implies $\mathscr{B}$ is to say that $\mathscr{A} \to \mathscr{B}$ is valid, not merely true; that is, that $\mathscr{A} \to \mathscr{B}$ isn't only true but true in every possible circumstance, or true *necessarily*. If we want to study the notion of validity within a symbolic language, therefore, we must study possibility and necessity.

We'll define a new logical connective, standing for *strict implication,* in terms of the classical truth-functional conditional and the concept of necessity. Although the classical conditional is truth-functional, necessity certainly isn't:

$\mathscr{A}$	Necessarily $\mathscr{A}$
T	?
F	F

If a sentence isn't true, then it can't be necessarily true. If a sentence is true, however, it may or may not be necessarily true. Let's say that 'John will wear a blue shirt tomorrow' is true. It is nevertheless not necessarily true; John could wear another shirt. In contrast, 'John is a creature with a soul,' if true, may well be necessarily true.

9.2 MODAL CONNECTIVES

To exploit this idea, we must supplement our sentential language with new connectives. First, we'll add the *necessity* connective, the *box,* $\Box$, to be read as "it is necessary that" or "necessarily." Many other English words express some sort of necessity; the box can sometimes symbolize *must, have to, got to,* and *need to.* As we'll see in a moment, however, these English phrases are ambiguous; symbolizing them requires caution.

All other connectives of standard modal logic can be defined in terms of truth-functional connectives and the box. Nevertheless, we'll add some other connectives for the sake of convenience. Our second connective is the *strict implication* or *strict conditional* connective, the *fishhook,* $\dashv$, to be read as "if . . . then," "only if," or "(strictly) implies." It bears a very close relation to $\rightarrow$. Implication holds whenever the corresponding conditional statement is necessarily true, or valid. So, it's easy to define the fishhook by using the box and the ordinary conditional:

DEFINITION. $(\mathscr{A} \dashv \mathscr{B}) \leftrightarrow \Box(\mathscr{A} \rightarrow \mathscr{B})$

The strict conditional is built on the truth-functional conditional. It provides, in many ways, a better approximation to natural language conditionals than $\rightarrow$ does. We'll consequently use the fishhook to translate all English sentential connective phrases for which we would previously have used $\rightarrow$. So, for the duration of this chapter, natural language conditionals will translate into formulas containing $\rightarrow$.

Third, we'll add the *possibility* connective, the *diamond*, $\diamond$, to be read as "it is possible that" or "possibly." Many other English words signal the notion of possibility: *can, could, may, might,* and, sometimes, the suffix *-ble.* We can define the diamond, too, in terms of the box and the familiar truth-functional connectives of sentential logic. To see how to do this, think about the meaning of *impossible.* If it's impossible for the Cubs to win the pennant, then it's necessarily true that they won't, and vice versa. So $\neg \diamond p$ amounts to $\square \neg p$. But something is possible just in case it isn't impossible. $\diamond p$, therefore, amounts to $\neg \square \neg p$:

DEFINITION. $\diamond \mathscr{A} \leftrightarrow \neg \square \neg \mathscr{A}$

It's possible for a sentence to be true, that is, just in case it's not necessarily false.

Fourth, we'll add the *strict equivalence* or *strict biconditional* connective, the *double fishhook,* $\Leftrightarrow$, to be read as "if and only if," "just in case," or " is equivalent to." We'll use the double fishhook to symbolize any sentence that we would formerly have symbolized with a double arrow. The double fishhook is a modalized version of the biconditional. We'll define it in terms of the box and the biconditional:

DEFINITION. $(A \Leftrightarrow B) \leftrightarrow \square(A \leftrightarrow B)$

The double fishhook bears the relation to the fishhook that the double arrow bears to the arrow. That is, a double fishhook is equivalent to a fishhook going in both directions:

$$(\mathscr{A} \Leftrightarrow \mathscr{B}) \leftrightarrow ((\mathscr{A} \rightarrow \mathscr{B}) \,\&\, (\mathscr{B} \rightarrow \mathscr{A}))$$

The double fishhook thus captures the intuition that '$\mathscr{A}$ if and only if $\mathscr{B}$' should be true if $\mathscr{A}$ and $\mathscr{B}$ agree in truth value, not just in the real world, but in any possible circumstance.

We've expanded SL by adding two singulary and two binary connectives. This is the vocabulary and grammar of our modal language ML:

Vocabulary

Sentence letters: p, q, r, p_1, q_1, r_1, etc.
Singulary connectives: $\neg, \square, \diamond$
Binary connectives: $\&, \vee, \rightarrow, \leftrightarrow, \rightarrow, \Leftrightarrow$
Grouping indicators: (,)

Formation Rules

Any sentence letter is a formula.
If $\mathscr{A}$ is a formula, $\neg\mathscr{A}$, $\Box\mathscr{A}$, and $\Diamond\mathscr{A}$ are formulas.
If $\mathscr{A}$ and $\mathscr{B}$ are formulas, $(\mathscr{A}\ \&\ \mathscr{B})$, $(\mathscr{A} \vee \mathscr{B})$, $(\mathscr{A} \rightarrow \mathscr{B})$, $(\mathscr{A} \leftrightarrow \mathscr{B})$,
$(\mathscr{A} \dashv \mathscr{B})$, and $(\mathscr{A} \boxminus \mathscr{B})$ are formulas.
Every formula can be constructed from a finite number of applications
of these rules.

Note that we haven't yet said anything, officially, about the semantics of this expanded language. A major issue in contemporary discussions of modal logic is what meanings ought to be associated with modal connectives.

Our sentential language now gives us the power to analyze some nontruth-functional connectives in natural language. Modal concepts, however, tend to be extremely complex. Logicians haven't settled on any single system as *the* logic of the modalities. In this chapter we'll develop S5, a commonly used modal logic, in detail.[2] We'll also try to indicate when this logic leads us into controversial positions. At the end of this chapter, furthermore, we'll see that some minor changes yield two other commonly used modal systems, S4 and M.

The semantics we're about to consider is controversial. It applies to S5, but not to other systems of modal logic. Its portrayal of the meanings of modal connectives relies on the concept of a *possible world*. An argument is valid when it's impossible for its premises all to be true while its conclusion is false; when, in other words, in every possible circumstance in which its premises are all true, its conclusion is true as well. Logicians often use the term *possible worlds* to refer to the actual and possible circumstances we must consider in evaluating arguments as valid or invalid. Possible worlds are ways the world might be or have been. They are possible circumstances or states of affairs. We can think of consistent works of fiction and flights of imagination as descriptions of possible worlds other than our own. In the context of sentential logic, we can think of a possible world as an assignment of truth values to all the letters in the language.

What does it mean for a sentence to be necessarily true? The answer in S5 was first formulated by the German philosopher and mathematician Gottfried Leibniz (1646–1716): A sentence is necessarily true if and only if it's true in all possible worlds. Similarly, a sentence is possibly true just in case it's true in some possible world. To say that God necessarily exists, then, is to say that God exists in *every* possible world. To say that I might have been a frog (except for the whims of fate, say) is to say that, in some possible world, I *am* a frog. On any given interpretation, these truth conditions apply:

Semantics

$\Box \mathscr{A}$ is true in a world iff $\mathscr{A}$ is true in every possible world.

$\Diamond \mathscr{A}$ is true in a world iff $\mathscr{A}$ is true in some possible world.

$(\mathscr{A} \dashv \mathscr{B})$ is true in a world iff $\mathscr{B}$ is true in every possible world in which $\mathscr{A}$ is true.

$(\mathscr{A} \mathrel{\boxminus} \mathscr{B})$ is true in a world iff $\mathscr{A}$ and $\mathscr{B}$ agree in truth value in every possible world.

S5 has the simplest semantics of any appealing system of modal logic. All possible worlds appear "on the same level," in the sense that judgments about modality involve the entire realm of possible worlds. So which world is actual—from which world, in particular, we make our judgments about possibility and necessity—makes no difference to those judgments. Notice that, although we're defining truth in a particular world, nothing referring to that world appears on the right side of any of the above clauses. If something is necessary in our world, it is necessary in any world. Similarly, anything that is possible in our world is also possible in any other world.

This is a natural way of thinking about the meanings of modal connectives. But it does have important consequences. S5, because of its approach to possible worlds, brings with it a commitment to five theses concerning the nature of modality.

Thesis 1: Every logical truth is necessary. With this principle, we assert that everything that is logically true is necessarily true. It's easy to see why S5 commits us to this claim. A logical truth, or valid sentence, is true in every possible world. But, according to S5's explication of necessity, that makes it necessarily true.

Thesis 2: Necessary truths are true. This principle, in effect, is that $\Box \mathscr{A}$ should imply $\mathscr{A}$. In S5, we commit ourselves to this thesis by saying that a necessary truth is true in all possible worlds, including the actual world.

Thesis 3: Necessary truths imply only necessary truths. This is the principle that, given some necessary truths, we can deduce only other necessary truths by valid logical means. Truths of logic, for example, when taken by themselves, should imply nothing but other truths of logic; we should not be able to conclude anything about the legitimacy of the Nicaraguan government or the prospects for the Pittsburgh Pirates next season from logically valid premises. Thesis 3 means that, if $\Box \mathscr{A}$ and $(\mathscr{A} \dashv \mathscr{B})$ are true, then $\Box \mathscr{B}$ must be true as well. In S5, we commit to the thesis in the following way: If $\Box \mathscr{A}$ is true, $\mathscr{A}$ is true in every possible world. But, if $(\mathscr{A} \dashv \mathscr{B})$ is true, $\mathscr{B}$

is true in every world in which $\mathscr{A}$ is true. Since $\mathscr{A}$ is true in every world, $\mathscr{B}$ must be also. So $\Box\mathscr{B}$ must be true.

Thesis 4: What is necessary is necessarily necessary. This is the principle that necessity is itself a matter of necessity. Necessary truths, in S5, *had* to be necessary. They can't be necessary simply by chance or happenstance. This thesis is thus simply the statement that, if $\Box\mathscr{A}$ is true, so is $\Box\Box\mathscr{A}$. It's easy to see why S5 commits us to this principle. If a sentence $\mathscr{A}$ is necessarily true in our world, it will be necessarily true in every other world, too. But then $\Box\mathscr{A}$ will be true in every world, and so $\Box\mathscr{A}$ will be necessarily true. Thus, $\Box\Box\mathscr{A}$ will be true.

Thesis 5: What is possible is necessarily possible. This is the principle that possibility, too, is a matter of necessity. Things can't just happen to be possible; if possible at all, they *had* to be possible. This thesis is thus a statement that, if $\Diamond\mathscr{A}$ is true, so is $\Box\Diamond\mathscr{A}$. Together with thesis 4, it implies that all judgments about modality are necessary. So, issues concerning possibility and necessity, according to these principles, look the same in every possible world. A sentence must have whatever modal status it has: possibly true, necessarily true, impossible, and so on. Its modal status couldn't be other than it is. Whether a sentence is necessarily true, necessarily false, or contingent is a matter that, in S5, isn't itself contingent. To see why S5 commits us to this thesis, consider that, if $\mathscr{A}$ is possible in one world, then $\mathscr{A}$ is true in some possible world. Thus, $\mathscr{A}$ is possible in every other world, too. So, $\Diamond\mathscr{A}$ will be true in every world; that makes $\Box\Diamond\mathscr{A}$ true.

Many systems of modal logic accept the first three theses. Theses 4 and 5, however, are highly controversial. In the last section of this chapter, we discuss some of the reasons for this controversy.

9.3 TRANSLATION

Many English expressions translate readily into our expanded language containing modal connectives. Nevertheless, translation is a complex matter. Almost every English expression capable of expressing modality is ambiguous. Modal phrases may express any of a variety of modalities.

One important kind of possibility is *technological* possibility. Something is technologically possible if we have the technical knowledge and ability to achieve it. It is now technologically possible to send astronauts to the moon. But space travel wasn't technologically possible in 1960. What is technologically possible, therefore, changes over time.

Something is *physically* possible if its occurrence is compatible with the physical laws of the universe. Traveling from New York to Tokyo in two hours is not technologically possible in 1986, but it is surely physically possible. If we assume that the theory of relativity is correct, however, travel at speeds faster than the speed of light is not physically possible. Anything that is technologically possible is physically possible. What is physically possible also changes over time, since some physical processes are irreversible.

Something is *epistemically* possible for a person if it is compatible with what that person knows. Since most of us have limited knowledge, much more is epistemically possible for us than is physically or technologically possible. What is epistemically possible for us changes as our knowledge grows. If, by adding a phrase of the form 'For all *a* knows' or 'given what *a* knows,' we do not alter the meaning of a sentence with a modal phrase, that phrase expresses epistemic modality.

Something is *logically* possible, finally, if it's compatible with the laws of logic. Everything technologically or physically possible is logically possible. Indeed, the only things that are logically impossible are those that involve a contradiction.

Modal phrases can express any of the above modalities. But expressions such as *must, might, can,* and *may* can also express varieties of obligation and permission. *Must* and *have to* can sometimes be replaced by *should* or *ought to* without a significant change in meaning. *Can* and *may,* in some cases, function as *is permitted to* and *is allowed to* do.

Phrases capable of expressing modality, then, tend to be multiply ambiguous. Further, we can represent only some of their meanings faithfully with our modal connectives. We can use $\Box, \Diamond$, $\dashv$, and $\boxminus$ to symbolize technological, physical, epistemic, or logical modality, provided that we maintain a single usage throughout the argument at hand. But we cannot use them to translate phrases that express any kind of obligation or permission. Connectives for that task are the subject of Chapter 11.

Consider expressions conveying some kind of necessity or obligation:

(6) a. Necessarily, John will go.
   b. It is necessary that John go.
   c. It is necessary for John to go.
   d. John has to go.
   e. John must go.
   f. John needs to go.
   g. John's got to go.

Each sentence in (6) is ambiguous. Without placing them in a context, we can't tell whether they involve logical, physical, epistemic, or technological necessity, or even a sort of obligation. It's hard to imagine it being logically necessary for John to go. But we can imagine it being physically necessary (if a bouncer has thrown him in the direction of the exit) or technologically necessary (if it's not technologically feasible for John to avoid going—because, say, he's decided to quit the mission but the rocket has already been launched). And John's departure might be epistemically necessary, if we know he's going or has gone. Construed in these ways, the sentences in (6) can translate as formulas having the structure $\Box \mathscr{A}$.

But there are at least two other ways of reading the sentences in (6). First, we might think that John *should* go, and say 'John must go' to indicate that we think he has a duty to go. We might also say 'John has to go' to indicate that he should go if he wants to achieve some goal. So interpreted, the sentences in (6) express a kind of obligation, and cannot be translated into S5.

Second, we might mean that John has to go in order to do something, or for something to happen. On such readings, the sentences in (6) convey the idea that John's departure is a *necessary condition* for something. $\mathscr{A}$ is a necessary condition for $\mathscr{B}$ just in case the truth of $\mathscr{A}$ is necessary for the truth of $\mathscr{B}$. If $\mathscr{A}$ is false, that is, $\mathscr{B}$ cannot be true. So, $\mathscr{A}$ is a necessary condition for $\mathscr{B}$ just in case $\mathscr{A}$ is true in every circumstance in which $\mathscr{B}$ is true; just in case, in other words, $\mathscr{B} \dashv \mathscr{A}$ is true. Conversely, $\mathscr{A}$ is a *sufficient condition* for $\mathscr{B}$ just in case the truth of $\mathscr{A}$ guarantees the truth of $\mathscr{B}$; just in case, that is, $\mathscr{A} \dashv \mathscr{B}$ is true. Sentences saying that one thing is a necessary or sufficient condition for another therefore translate into formulas containing a fishhook. If we construe the sentences in (6) as expressing that John's departure is a necessary condition for something, we can symbolize them in formulas having the structure $\mathscr{B} \dashv \mathscr{A}$, although the sentences themselves do not say anything about $\mathscr{B}$. We must depend on the context to clarify the reason that John's going is necessary.

The ambiguity of modal expressions also afflicts sentences that convey necessary and sufficient conditions. When we say that $\mathscr{A}$ is a necessary condition for $\mathscr{B}$, we may mean that $\mathscr{A}$ is logically necessary for $\mathscr{B}$ (as the validity of an argument, for instance, is a necessary condition for its soundness); that $\mathscr{A}$ is physically necessary for $\mathscr{B}$ (as a temperature of 100 degrees Centigrade is necessary for water to boil at standard pressure); that $\mathscr{A}$ is technologically necessary for $\mathscr{B}$ (as rockets carrying their own oxygen are necessary for flight outside the atmosphere); or even that $\mathscr{A}$ is epistemically necessary for $\mathscr{B}$.

There is a further source of ambiguity in sentences with modal expressions: Some conditional sentences have phrases expressing necessity in their consequents. These necessity phrases may express that the truth of the consequent is necessary for the truth of the antecedent; in such cases, the modality properly applies to the conditional statement as a whole, not just the consequent portion. In other cases, the modality applies solely to the consequent. Fortunately, one interpretation is usually far more plausible than the other in particular cases. These sentences illustrate the difference.

(7) a. If the solution has turned blue, a ferric sulfide must be present.
   b. If this is Tuesday, this must be Belgium.
   c. If a mathematical statement is true, it's necessarily true.

In the most plausible interpretations of (7)a and b, the modality conveyed by *must* applies to the entire conditional. (7)a draws our attention to a physically necessary connection between the color of a solution and its chemical analysis; (7)b points toward an epistemically necessary connection between the day of the week and a location. (7)c, in contrast, contains a modal phrase that applies to its consequent alone. (7)c says that mathematical statements are necessary truths if they are true at all. Correct or incorrect, this thesis is not trivial. If we take *necessarily* as applying to the entire conditional, however, we have to read (7)c as saying merely that it's a necessary truth that a mathematical statement is true if it's true.

Just as a phrase expressing necessity may appear in the consequent of a conditional even though it properly applies to the entire conditional, so such a phrase may appear in the conclusion of an argument. In some cases, the phrase applies to the entire argument and, so, expresses the necessity of the link between premises and conclusion. Consider, for example, the argument:

(8) John is a detective. All detectives are accustomed to danger. So, John must be accustomed to danger.

The argument does not mean to show that 'John is accustomed to danger' is a necessary truth in any absolute sense. The word *must* signals that a conclusion is being drawn. The truth of the sentence 'John is accustomed to danger' is not necessary per se; it's a necessary condition for the truth of the premises. In this circumstance, the word *must* need not be translated at all.

Another oddity of conditional sentences containing modal expressions is that phrases expressing possibility may also convey information about necessary or sufficient conditions. The following sentences seem alike in meaning:

(9) a. Joe can pass this course only if he makes an A on the test.
b. To pass this course, Joe must make an A on the test.
c. Making an A on the test is a necessary condition for Joe's passing the course.

Actually, (9)a seems ambiguous; making an A on the test may be necessary for Joe to pass, or necessary for Joe to be capable of passing. In general, the use of *can* or *possibly* in the antecedent of a conditional produces the same ambiguity that *must* produces in the consequent.

Under most circumstances, nevertheless, sentences involving *possible, possibly,* or *can* translate into formulas containing the diamond. Each of the following sentences translates into $\diamond p$:

(10) a. It's possible that there is life elsewhere in the universe.
b. I might go to the Whalers game on Friday.
c. Dr. Green might not be at the hospital today.
d. You could have been a great theoretical physicist.
e. The president may get the bill through the Senate intact.
f. Professor Asher's lecture was comprehensible.
g. It's possible for Fred to win the nomination.

Just as we'll translate conditionals and sentences specifying necessary or sufficient conditions into formulas containing fishhooks, we'll symbolize *if and only if, just in case, is a necessary and sufficient condition for,* and *is equivalent to* (or *are equivalent*) as the double fishhook. Each of these sentences thus translates into a formula having the structure $\mathcal{A} \boxminus \mathcal{B}$:

(11) a. I'll come to your party if and only if Jane does.
    b. This sentence is valid just in case it's necessarily true.
    c. Ralph's advocating investment in silver is equivalent to his encouraging speculation.
    d. John's selling the stock is necessary and sufficient for his being convicted.

We can also translate sentences containing *is compatible with* or *is consistent with,* and their opposites, *is incompatible with, is inconsistent with,* or *contradicts.* Two sentences are consistent if they can both be true. This fact suggests a natural analysis of compatibility: $\mathcal{A}$ is compatible with $\mathcal{B}$ just in case it's possible that $\mathcal{A}$ and $\mathcal{B}$ are both true. Incompatibility, of course, is just the opposite of compatibility. So we can render $\mathcal{A}$ *is compatible with* $\mathcal{B}$ as $\Diamond(\mathcal{A} \& \mathcal{B})$, and $\mathcal{A}$ *contradicts* $\mathcal{B}$ as $\Box\neg(\mathcal{A} \& \mathcal{B})$ (or, equivalently, as $\mathcal{A} \dashv3 \neg\mathcal{B}$). These sentences translate as formulas having the structure $\Diamond(\mathcal{A} \& \mathcal{B})$:

(12) a. That this vase was produced before 1000 B.C. is compatible with the assumption that the Hittites had iron weapons.
    b. That $2 + 3 = 6$ is consistent with Dale's theory of arithmetic.
    c. Leela's savagery and innocence are compatible.

These sentences translate as formulas having the structure $\Box\neg(\mathcal{A} \& \mathcal{B})$:

(13) a. That it was raining that night contradicts the allegation that the full moon would have provided enough light.
    b. The demand that you be given higher pay is incompatible with your insistence on job security.

Problems

Translate the following into sentential modal logic:

1. It's possible for John to get what he wants.

2. You have to understand this material well.

3. Sarah can go to law school only by taking out loans.

4. To prove the theorem, Ralph will have to use an indirect proof.

▶ **5.** Rhonda's perceptiveness is inconsistent with the report that she made the error.

6. Leigh's analyzing the data is necessary for a realistic forecast.

7. John's refusing to accept the thesis is equivalent to his denying Ed the degree.

8. If Beth knows what Jerry is up to, she can stop him.

9. Only if she is able to defeat the challenge can Lynn persuade the board.

▶ 10. If computing the area is equivalent to taking an integral, then the software can do it only by incorporating certain complex approximation techniques.

11. If you build a castle in the air, you won't need a mortgage (Philip Lazarus).

12. You can be sincere and still be stupid (Charles Kettering).

13. Sweetness and light are not necessarily consistent with productivity (Robert L. Katz).

14. If you are patient in one moment of anger, you will escape 100 days of sorrow (Chinese proverb).

▶ 15. It is only by painful effort, by grim energy and resolute courage, that we move on to better things (Theodore Roosevelt).

16. If you don't get what you want, it is a sign either that you did not seriously want it, or that you tried to bargain over the price (Rudyard Kipling).

17. If we are to preserve civilization, we must first remain civilized (Louis St. Laurent).

18. So long as man is capable of self-renewal he is a living being (Henri Frederic Amiel).

19. If the mind of the teacher is not in love with the mind of the student, he is simply practicing rape, and deserves at best our pity (Adrienne Rich).

▶ 20. I could not love thee, dear, so much/Loved I not honor more (Richard Lovelace).

9.4 Modal Semantic Tableaux

S5 is an extension of classical sentential logic. In S5, the logic of the truth-functional connectives is not revised; a theory about other connectives is simply added. Consequently, in S5, all the usual sentential semantic tableau rules are taken for granted. For a detailed presentation of those rules, see Chapter 3.

Recall that a tableau branch closes just in case some "live" (usually, undispatched) formula appears on both sides of the branch. Since the left side of a tableau represents truth and the right represents falsehood, a live formula on both sides of a branch is tantamount to a contradiction on that branch. A tableau closes if and only if every branch of the tableau closes.

We use a tableau, in effect, to try to find interpretations of the formulas at the tops that make those on the left true and those on the right false. Interpretations, we've said, correspond to circumstances that make the sentences those formulas symbolize true or false. An argument is valid if and only if no possible circumstance makes its premises all true and its conclusion false. In checking for validity, therefore, we are, in effect, trying to find such a possible world. We assume that all the premises are true and the conclusion false. If all branches close, there is no such world; the argument is valid.

Since we could translate all the other modal connectives, we need tableau rules only for the box. Nevertheless, we'll have rules for each connective; the added rules allow tableau construction to proceed more simply and more naturally than it would if we had to translate formulas into others containing only boxes and sentential connectives.

Suppose we find □𝒜 on the left of a branch. If □𝒜 is true, what can we conclude about 𝒜? That it's true. We'll thus formulate the rule □L as follows:

Necessity Left (□L)

$$\begin{array}{c|} \Box\mathscr{A} \\ \mathscr{A} \end{array}$$

If □𝒜 is true, 𝒜 must be true. This rule thus expresses thesis 2. In applying □L, we begin with the information that something is necessarily true and end up only with the information that it's true. For this reason, we don't need to dispatch the formula to which we apply □L. Even if we've deduced from □L that a formula is true, that formula is still true necessarily.

To see how this rule works, let's show that the second modal principle of S5—that all necessary truths are true—is valid. A particular case of the principle is □p → p; nothing in the tableau, however, will depend on p. Any formula could play its role. To prove the formula valid, we try to find a world or interpretation in which it's false. So, we begin by placing the formula on the right and applying →R to its main connective, →:

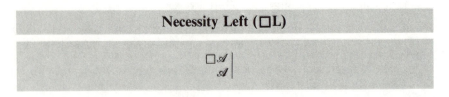

At this point we can close the tableau by applying □L:

$$\begin{array}{c|c} & \Box p \to p \quad \checkmark \\ \Box p & p \\ p & \\ & \text{Cl} \end{array}$$

The rule □R poses a trickier problem. Suppose that □𝒜 is false. What does that fact tell us about 𝒜? Nothing, directly; that 𝒜 isn't a necessary truth seems to tell us nothing about whether 𝒜 is actually true or false. 𝒜 might be false, but it also might be contingently true. Recall, however, that necessary truth is truth in all possible worlds. If 𝒜 isn't necessarily true, then 𝒜 must be false in some possible world. So, we need a way of saying that, if □𝒜 is false, then 𝒜 is false in some world, though not necessarily the world we first envisioned. We'll formulate the rule □R in this way:

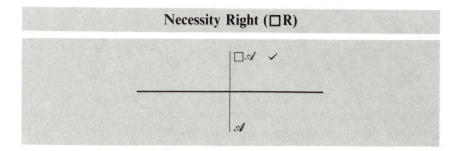

Necessity Right (□R)

We'll call the line separating these formulas a *world shift line*. It indicates that the world in which 𝒜 is false isn't necessarily the same world, described earlier in the tableau, in which □𝒜 is false. Tableaux are, in effect, diagrams of possible worlds. With world shift lines, we can use tableaux to diagram several worlds by switching at certain points from one world to another. In one tableau, we can shift from speaking of truth and falsehood in one world to speaking about truth and falsehood in another.

To see how □R works, let's show that tableaux demonstrate that the first principle of S5—that all logical truths are necessary—is valid. Actually, we'll show this only for one particular logical truth, $p \lor \neg p$. To show that $\Box(p \lor \neg p)$ is valid:

$$
\begin{array}{c}
\Box(p \lor \neg p) \quad \checkmark \\
\hline
p \lor \neg p \quad \checkmark \\
p \\
\neg p \quad \checkmark \\
p \\
\text{Cl}
\end{array}
$$

Notice that a logical truth could fail to be necessary only by being false in some world. But no world can make a logical truth false. Valid sentences, after all, are those true in all possible worlds. The result of placing a box before a valid formula is thus always another valid formula.

Problems

Show that the following are valid in S5.

1. $\Box(p \to q)$; p; $\therefore q$

2. $\Box(p \to q)$; $\neg q$; $\therefore \neg p$

3. $\Box(p \leftrightarrow q)$; q; $\therefore p$

4. $\Box(p \leftrightarrow q)$; $\neg p$; $\therefore \neg q$

▶ **5.** $\Box(p \leftrightarrow \neg p)$; $\therefore q$

6. $\Box(p \to q)$; $p \lor q$; $\therefore q$

7. $\Box\neg(p \,\&\, q)$; q; $\therefore \neg p$

8. $\Box(p \lor q)$; $\neg q$; $\therefore p$

9. $\neg(\neg\Box p \,\&\, \neg\Box q)$; $\therefore p \lor q$

▶ **10.** $\Box(\Box p \to \Box q)$; $\Box\neg q$; $\therefore \neg\Box p$

9.5 OTHER TABLEAU RULES

Given our rules for necessity, and the definitions of other modal connectives in terms of the box, we can derive tableau rules for possibility, strict implication, and strict equivalence. Let's begin with possibility. If $\Diamond\mathcal{A}$ appears on the left side of a tableau, we're assuming that it's true. What can we conclude about $\mathcal{A}$? If $\Diamond\mathcal{A}$ is true, then $\mathcal{A}$ is true in some possible world, though not necessarily the actual one. So we can write $\mathcal{A}$ on the left, after drawing a world shift line.

Possibility Left (◇L)

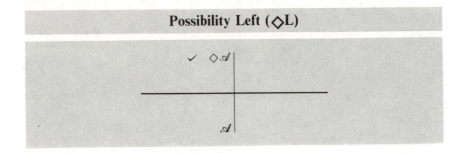

This rule allows us to show that contradictions such as $p \,\&\, \neg p$ are impossible:

$$
\begin{array}{c|c}
& \neg\Diamond(p \,\&\, \neg p) \;\checkmark \\
\checkmark \;\; \Diamond(p \,\&\, \neg p) & \\
\hline
\checkmark \;\; p \,\&\, \neg p & \\
p & \\
\checkmark \;\; \neg p & \\
& p \\
\mathrm{C1} &
\end{array}
$$

If $\Diamond \mathcal{A}$ appears on the right side of a tableau branch, we're assuming that it's false. So, if $\Diamond \mathcal{A}$ is false, what can we conclude about $\mathcal{A}$? That it is false. If something isn't even possible, then it certainly isn't true. So we can write $\mathcal{A}$ on the right if we find $\Diamond \mathcal{A}$ there.

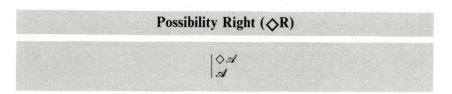

Possibility Right ($\Diamond$R)

This rule allows us to show that whatever is actual is possible:

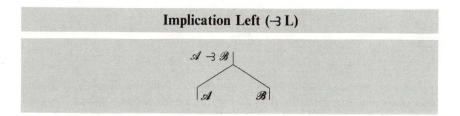

Rules for the fishhook are even easier to formulate. If $\mathcal{A} \dashv \mathcal{B}$ appears on the left, we're assuming that it's true. Since $\mathcal{A} \dashv \mathcal{B}$ amounts to $\Box(\mathcal{A} \to \mathcal{B})$, if $\mathcal{A} \dashv \mathcal{B}$ is true, we can conclude that $\mathcal{A} \to \mathcal{B}$ is true. Our rule for fishhooks on the left, then, is the same as the rule for conditionals on the left.

Implication Left ($\dashv$ L)

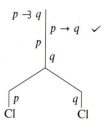

This rule allows us to show that strict conditionals imply ordinary, truth-functional conditionals:

If a fishhook appears on the right, we're assuming that it's false. $\mathscr{A} \dashv \mathscr{B}$ can be false only if there is a possible world in which $\mathscr{A}$ is true and $\mathscr{B}$ is false. So, if $\mathscr{A} \dashv \mathscr{B}$ appears on the right, we can draw a world shift line, writing below it $\mathscr{A}$ on the left and $\mathscr{B}$ on the right.

Implication Right (⊰R)

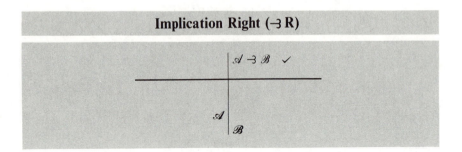

⊰R allows us to show that $(p \& q) \dashv q$ is valid.

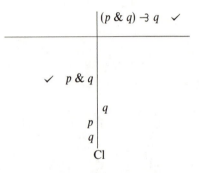

Finally, we can develop rules for the double fishhook. If $\mathscr{A} \; \mathrm{\upshape ε3} \; \mathscr{B}$ is true, then $\mathscr{A} \leftrightarrow \mathscr{B}$ is true also. So, the rule for double fishhooks on the left is similar to that for biconditionals on the left.

Equivalence Left (ε3L)

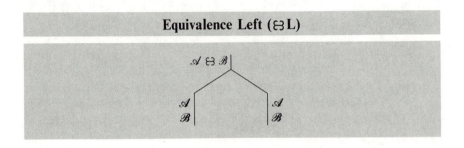

Using this rule, we can show that $p \; \mathrm{ε3} \; q$ implies $p \leftrightarrow q$.

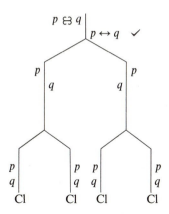

If a double fishhook appears on the right, we're assuming that it's false. 𝒜 ⇌ ℬ can be false only if 𝒜 and ℬ disagree in truth value in some possible world. So, there is a world in which 𝒜 is true and ℬ is false, or ℬ is true and 𝒜 is false.

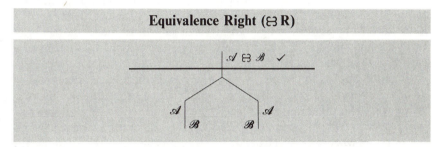

By means of ⇌R, it's possible to show that $(p \lor q) \between (\neg p \to q)$ is valid:

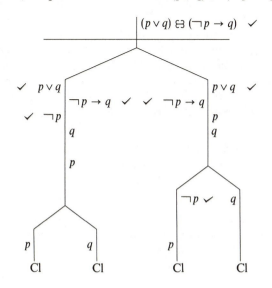

Problems

Show that the following are valid in S5.

1. $p \rightthreetimes q$; p; $\therefore q$

2. $p \rightthreetimes q$; $\neg q$; $\therefore \neg p$

3. $p \mathbin{\text{\raisebox{0pt}{⊖}}} q$; q; $\therefore p$

4. $p \mathbin{\text{\raisebox{0pt}{⊖}}} q$; $\neg p$; $\therefore \neg q$

5. $p \mathbin{\text{\raisebox{0pt}{⊖}}} \neg p$; $\therefore q$

6. $p \rightthreetimes q$; $p \vee q$; $\therefore q$

7. $\neg \Diamond (p \mathbin{\&} q)$; q; $\therefore \neg p$

8. $\neg \Diamond \neg (p \vee q)$; $\neg q$; $\therefore p$

9. $\Diamond (\Box p \mathbin{\&} \Box \neg p)$; $\therefore p \mathbin{\&} \neg p$

10. $\Diamond p \rightthreetimes \Box q$; $\Box \neg q$; $\therefore \Diamond \neg p$

9.6 WORLD TRAVELING

These rules for the box and the other modal connectives are crucial to S5 and, indeed, to many systems of modal logic. But they are, in themselves, too weak to capture S5. When we shift from world to world, we need to know something about the new world. In other words, modal systems generally assume that our "world traveling" isn't completely blind; some of what we know about the first world will apply to the second world as well. *What* information applies to the new world, however, is the source of controversy among proponents of different systems; many systems that agree on the rules we've seen so far differ in this matter. We need a rule to specify what information crosses world shift lines in S5. Because the rule indicates what information survives when we travel from one world to the next, we'll call it a *survival rule*.

The term *survival* is especially apt. The issue is: What formulas on the tableau should we count as "live"? In ordinary, truth-functional tableaux, we regard any undispatched formula as live. Once we incorporate world shift lines into our system, however, we can move from world to world. Certain formulas giving us information about a particular world prove unhelpful once we move to another world. World shifts, that is, kill formulas. We can't assume that we can transfer the information about the first world to the second. Thus, we need to alter our definition of live formulas.

> **DEFINITION.** A formula is *live* on a tableau branch iff it is undispatched and appears between the last world shift line and the tip of the branch.

When we draw a world shift line, we threaten to kill all the information we have. So, we must answer the question: What formulas should survive world shifts? The answer is provided by the survival rule.

S5's survival rule should reflect S5's five principles about modality. In particular, all modal formulas—formulas whose main connectives are modal—should themselves be necessary if they are true at all. This is so because S5 treats modal judgments as invariant across possible worlds. How can we construct the rule so it has this effect? Clearly modal formulas, since they are all necessarily true or necessarily false, should survive world shifts. If they hold in one world, they hold in all. Nonmodal formulas, in contrast, should not survive. That it is raining in this world is no guarantee that it will be raining in the world we visit next in the tableau.

To formulate S5's survival rule, we need an additional definition.

> **Definition.** A formula is *modally closed* iff each of its atomic subformulas is within the scope of a modal connective.

Consequently, all formulas whose main connectives are modal are modally closed. In addition, formulas such as $\neg\Box p$, $\Diamond p \rightarrow \Box q$, $\neg(p \mathbin{\mathit{8}} q)$, and $\Box p \leftrightarrow (p \mathbin{-\!3} q)$ are modally closed. Although their main connectives are truth-functional, each sentence letter occurrence within them falls inside the scope of $\Box$, $\Diamond$, $-\!3$, or $\mathit{8}$. In contrast, p, $p \rightarrow \Box q$, and $\Diamond(p \mathbin{\&} q) \vee r$ are modally open.

With the help of this definition, we can formulate the rule specifying what information transfers into a new possible world on a tableau as follows:

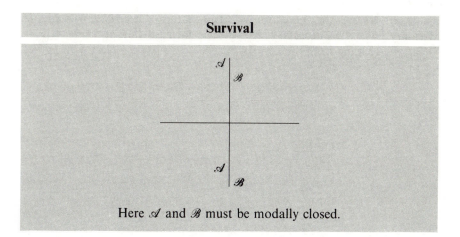

Survival

Here $\mathcal{A}$ and $\mathcal{B}$ must be modally closed.

Every modally closed formula survives a world shift. It makes no difference on which side of the branch the formula appears. Notice that the formula that survives need not be dispatched. Notice also that the shift line here is in regular type; drawing a shift line is not part of the rule. The line must already be present for the rule to be applied.

We can demonstrate these rules by showing that modal theses 3, 4, and 5 are valid in S5. Thesis 3 is that necessary truths imply only necessary truths: that, if $\mathscr{A}$ is necessary, and $\mathscr{A}$ implies $\mathscr{B}$, then $\mathscr{B}$ is necessary too. A typical symbolic instance of this principle is $(\Box p \mathbin{\&} (p \mathbin{-3} q)) \to \Box q$.

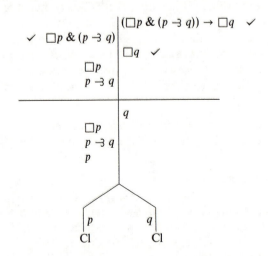

Notice that the world shift is a consequence of applying $\Box$R to $\Box q$. $\Box p$ and $p \mathbin{-3} q$ cross the shift line by the survival rule, since both are modally closed.

Thesis 4 states that necessity itself is a matter of necessity; whatever is necessary has to be necessary. A symbolic form of this principle, $\Box p \to \Box\Box p$, is also valid in S5.

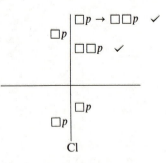

The $\Box p$ on the left, below the world shift line, came from the earlier occurrence of $\Box p$ above by use of the survival rule. The same formula on the right came from $\Box\Box p$ by an application of $\Box$R.

Finally, thesis 5 says that whatever is possible is so necessarily. This thesis, sometimes called the "characteristic principle" of S5, states that, if a formula, $\mathscr{A}$, is possible, it is necessarily possible. In symbols, this thesis amounts to $\Diamond p \to \Box\Diamond p$.

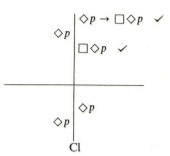

Cl

Finally, S5 requires what we'll call a *revival rule*. Normally, after we dispatch or kill a formula by a world shift, the formula remains dead throughout the rest of the tableau. S5, however, allows us to revive some dead formulas. We need such a rule because of formulas such as $p \rightarrow \Box \Diamond p$, which we can read as "if p is true, p has to be possible." We can't prove that this formula is valid using just $\Box$L, $\Box$R, and the survival rule:

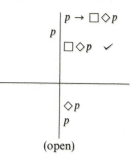

(open)

There is a p on both sides of the branch, but p is a live formula only on the right. A branch closes only if the same formula appears *live* on both sides of the branch. The world shift line prevents us from closing the tableau.

Nevertheless, it's easy to see that the principle should be valid in S5. If something is true, after all, it is surely possible. And, in S5, whatever is possible is necessarily possible. So, whatever is true should also be necessarily possible.

Let's call the world described at the top of the tableau world 1, and the world described after the shift world 2. In world 2, it's true that $\neg \Diamond p$. Since all worlds share the same modal truths, $\neg \Diamond p$ should also be true in world 1. We have no rule, however, that allows us to transfer this information back to world 1. In other words, the survival rule allows us to apply modal formulas to worlds *later* in the tableau; no rule allows us to apply modal formulas to worlds *earlier* in the tableau.

We need a rule that will allow us to bring the information we derive from later worlds back to worlds we've already visited. We could allow ourselves to cross back over world shift lines, but it would be very hard to separate legitimate from illegitimate crossings. Instead, therefore, we'll formulate a rule that allows us to revive a group of dead formulas. This rule has the effect of permitting us to return to worlds we've left in our earlier travels.

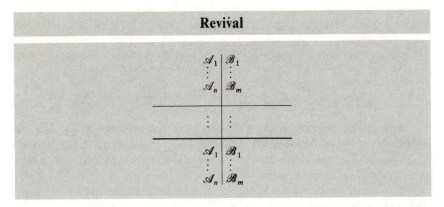

Revival

Note that all the transferred formulas must be undispatched. This rule says that we can draw a world shift line and write down every undispatched formula from a previous portion of the branch lying between successive world shift lines (or between a world shift line and the origin). These formulas need not be modally closed. In effect, we can use the revival rule to return to a world we previously visited. So, after drawing a world shift line to mark our return, we revive the information we had in that world, modally closed or not.

It's vital that the second shift line in the statement of the rule be boldface; we must draw the line in applying the rule. In constructing modal tableaux, we assume that the formulas lying between any two successive world shift lines all describe the same world. The revival rule means that S5 allows us to revive entire worlds. But we must draw another world shift line to indicate that we are going to be talking about a world different from the one we've been discussing.

Though complicated to state, this rule is easy to use. Consider $p \rightarrow \Box \Diamond p$, which we may now show to be valid:

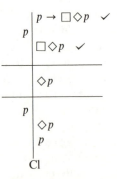

The first world shift line of this tableau results from an application of $\Box$R to the formula $\Box \Diamond p$. Having derived $\Diamond p$ on the right, we don't use $\Diamond$R, as we did before, but instead move back to the first world we talked about. The second world shift line, therefore, results from applying the revival rule. The formula p appears on the left because p was the only undispatched formula appearing between the origin of the tableau and the first shift line; p represents all the information from the earlier part of the tableau we have available concerning this world. Of course, $\Diamond p$ survives, and so we can transfer that information to the world as well. This transfer is enough to create a contradiction.

The best strategy for using the revival rule is this:

1. Begin by applying the rules for the connectives.

2. Wait until you have formulas with boxes as main connectives, which will survive the shift into a "revived" world.

3. Revive the world with the information you hope to use.

As the example above suggests, modal tableaux are more complex than standard sentential semantic tableaux. Instead of charting possible worlds, we are now, in effect, charting systems of possible worlds. We still begin by trying to describe systematically a world in which, for example, a formula is false. But, to describe that world, we may have to describe a variety of others.

Just as open truth-functional tableaux specify interpretations assigning the formulas at their tops appropriate truth values, open modal tableaux do the same. The method for constructing an interpretation from a tableau branch is the same, except that world shift lines move us from one world to another.

For example: in S5, $\Box(p \lor q)$ does not imply $\Box p \lor \Box q$.

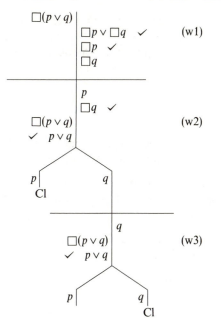

The tableau remains open, because *p* is live only on the left. The information the tableau gives us is:

World	True	False
w1	$\Box(p \vee q)$	$\Box q$
w2	$\Box(p \vee q), q$	p
w3	$\Box(p \vee q), p$	q

The list above takes account of all undispatched formulas. We can write an interpretation by focusing on the sentence letters:

	p	*q*
w1		
w2	F	T
w3	T	F

The table above is not an interpretation, but a *kind* of interpretation: We haven't specified values of *p* and *q* in w1. Indeed, our treatment of world w1 is incomplete in the tableau: We know that $\Box(p \vee q)$ is true in w1. It follows that $p \vee q$ is also true there. So the tableau allows us to specify three interpretations making $\Box(p \vee q)$ true but $\Box p \vee \Box q$ false:

	p	*q*	*p*	*q*	*p*	*q*
w1	T	F	F	T	T	T
w2	F	T	F	T	F	T
w3	T	F	T	F	T	F

As we've seen above, there is no simple way of knowing when to apply the revival rule. An argument form will be valid in S5 just in case *some* appropriate tableau closes. As in natural deduction systems, therefore, success (here, in closing a tableau) does show validity, but failure does not establish invalidity.

Problems

These arguments and sentences appeared earlier in this chapter as examples of bizarre validities in classical logic. Are any of these valid in S5?

1. The Colorado River is good for white-water canoeing. So, if a nuclear bomb just exploded over the Rockies, the Colorado River is good for white-water canoeing.

2. Not many Americans eat Thai food. If many Americans eat Thai food, therefore, sales of antacids will soar.

3. It isn't true that, if God is dead, then everything is permitted. So God is dead.

4. This gun will fire if you load it and pull the trigger. Hence, either this gun will fire if you load it, or it will fire if you pull the trigger.

▸ 5. Either you'll die if you eat tomatoes, or you'll die only if you eat tomatoes.

6. Either the patient will die if we operate, or he will die if we don't operate.

Evaluate these arguments as valid or invalid in S5:

7. If it's possible that God exists, it's possible that this is the best of all possible worlds. So, God's existence is compatible with this world's being the best possible world.

8. My willingness to compromise is compatible with the fact that I strive to do what is right. So, if it's possible for me to strive to do right, and be a good person, I can be willing to compromise.

9. If you go to college, you'll have to work hard. But working hard and having fun are compatible. So, you can go to college and have fun too.

▸ 10. Advocating tax cuts is consistent with abhorring deficits. If one advocates tax cuts, however, one must be willing to shrink revenues in the short term. Thus, abhorring deficits is compatible with a willingness to shrink revenues in the short term.

11. If we have any knowledge at all, we know that we exist. So, knowledge is possible only if we know we exist.

12. The Warren Commission account of the Kennedy assassination is compatible with many descriptions of the shooting given by eyewitnesses in Dealey Plaza. But the Commission's report is inconsistent with the film of the incident made by Abraham Zapruder. Therefore, the Zapruder film is incompatible with the descriptions given by many observers who were in Dealey Plaza that day.

13. Whether we choose to adopt different social and linguistic conventions and, so, change our language in fundamental ways is a contingent matter (that is, it's possible but not necessary). The laws of logic and mathematics, by contrast, are necessarily true. The assumption that the laws of logic and mathematics are true is thus consistent even with the adoption of different social and linguistic conventions and with fundamental changes in our language.

14. The conjecture that $2 + 2 = 3$ is necessarily false. So, if $2 + 2 = 3$, everything anyone has ever learned is an outrage.

15. If the state university becomes more selective in its admissions process, the legislature might become angry. But if the university doesn't tighten admissions standards, there won't be enough classroom space. So, it's necessarily true that, if the state university has enough classroom space, the legislature might become angry.

Each pair of formulas below appears to be equivalent. Are any really equivalent in S5?

16. □(p & q)   □p & □q

17. □(p ∨ q)   □p ∨ □q

18. □(p → q)   □p → □q

19. □(p ↔ q)   □p ↔ □q

20. □¬p   ¬□p

21. □(p ⊰ q)   □p ⊰ □q

22. □(p ⪕ q)   □p ⪕ □q

23. □◇(p & q)   ◇(□p & □q)

24. □¬◇(p & q)   ¬◇(□p & □q)

▶ 25. □◇p   ◇□p

26. ◇(p & q)   ◇p & ◇q

27. ◇(p ∨ q)   ◇p ∨ ◇q

28. ◇(p → q)   ◇p → ◇q

29. ◇(p ↔ q)   ◇p ↔ ◇q

30. ◇¬p   ¬◇p

31. ◇(p ⊰ q)   ◇p ⊰ ◇q

32. ◇(p ⪕ q)   ◇p ⪕ ◇q

33. ◇◇(p & q)   ◇(◇p & ◇q)

34. ◇¬◇(p & q)   ¬◇(◇p & ◇q)

Various authors have proposed these principles as plausible theses of modal logic. (The column to the right lists a system in which the principle counts as valid.) Are any of these principles valid in S5? Contradictory in S5?

▶ 35. p ⊰ (◇□p ⊰ p)   S4.04

36. (□p ⊰ q) ∨ (□q ⊰ p)   S4.3

37. (□p ⊰ □q) ∨ (□q ⊰ □p)   S4.3

38. □p ∨ ((p ⊰ q) ∨ (q ⊰ p))   V1

39. p ⊰ (◇p ⊰ p)   K1.2

40. ((p ⊰ □p) ⊰ p) ⊰ p   K1.1

41. ((p ⊰ □p) ⊰ □p) ⊰ □p   K1.1

42. $((p \rightthreetimes q) \rightthreetimes p) \rightthreetimes \Diamond \Box p$ \hfill K1

43. $((p \rightthreetimes \Box p) \rightthreetimes p) \rightthreetimes (\Diamond \Box p \rightthreetimes p)$ \hfill S4.3.1

44. $((\Diamond \Box p \rightthreetimes p) \vee \Box q) \vee ((q \rightthreetimes p) \vee (p \rightthreetimes q))$ \hfill S4.5

Many differences of opinion among modal logicians find expression in attitudes about compatibility and incompatibility. Determine whether each of these principles about compatibility is valid in S5.

▶ **45.** $\Diamond(\Diamond p \ \& \ \neg \Diamond q) \rightarrow \Diamond(p \ \& \ \neg q)$ \hfill Q

46. $\neg \Diamond(p \ \& \ \neg \Diamond p)$ \hfill S1

47. $\Box \neg (\Diamond(p \ \& \ q) \ \& \ \neg \Diamond p)$ \hfill S2

48. $\Diamond(p \ \& \ q) \rightthreetimes \Diamond p$ \hfill S2

49. $\Diamond(p \ \& \ \neg p) \rightthreetimes p$ \hfill M

50. $\neg \Diamond \Diamond(p \ \& \ \neg p)$ \hfill S4

51. $(\Box \Diamond p \ \& \ \Box \Diamond q) \rightthreetimes \Diamond(p \ \& \ q)$ \hfill S4.1

52. $\Diamond(p \ \& \ q) \rightthreetimes (\Diamond(p \ \& \ \neg q) \rightarrow \Box p)$ \hfill V1

The modal logics S6, S7, S8, and S9, together with some offshoots, were developed during the 1940s to reflect rather different intuitions about modality from those embodied in S5.[3] Here are some principles they assert. Are any of these valid in S5? Are any contradictory? Do any become contradictory if $\mathscr{A}$ is a contradiction, say, $p \ \& \ \neg p$?

53. $\Diamond \Diamond \mathscr{A}$ \hfill S6

54. $\Diamond \Diamond (\mathscr{A} \ \& \ \neg \mathscr{A})$ \hfill S6

55. $\Diamond \neg \Diamond \Diamond (\mathscr{A} \ \& \ \neg \mathscr{A})$ \hfill S7.1

56. $\Diamond \Box \Diamond \Diamond \mathscr{A}$ \hfill S7.5

▶ **57.** $\Box \Diamond \Diamond \mathscr{A}$ \hfill S8

58. $\neg \Diamond \neg \Diamond \Diamond (\mathscr{A} \ \& \ \neg \mathscr{A})$ \hfill S8

59. $\neg \Box \Box \mathscr{A}$ \hfill S9

60. $(\mathscr{A} \rightthreetimes \mathscr{A}) \rightthreetimes \neg(\neg(\mathscr{A} \rightthreetimes \mathscr{A}) \rightthreetimes (\mathscr{A} \rightthreetimes \mathscr{A}))$ \hfill S9

61. $\Diamond \Box \mathscr{A}$ \hfill Tx

62. In S5, are 53 and 56 equivalent?

63. In S5, are 56 and 57 equivalent?

64. In S5, are 57 and 59 equivalent?

65. In S5, are 59 and 61 equivalent?

Boethius's thesis about conditionals is that 'If $\mathscr{A}$, then $\mathscr{B}$' and 'If $\mathscr{A}$, then not $\mathscr{B}$' are contradictory. Aristotle's thesis is that no formula is implied by its own negation.

66. Is Boethius's thesis valid in S5?

67. Is Aristotle's thesis valid in S5?

68. Does Boethius's thesis imply Aristotle's in S5?

69. Does Aristotle's thesis, in S5, imply Boethius's thesis?

70. Suppose that we were to restrict Boethius's thesis to contingent formulas, to obtain the principle $(\neg\Box\mathscr{A} \,\&\, \neg\Box\neg\mathscr{A}) \dashv3 \Box\neg((\mathscr{A} \dashv3 \mathscr{B}) \,\&\, (\mathscr{A} \dashv3 \neg\mathscr{B}))$. Is this principle valid in S5?

71. Suppose we were to restrict Aristotle's thesis to contingent formulas, obtaining $(\neg\Box\mathscr{A} \,\&\, \neg\Box\neg\mathscr{A}) \dashv3 \neg(\neg\mathscr{A} \dashv3 \mathscr{A})$. Is this principle valid in S5?

In the system K2, $\Box\Diamond\mathscr{A}$ is equivalent to $\Diamond\Box\mathscr{A}$.[4]

72. Show that they are not equivalent in S5.

73. Show that, if this K2 equivalence were adopted in S5, $\mathscr{A}$ would be equivalent to both $\Box\mathscr{A}$ and $\Diamond\mathscr{A}$. (This equivalence, of course, would collapse all modalities in S5.)

74. George Lakoff holds that these two sentences are equivalent in natural language: (a) 'It is possible that Sam will smoke pot, if he can get it cheap'; (b) 'If he can get it cheap, then it is possible that Sam will smoke pot'.[5] Lakoff thinks these should translate into modal logic as $\Diamond(\Diamond p \to q)$. Here are some other possibilities. Are any equivalent to Lakoff's suggestion? (c) $\Diamond(\Diamond p \dashv3 q)$; (d) $\Diamond p \to \Diamond q$; and (e) $\Diamond p \dashv3 \Diamond\Diamond q$.*

75. Surely this argument should be valid: 'It is possible that Sam will smoke pot, if he can get it cheap. Sam can get pot cheap. So it's possible that he'll smoke it.' Is this argument valid, given each of the four translations for the first premise given in problem 74?

9.7 MODAL DEDUCTION

Natural deduction in the modal system S5 begins with the standard deduction rules for sentential logic. Rules for modal connectives and some special deduction methods are then added. Recall that, in classical sentential logic, proofs fall into three types. Direct proofs begin with premises or lines deduced from earlier lines and proceed to the formulas to be shown. Conditional proofs begin with the assumption of the antecedent of the conditional to be established and proceed to demonstrate the consequent. Indirect proofs begin by assuming formulas whose negations are to be shown and proceed to derive a contradiction. Conditional proof acts like a conditional introduction rule; indirect proof, like a negation introduction rule.

To go any further, we need ways of introducing and eliminating □. Formulating a necessity elimination rule is very easy. What can we deduce from □p? Obviously, p. We can thus establish the rule that we can drop boxes as main connectives whenever we like:

Necessity Exploitation (□E)

n.	□𝒜
n + p.	𝒜 □E, n

This rule corresponds closely to thesis 1: Necessary truths are true. In fact, having this rule allows us to show the validity of that thesis in a symbolic form:

1. Show □p → p
2. ⌈ □p   ACP
3. ⌊ p   □E, 2

To introduce boxes, we need a complex rule tantamount to a new kind of proof. Instead of having a necessity introduction rule, we'll define a *modal proof*:

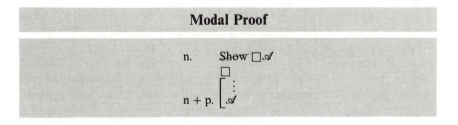

Modal Proof

n.	Show □𝒜
	□
n + p.	⌊ 𝒜

To prove □𝒜, we must try to prove 𝒜. Obviously, this rule requires us to make some further adjustments in our system of deduction, since we don't want to be able to deduce □𝒜 from 𝒜 in general. The idea is that we can prove □𝒜 by proving 𝒜 *under certain special conditions*. To emphasize that special restrictions apply to modal proofs (and subproofs), we'll write a □ immediately below any line we are trying to establish using such a proof. This signals that we cannot reiterate or even use formulas above the box unless they meet certain conditions.

Before considering the nature of those conditions, we can observe that modal proof itself allows us to prove that logical truths are necessary. It's easy to demonstrate the necessity of a simple logical truth, say, $p \to p$.

1. ~~Show~~ $\Box(p \to p)$
   $\Box$
2. $\lceil$ ~~Show~~ $p \to p$
3. $\lfloor \lceil p$ ACP

In general, if $\mathscr{A}$ is valid, then so is $\Box\mathscr{A}$. We can construct a proof of $\Box\mathscr{A}$ simply by adding the lines

1. Show $\Box\mathscr{A}$
   $\Box$

to a proof of $\mathscr{A}$. Our proof system thus respects thesis 2. Furthermore, we can show that logical truths are necessary without worrying about the conditions for using or reiterating formulas in a modal proof.

What restrictions are appropriate to modal proofs? To answer this question, recall that, in S5, necessary truths are true in every possible world. To show that something is necessarily true, then, we have to do the equivalent of showing it true in every possible world. We can do this by showing that it is true in some purely arbitrary world. We want to show that $\mathscr{A}$, say, is true no matter what world we are talking about. So, we want to assume about this arbitrary world only those things we can assume about every possible world.

S5 holds that modality, in general, is a matter of necessity. Whatever modal status a statement has, it has necessarily, that is, in every possible world. Any formula with a modal main connective has the same truth value in every possible world; negations and other truth-functional combinations of these formulas, too, have this property. If $\Box\mathscr{A}$ is true (or false) in some world, it's likewise true (or false) in all other worlds. Similarly, a formula such as $\Box p \to \Box q$ has the same truth value in every world, since neither $\Box p$ nor $\Box q$ varies in truth value across possible worlds, and since the conditional combines these values in the same way in every case. As above, we'll call a formula that has the same truth value in every possible world *modally closed:*

> **DEFINITION.** A formula is *modally closed* iff each of its atomic sub-formulas lies within the scope of a modal connective.

The restriction we need, then, is this: Only modally closed formulas may reiterate into modal proofs. Furthermore, only modally closed formulas are free to lines within a modal proof. So, we need to revise (or, more precisely, extend) the reiteration rule. As it stands, reiteration allows us to write in a subproof a formula that has appeared above in a superproof, that is, a proof to which the subproof is subordinate. This formula, of course, must also be "live" in the sense that it is not imprisoned within a completed proof or part of an uncancelled *Show* line. But we've talked, until now, only about direct, indirect, and conditional subproofs. When dealing with modal subproofs, we must limit this rule. We should be able to write and subsequently use in a modal subproof only those live formulas that have appeared above that are modally closed. We'll formulate the reiteration rule for our modal system, then, in this way:

Reiteration into Modal Proofs (⌐R)

n.　　$\mathscr{A}$

□

n + p.　$\mathscr{A}$　　　□R, n

$\mathscr{A}$ must be modally closed.

Since all worlds, in S5, share the same modal judgments, it makes no difference how many boxes we cross in applying reiteration.

To see how modal proof works, let's show that S5's fifth thesis—that whatever is possible is necessarily possible—holds. The proof doesn't even make use of any rules for the diamond.

1. ~~Show~~ $\Diamond p \to \Box \Diamond p$
2. $\Diamond p$　　　　　　　ACP
3. ~~Show~~ $\Box \Diamond p$
   □
4. $\Diamond p$　　　　　□R, 2

On line 3, we begin a modal proof. We try to show that p is necessarily possible by showing that p is possible in some arbitrary world; hence line 4. But we already know that p is possible in another world. And, since this judgment is modal (that is, since $\Diamond p$ is modally closed), that information reiterates into the arbitrary world we're talking about.

We can similarly demonstrate the validity of thesis 4, that whatever is necessary is necessarily necessary.

1. ~~Show~~ $\Box p \to \Box \Box p$
2. $\Box p$　　　　　　　ACP
3. ~~Show~~ $\Box \Box p$
   □
4. $\Box p$　　　　　□R, 2

We can reiterate $\Box p$ into our modal proof, since $\Box p$ is modally closed.

Rules for possibility are fairly straightforward. The simpler rule is possibility introduction. Under what circumstances are we justified in introducing $\Diamond\mathscr{A}$ into a proof? Clearly, if we know that $\mathscr{A}$ is true, we know that it's possible. So we can justify writing $\Diamond\mathscr{A}$ if we already have $\mathscr{A}$.

Possibility Introduction ($\Diamond$I)

$$
\begin{array}{lll}
\text{n.} & \mathscr{A} & \\
\text{n + p.} & \overline{\Diamond\mathscr{A}} & \Diamond\text{I, n}
\end{array}
$$

It's always acceptable, then, to add a diamond to any formula already established in a proof. This rule allows us to derive $\Diamond p$ from $\Box p$:

$$
\begin{array}{lll}
\text{1.} & \Box p & \text{A} \\
\text{2.} & \text{Show } \Diamond p & \\
\text{3.} & \lceil p & \Box\text{E, 1} \\
\text{4.} & \lfloor \Diamond p & \Diamond\text{I, 3}
\end{array}
$$

The rule for exploiting possibility statements is slightly more complex. What can we conclude from the information that something is possible? We certainly can't deduce that it's true. All we know is that the sentence is true in some possible world. If the sentence implies some consequence, we can conclude that the consequence is true in some world. But if what the sentence implies is modally closed, the truth of the consequence in one world entails its truth in every world. So, if something we know to be possible implies something modally closed, we can deduce that the modally closed formula is true in our world.

Possibility Exploitation ($\Diamond$E)

$$
\begin{array}{lll}
\text{n.} & \Diamond\mathscr{A} & \\
\text{m.} & \underline{\mathscr{A} \dashv \mathscr{B}} & \\
\text{p.} & \mathscr{B} & \Diamond\text{E, n, m}
\end{array}
$$

Here $\mathscr{B}$ must be modally closed.

Modally closed formulas, under any given interpretation, have the same truth value in every world. So, proving the truth of such a formula in one world allows us to deduce its truth in any world whatever.

This rule is extremely useful. To see how it works, however, we must first develop rules for $\dashv$, since that connective plays an important role in $\Diamond$E.

Suppose we've established that $\mathscr{A} \dashv \mathscr{B}$. What can we deduce from that? If we have established $\mathscr{A}$, then we can deduce $\mathscr{B}$.

Strict Conditional Exploitation ($\dashv$ E)

> n. $\mathscr{A} \dashv \mathscr{B}$
> m. $\mathscr{A}$
> p. $\mathscr{B}$ $\dashv$E, n, m

This rule permits strict conditionals to operate as ordinary truth-functional conditionals do, once we've established them in the proof.

We can use this rule to demonstrate thesis 3, that necessary truths imply only necessary truths.

> 1. $\Box p$ A
> 2. $p \dashv q$ A
> 3. Show $\Box q$
> $\Box$
> 4. $\Box p$ $\Box$R, 1
> 5. $p \dashv q$ $\Box$R, 2
> 6. p $\Box$E, 4
> 7. q $\dashv$E, 5, 6

To introduce a strict conditional into a proof, we need to deduce its consequent from its antecedent. This idea is central to conditional proof. But, to justify a strict conditional, we need to do so within a modal proof. That is, the derivation of the consequent from the antecedent must be justified in any possible world, not just the world we begin describing in the proof. So, we need to restrict the information available to us in deriving the consequent from the antecedent. Introducing strict conditionals thus requires a new form of modal proof.

Modal Proof for Strict Conditionals

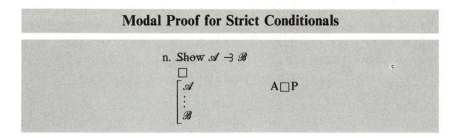

> n. Show $\mathscr{A} \dashv \mathscr{B}$
> $\Box$
> $\mathscr{A}$ A$\Box$P
> $\vdots$
> $\mathscr{B}$

This proof method is essentially conditional proof. The only difference is the box and the restrictions it imposes. These restrictions are precisely those for other modal proofs. Our rule for reiteration into modal proofs thus suffices for reiteration into modal proofs of strict conditionals.

We can use our new version of modal proof to show that $\Box(p \& q)$ implies $\Box p \dashv 3 \Box q$.

$$
\begin{array}{lll}
1. & \Box(p \& q) & A \\
2. & \text{Show } \Box p \dashv 3 \Box q & \\
 & \quad \Box & \\
3. & \quad \Box p & A\Box P \\
4. & \quad \Box(p \& q) & \Box R, 1 \\
5. & \quad \text{Show } \Box q & \\
 & \quad \quad \Box & \\
6. & \quad \quad \Box(p \& q) & \Box R, 4 \\
7. & \quad \quad p \& q & \Box E, 6 \\
8. & \quad \quad q & \&E, 7 \\
\end{array}
$$

We can now consider a simple example of the use of $\Diamond E$, the inference from 'p might be necessary' ($\Diamond \Box p$) to 'p is necessary' ($\Box p$).

$$
\begin{array}{lll}
1. & \Diamond \Box p & \\
2. & \text{Show } \Box p & \\
3. & \quad \text{Show } \Box p \dashv 3 \Box p & \\
 & \quad \Box & \\
4. & \quad \Box p & A\Box P \\
5. & \quad \Box p & \Diamond E, 1, 3 \\
\end{array}
$$

We know, from line 1, that $\Box p$ is possible. Because $\Box p$ strictly implies $\Box p$, we can conclude $\Box p$ by possibility exploitation.

As another example, let's show that the possible implies only the possible: that $\Diamond p$ and $p \dashv 3 q$ imply $\Diamond q$.

$$
\begin{array}{lll}
1. & \Diamond p & A \\
2. & p \dashv 3 q & A \\
3. & \text{Show } \Diamond q & \\
4. & \quad \text{Show } p \dashv 3 \Diamond q & \\
 & \quad \Box & \\
5. & \quad p & A\Box P \\
6. & \quad p \dashv 3 q & \Box R, 2 \\
7. & \quad q & \dashv 3 E, 5, 6 \\
8. & \quad \Diamond q & \Diamond I, 7 \\
9. & \quad \Diamond q & \Diamond E, 1, 4 \\
\end{array}
$$

As a further example, let's look at the formula $p \to \Box \Diamond p$. It says that whatever is true is necessarily possible. In S5, this formula is valid. But the proof requires us to go first from p to $\Diamond p$, and then from $\Diamond p$ to $\Box \Diamond p$.

1. S̶h̶o̶w̶ $p \rightarrow \Box \Diamond p$
2. ⌈ p ACP
3. | $\Diamond p$ $\Diamond$I, 2
4. | S̶h̶o̶w̶ $\Box \Diamond p$
     | $\Box$
5. ⌊⌈ $\Diamond p$ $\Box$R, 3

This way of pursuing the proof may seem roundabout. But, had we tried to show $\Box \Diamond p$ immediately, we would have run into trouble. The formula p is not modally closed, and so we could not have reiterated it. To get the information we need within the modal proof, we must derive a modally closed formula, such as $\Diamond p$, that can reiterate.

It remains to develop rules for the double fishhook. If we've established a strict biconditional $\mathscr{A} \otimes \mathscr{B}$, then we've established that $\mathscr{A}$ and $\mathscr{B}$ are strictly equivalent. So, we should be able to derive $\mathscr{A}$ from $\mathscr{B}$, and $\mathscr{B}$ from $\mathscr{A}$.

Strict Biconditional Exploitation ($\otimes$E)

n. $\mathscr{A} \otimes \mathscr{B}$
m. $\underline{\mathscr{A}}$ (or $\mathscr{B}$)
p. $\mathscr{B}$ (or $\mathscr{A}$) $\otimes$E, n, m

Introducing strict biconditionals is very similar to introducing truth-functional biconditionals. A truth-functional biconditional amounts to truth-functional conditionals going in both directions. A strict biconditional, likewise, amounts to strict conditionals going in both directions.

Strict Biconditional Introduction ($\otimes$I)

n. $\mathscr{A} \dashv \mathscr{B}$
m. $\mathscr{B} \dashv \mathscr{A}$
p. $\overline{\mathscr{A} \otimes \mathscr{B}}$ $\otimes$I, n, m

The strategy for using these rules is the same as for the corresponding truth-functional rules.

The rules we've presented in this chapter are complete: Every argument form valid in S5 can be established as valid using only these rules and the ordinary truth-functional rules. The rule of possibility exploitation is one of the most controversial features of S5. Adopting the rule amounts to asserting the validity of a version of the "ontological proof" of God's existence:

> If God exists at all, He exists necessarily. It is at least possible that God exists. Therefore, it is necessarily true that God exists.

This inference is highly controversial; the $\Diamond$E rule is, too. Whether it ought to be counted a legitimate rule has a great deal of philosophical and practical significance.

Problems

Use natural deduction to demonstrate the validity of each of the following formulas in S5.

1. $p \dashv3 (p \lor q)$

2. $\Box p \dashv3 \Box(p \lor q)$

3. $\neg \Diamond(p \ \& \ \neg p)$

4. $\Box((p \ \& \ q) \to (p \lor q))$

▸ 5. $\Diamond p \lor \Diamond \neg p$

6. $(p \ \& \ q) \dashv3 \neg(\neg p \lor (q \to \neg q))$

7. $(p \dashv3 q) \to (p \to q)$

8. $(p \ \&3 \ q) \to (p \leftrightarrow q)$

9. $\neg \Box p \dashv3 \Diamond \neg p$

▸ 10. $\Diamond \neg p \dashv3 \neg \Box p$

11. $\neg \Diamond p \dashv3 \Box \neg p$

12. $\Box \neg p \dashv3 \neg \Diamond p$

13. $\Box(p \ \& \ q) \dashv3 (\Box p \ \& \ \Box q)$

14. $(\Box q \lor \Box p) \dashv3 \Box(q \lor p)$

15. $\Box(p \lor q) \dashv3 (\neg \Diamond q \dashv3 \Box p)$

16. $(p \dashv3 q) \dashv3 \Box(p \dashv3 q)$

17. $\Diamond(p \dashv3 q) \dashv3 (p \dashv3 q)$

18. $(p \dashv3 q) \dashv3 \Box(p \to q)$

19. $\Box(p \to q) \dashv3 (p \dashv3 q)$

▸ 20. $(p \ \&3 \ q) \dashv3 \Box(p \leftrightarrow q)$

21. $\Box(p \leftrightarrow q) \dashv3 (p \ \&3 \ q)$

22. $\Box(p \lor q) \dashv3 (\neg q \dashv3 p)$

23. $(p \dashv3 q) \dashv3 \Box(q \lor \neg p)$

24. $(p \dashv3 q) \dashv3 (p \dashv3 (p \ \& \ q))$

25. $(p \dashv3 (p \ \& \ q)) \dashv3 (p \dashv3 q)$

26. $(p \ \& \ \neg q) \dashv3 \neg(p \dashv3 q)$

27. $((p \dashv3 q) \ \& \ \neg q) \dashv3 \neg p$ (This justifies a derivable rule, $\dashv3$E*, parallel to $\to$E*.)

28. $((p \dashv3 q) \ \& \ \Diamond \neg q) \dashv3 \neg \Box p$

29. $((p \dashv3 q) \ \& \ \neg \Diamond q) \dashv3 \neg \Diamond p$

30. $(p \dashv3 q) \dashv3 (\neg q \dashv3 \neg p)$

31. $(p \ \&3 \ q) \dashv3 (\Box p \ \&3 \ \Box q)$

32. $(p \ \&3 \ q) \dashv3 (\Diamond p \ \&3 \ \Diamond q)$

33. $(p \dashv3 q) \dashv3 (r \dashv3 (p \dashv3 q))$

34. $(p \ \&3 \ q) \ \&3 \ \Box(p \ \&3 \ q)$

▸ 35. $\Diamond(p \ \&3 \ q) \ \&3 \ (p \ \&3 \ q)$

36. $((p \dashv3 q) \ \& \ (p \dashv3 \neg q)) \dashv3 \neg \Diamond p$

37. $((p \dashv3 q) \ \& \ (\neg p \dashv3 q)) \dashv3 \Box q$

38. $(\Box p \ \& \ \Box q) \dashv3 (p \ \&3 \ q)$

39. $(\neg \Diamond p \ \& \ \neg \Diamond q) \dashv3 (p \ \&3 \ q)$

40. $((p \dashv3 q) \ \& \ (q \dashv3 p)) \dashv3 (p \ \&3 \ q)$

41. $(p \mathbin{\&} q) \mathbin{-3} ((p \mathbin{-3} q) \mathbin{\&} (q \mathbin{-3} p))$

42. $((p \mathbin{-3} (q \mathbin{\&} r)) \mathbin{\&} ((p \mathbin{-3} q) \mathbin{\&} (p \mathbin{-3} r))$

43. $((p \vee q) \mathbin{-3} r) \mathbin{\&} ((p \mathbin{-3} r) \mathbin{\&} (q \mathbin{-3} r))$

44. $(p \mathbin{\&} q) \mathbin{-3} ((p \mathbin{-3} r) \mathbin{\&} (q \mathbin{-3} r))$

▶ **45.** $(p \mathbin{-3} q) \mathbin{-3} ((r \mathbin{-3} p) \mathbin{-3} (r \mathbin{-3} q))$

46. $(p \mathbin{\&} q) \mathbin{-3} ((r \mathbin{-3} q) \mathbin{\&} (r \mathbin{-3} p))$

47. $(p \mathbin{\&} q) \mathbin{-3} ((p \mathbin{\&} r) \mathbin{\&} (r \mathbin{\&} q))$

48. $(p \mathbin{-3} q) \mathbin{-3} (p \mathbin{-3} \Box \Diamond q)$

49. $((p \mathbin{\&} q) \mathbin{-3} r) \mathbin{\&} (p \mathbin{-3} (q \mathbin{-3} r))$

▶ **50.** $(p \mathbin{\&} (q \mathbin{-3} r)) \mathbin{-3} (\Box p \vee \Box \neg p)$*

Use natural deduction to demonstrate the validity of each of the following argument forms.

51. $\Box p; \therefore p \vee q$ **52.** $\Box \neg p; \therefore \neg \Box p$

53. $\neg \Diamond p; \therefore \Diamond \neg p$ **54.** $\Box p; \therefore \Diamond \Diamond \Box p$

▶ **55.** $\Diamond p \mathbin{-3} q; \therefore p \mathbin{-3} q$ **56.** $p \mathbin{-3} \Box q; \therefore p \mathbin{-3} q$

57. $\Diamond p \mathbin{-3} \Box q; \therefore p \mathbin{-3} q$ **58.** $p \mathbin{-3} q; \therefore \Box p \mathbin{-3} q$

59. $p \mathbin{-3} q; \therefore p \mathbin{-3} \Diamond q$ **60.** $p \mathbin{-3} q; \therefore \Box p \mathbin{-3} \Diamond q$

61. $\Diamond (p \mathbin{\&} q); \therefore p \mathbin{\&} q$ **62.** $\Box p; \therefore \Box \Box \Box p$

63. $\Box \Box \Box p; \therefore \Box p$ **64.** $\Diamond \Box p; \therefore \Box p$

▶ **65.** $\Diamond \Diamond \Box p; \therefore \Box p$ **66.** $\Box \Box \Diamond p; \therefore \Diamond p$

67. $\Diamond \Diamond p; \therefore \Diamond p$ **68.** $\Diamond p; \therefore \Box \Box \Diamond p$

69. $\Diamond p; \therefore \Diamond \Diamond \Diamond p$ **70.** $\Box \Box p; \therefore \Diamond \Box p$

71. $\Diamond \Box p; \therefore \Box \Box p$ **72.** $\Box \Diamond p; \therefore \Diamond \Diamond p$

73. $\Diamond \Diamond p; \therefore \Box \Diamond p$ **74.** $\Diamond \Box \Diamond p; \therefore \Box \Diamond \Diamond p$

▶ **75.** $\Box \Diamond \Box p; \therefore \Diamond \Box \Box p$ **76.** $\Box \Box \Diamond \Box p; \therefore \Box \Diamond \Box \Diamond p$

77. $\Box \Box \Diamond \Box p; \therefore \Box \Diamond \Box \Diamond \Box p$ **78.** $p \vee q; \therefore \Diamond p \vee \Diamond q$

79. $\Box (p \mathbin{\&} q); \therefore \Box p \mathbin{\&} \Box q$ **80.** $\Box p \mathbin{\&} \Box q; \therefore \Box (p \mathbin{\&} q)$

81. $\Diamond (p \mathbin{\&} q); \therefore \Diamond p \mathbin{\&} \Diamond q$ **82.** $\Diamond (p \vee q); \therefore \Diamond p \vee \Diamond q$

83. $\Diamond p \vee \Diamond q; \therefore \Diamond (p \vee q)$ **84.** $\Box p \vee \Box q; \therefore \Box (p \vee q)$

▶ **85.** $\Box (p \rightarrow q); \therefore \Box p \rightarrow \Box q$ **86.** $\Box (p \leftrightarrow q); \therefore \Box p \leftrightarrow \Box q$

87. ◇(p → q); ∴ □p → ◇q   88. ◇(p ↔ q); ∴ □q → ◇p

89. ¬(p ⥽ q); ∴ ◇p   ▶ 90. ¬(p ⥽ q); ∴ ◇¬q

91. ¬(p ⫣ q); ∴ ◇p ∨ ◇q   92. ¬(p ⫣ q); ∴ ◇(p ∨ q)

93. ¬(p ⫣ q); ∴ □p → ◇¬q   94. ¬(p ⫣ q); ∴ ◇¬(p & q)

95. p ⥽ q; q ⥽ r; ∴ p ⥽ r   96. p ⥽ q; ∴ (p & r) ⥽ q

97. (p → q) ⥽ r; ∴ q ⥽ r   98. □p ⥽ q; ∴ □p ⥽ □q

99. ◇p ⥽ q; ∴ p ⥽ □q   ▶ 100. ◇p ⥽ q; ∴ ◇p ⥽ □q

101. p ⥽ □q; ∴ ◇p ⥽ q   102. p ⥽ □q; ∴ ◇p ⥽ □q

103. p ⥽ ◇q; ∴ ◇p ⥽ ◇q   104. p ⫣ ◇q; ∴ ◇p ⫣ ◇q

105. p ⫣ ◇q; ∴ □p ⫣ ◇q   106. p ⫣ □q; ∴ ◇p ⫣ □p

107. p ⥽ q; r ⥽ p; ◇q ⥽ s; ∴ r ⥽ □s

108. p ⥽ (q & r); ◇r ⥽ t; q ⥽ s; ∴ p ⥽ (s & □t)

109. ¬s ⥽ ¬q; r ⥽ ¬p; ◇p ⥽ (q ∨ r); ∴ p ⥽ □s

▶ 110. (p ⥽ s) ∨ q; □r → ◇(p & ¬s); ∴ □r → q

111. ¬p ⥽ ¬q; p ⥽ (r & s); ∴ (q ⥽ r) & (q ⥽ s)

112. p ⥽ (q & r); (r ∨ s) ⥽ t; ∴ (p ∨ s) ⥽ t

113. p ⥽ q; p ⥽ ¬r; (q ∨ r) ⥽ s; ∴ (s ⥽ p) ⥽ (q ⥽ ¬r)

114. (p ∨ q) ⥽ r; ¬s ⥽ ¬q; ¬◇(r & s); ∴ (p ⥽ r) & (q ⥽ t)

▶ 115. (p → (q → r)) ⫣ ((p → q) → r); ∴ □(p ∨ r)

Use deduction to show that each of these arguments is valid in S5.

116. If there's a God, we're all His creations. Since it's possible that God exists, it's possible that we're all His creations.

117. If it's possible for Frank to do good work, they'll hire him. It's not necessarily true that they will hire him. So, Frank can't do good work.

118. Reasoning must proceed according to discernible laws. So, if you drink enough tequila, reasoning proceeds according to discernible laws.

119. If the Romanians either cut military spending or relax internal regulations, they'll keep "most-favored nation" status. It's possible that Romania will cut military spending. Thus, it's not necessarily true that Romania will lose its "most-favored nation" status.

120. It's necessarily true that the laws of logic are valid, and it's necessary that the theorems of mathematics are true. Hence, the validity of the laws of logic implies the truth of the theorems of mathematics.

121. Unless we take drastic measures, it will be necessary for us to take out a loan. It's possible that we won't take out a loan, but if and only if sales suddenly improve. We'll take drastic measures if sales don't suddenly improve. Therefore, we're going to take drastic measures.

122. If all our knowledge comes from experience, it must be that we know nothing beyond what we perceive. But the fact that all our knowledge comes from experience doesn't imply that all our knowledge arises out of experience alone. So, the fact that some knowledge arises out of something other than experience is compatible with our knowing nothing beyond what we perceive.

123. The claim that Alfred is guilty of the murder might be consistent with Boris's testimony that Alfred was in Normandy. So it's possible that Alfred is guilty of the murder.

124. If Osmond's testimony is true, our client might have committed the burglary. If he did commit the burglary, then he has perjured himself. What Osmond said could be true. Thus, our client may have perjured himself.

▶ 125. A failure in our attempt to prevent a leftist takeover would be incompatible with a reassertion of our leadership role in the world. It is quite possible, unfortunately, that it is impossible for our attempt to prevent a leftist takeover to succeed. Therefore, we can't reassert our role as world leaders.

126. Oak trees will grow tall only if it's possible for them to obtain a reasonable amount of moisture. They can't get a reasonable amount of moisture if it's possible for you to plant them only in less than two feet of soil. So, if you can plant oaks only in less than two feet of soil, it won't be possible for the trees to grow tall.

127. Experience of the world is possible only if minds impose a structure on what is outside them. It's obviously possible to experience the world. Therefore, minds must necessarily impose a structure on what lies outside them.

128. The witness, Jones, must be honest. But it's possible that it wasn't necessary for the suspect to leave town at 7:00 P.M. on the night of the murder. Hence it isn't true that, if Jones is being honest, the suspect left town before 7:00 P.M. on the night of the murder.

129. The assumption that everything is an illusion caused by an evil demon is consistent with everything that we think we know being false. Therefore, that everything could be an illusion that an evil demon causes implies that everything that we think we know might be false.

▶ 130. If a recent Supreme Court decision is correct, then the demand that men and women be treated equally under the law is compatible with our permitting the criminal status of an act to depend on the sex of the agent (specifically, in cases of statutory rape). We can permit the criminal status of an action to depend on the sex of the agent, however, only if we are willing to accept any sexual discrimination that might benefit society. Thus, if a recent Supreme Court decision is correct, the demand that men and women be treated equally under the law and a willingness to accept any sexual discrimination that may benefit society are not contradictory.

In 1959, Lemmon and Gjertsen formulated several axioms for contingency (∇).[6] A statement is contingent iff it's possible, but not necessary. That is, $\nabla\mathscr{A}$ is equivalent to $\Diamond\mathscr{A} \,\&\, \neg\Box\mathscr{A}$. Show that each axiom is valid in S5.

131. $\nabla p \leftrightarrow \nabla\neg p$ **132.** $\neg\nabla(p \leftrightarrow q) \rightarrow (\nabla p \rightarrow \nabla q)$

133. $\nabla(p \rightarrow q) \rightarrow (\neg\nabla p \rightarrow p)$ **134.** $\neg\nabla(p \rightarrow \nabla q)$

▶ **135.** $\Box p \leftrightarrow (p \,\&\, \neg\nabla p)$

In spite of our desire to avoid paradoxes of implication, S5 falls prey to paradoxes of strict implication. Show that each of these argument forms is valid.

136. $\Box p$; $\therefore q \dashv 3\ p$ **137.** $\Box\neg p$; $\therefore p \dashv 3\ q$

138. A formula is *Clavian* just in case it's implied by its own negation. Show that, in S5, a formula is Clavian iff it's necessarily true.

139. A formula is *E-necessary* iff it follows from the statement that it implies itself.[7] Show that, in S5, a formula is E-necessary just in case it's necessarily true.

▶ **140.** Suppose that a formula is *irrefutable* just in case its negation entails a contradiction, say, $p \,\&\, \neg p$. Show that, in S5, all and only necessarily true formulas are irrefutable.

9.8 OTHER MODAL SYSTEMS

The framework we've established in this chapter allows us to formulate not only S5 but several other modal logics. The first alternative to S5 we'll consider is S4. S4 accepts the first four theses that characterize the perspective of S5. It rejects only the contention that whatever is possible is necessarily possible. The second alternative, M, also rejects the thesis that whatever is necessary is necessarily necessary. Thus M advocates theses 1–3; S4, theses 1–4; and S5, theses 1–5.[8]

M and S4 also use the language ML, but give it a slightly different semantics. They too build on Leibniz's idea that necessity is truth in all possible worlds. But they introduce an *accessibility* relation: Necessity in a world, w, is truth in all possible worlds accessible from w. On any interpretation,

> □𝒜 is true in w iff 𝒜 is true in all worlds accessible from w.
>
> ◇𝒜 is true in w iff 𝒜 is true in some world accessible from w.
>
> 𝒜 ⊰ ℬ is true in w iff ℬ is true in every world accessible from w where 𝒜 is true.
>
> 𝒜 ⊟ ℬ is true in w iff 𝒜 and ℬ agree in truth value in every world accessible from w.

Inherent in S5 are the assumptions that what is necessary in one world is necessary in all and that what is possible in one world is possible in all. These assumptions are plausible if we think about logical modality. But they seem less plausible if we think about physical or technological modality. If Jack were to borrow Jill's BMW and drive it off a cliff, then Jill can no longer sell it to pay off her gambling debts; what is possible in our world would be impossible in that world. So, we might want to say that something is possible in a world just in case there is a world we can reach from that world in which it occurs. Roughly, if we can reach one world from another, we say that the former is accessible from the latter.

Once we introduce an accessibility relation, however, we must say what it is like. In our earlier approach to S5, we in effect assumed that every world is accessible from every other. In M, the assumption is only that every world is accessible from itself; M thereby adopts thesis 1, that every necessary truth is true. In S4, in addition, the assumption is that accessibility is transitive: that, whenever w is accessible from w1 and w1 is accessible from w2, w is accessible from w2. So, no world accessible from our world has a broader range of accessible worlds than ours has. Our world, in other words, allows as broad a range of possibilities as any other world accessible from it. The same holds of other worlds. This is responsible for S4's advocacy of thesis 4, that necessary truths are necessarily necessary.

We can use tableaux to evaluate validity in M and S4. We need, however, to change the rules appropriate for S5. To use tableaux in M and S4:

1. Eliminate the revival rule; it holds in neither M nor S4.

2. Change the survival rule. Not all modally closed formulas survive world shifts in M or S4.

When we move from a world to another accessible from it, what information about the "old" world remains true? Because S4 accepts thesis 4, necessary truths remain necessary in the new world. But, because S4 rejects thesis 5, there is no similar guarantee for statements of possibility. The survival rule for S4 thus restricts survival to necessary formulas:

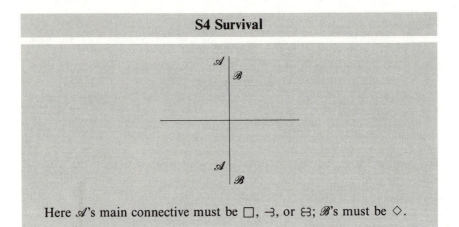

S4 Survival

Here $\mathscr{A}$'s main connective must be □, ⫞, or ⊰; $\mathscr{B}$'s must be ◇.

To illustrate the use of this rule, let's show that S4 accepts thesis 4.

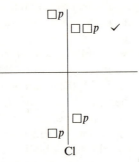

Note, however, that S4 rejects thesis 5.

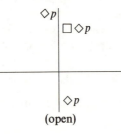

We cannot write ◇p on the left, as we could in S5; formulas with diamonds as main connectives survive only on the right.

M rejects thesis 4 and thesis 5. So, neither statements of necessity nor those of possibility survive a world shift intact. But we do know that, if something is necessarily true in world w, it's true in all the worlds accessible from w. It might not, however, be *necessarily* true in those worlds. We may formulate M's survival rule, then, as follows. Assume that formulas having the structures on the right below are *demodalized versions* of those having the structures on the left.

Formula	Demodalized Version
$\Box\mathscr{A}$	$\mathscr{A}$
$\mathscr{A} \rightarrow\!\!\!\!\! \cdot\ \mathscr{B}$	$\mathscr{A} \rightarrow \mathscr{B}$
$\mathscr{A} \,\epsilon\!\!\!\ni\, \mathscr{B}$	$\mathscr{A} \leftrightarrow \mathscr{B}$
$\Diamond\mathscr{A}$	$\mathscr{A}$

When a modal formula survives a world shift in M, it survives only in its demodalized version.

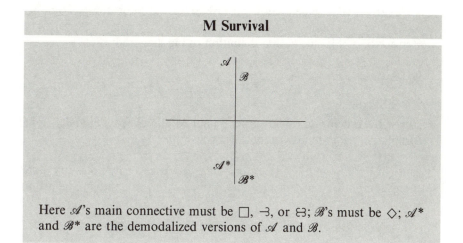

M Survival

Here $\mathscr{A}$'s main connective must be $\Box$, $\rightarrow\!\!\!\!\! \cdot$, or $\epsilon\!\!\!\ni$; $\mathscr{B}$'s must be $\Diamond$; $\mathscr{A}^*$ and $\mathscr{B}^*$ are the demodalized versions of $\mathscr{A}$ and $\mathscr{B}$.

It's easy to see that M must reject thesis 4.

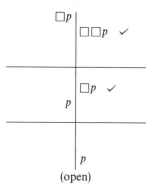

□*p* survives the first world shift as *p*; for this reason, it cannot survive the second world shift.

M still accepts thesis 3, however; □*p* and (*p* ⥽ *q*) still imply □*q*.

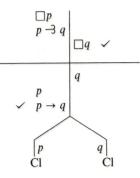

We can also devise deduction systems for S4 and M by doing two things:

1. Replace S5's ◇E rule with a definition of the diamond:

$$
\begin{array}{ll}
\text{n.} & ◇\mathscr{A} \\
\hline
\text{n + p.} & ¬□¬\mathscr{A} \quad \text{Df}◇, \text{n}
\end{array}
$$

2. Revise the reiteration rule. Neither S4 nor M allows every modally closed formula to reiterate into modal proofs.

Because in both S4 and S5 necessary truths are necessarily necessary, in S4 formulas symbolizing statements of necessity can reiterate into modal proofs.

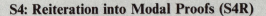

S4: Reiteration into Modal Proofs (S4R)

$$
\begin{array}{ll}
\text{n} & \mathscr{A} \\
 & □ \\
\text{n + p.} & \mathscr{A} \quad \text{S4R, n}
\end{array}
$$

$\mathscr{A}$'s main connective must be □, ⥽, or ⧟.

M allows such formulas to reiterate, but only in their demodalized versions.

M: Reiteration into Modal Proofs (MR)

$$
\begin{array}{ll}
\text{n} & \mathscr{A} \\
 & □ \\
\text{n + p.} & \mathscr{A}* \quad \text{MR, n}
\end{array}
$$

𝒜's main connective must be □, ⊰, or ⋶; 𝒜* must be the demodalized version of 𝒜.

To see the difference between these two rules, compare the S4 and M proofs of thesis 3:

	S4				M	
1.	□p	A		1.	□p	A
2.	p ⊰ q	A		2.	p ⊰ q	A
3.	Show □q			3.	Show □q	
	□				□	
4.	□p	S4R, 1		4.	p	MR, 1
5.	p ⊰ q	S4R, 2		5.	p → q	MR, 2
6.	p	□E, 4		6.	q	→E, 4, 5
7.	q	⊰E, 5, 6				

As we've observed, there is no consensus on which modal logic best captures our use of modal expressions in English. Indeed, it may be that different logics capture different senses of necessity. Of the logics we've developed in this chapter, S5 is strongest; it counts the most formulas and argument forms valid. Anything valid in M is valid in S4; anything valid in S4 is valid in S5.

In S5, there are six modalities or modes of truth: A sentence may be (a) (actually) true, (b) (actually) false, (c) necessary, (d) possible, (e) impossible, and (f) unnecessary. That is, S5 distinguishes among 𝒜, ¬𝒜, □𝒜, ◇𝒜, ¬◇𝒜 (or □¬𝒜), and ◇¬𝒜 (or ¬□𝒜). All combinations of boxes and diamonds—such as □□, ◇◇, □◇, and ◇□—are identified with one of the basic six modalities. In fact, in S5 we can replace a string of boxes and diamonds with the last connective in the string to obtain an equivalent formula. ◇◇□◇p is equivalent, in S5, to ◇p; □◇□◇◇□p is equivalent to □p.

S4 distinguishes fourteen modalities. M distinguishes infinitely many modalities. Choices between modal systems thus become difficult. Nobody has clear intuitions about whether, for example, being necessary is the same as being necessarily possibly necessary. But the disagreements between M, S4, and S5 are confined to formulas with *nested* modalities, that is, formulas having modal connectives within the scope of other modal connectives. On simpler formulas, the logics concur.

Problems

1. Use semantic tableaux to evaluate the arguments, argument forms, and formulas at the end of section 9.6 in either M or S4.

2. Use natural deduction to show valid as many of the arguments, argument forms, and formulas at the end of section 9.7 as you can, in either M or

S4. (Some will not be valid; but trying to prove them valid will show where M and S4 are weaker than S5.)

Notes

[1] From James D. McCawley, *What Linguists Have Always Wanted to Know About Logic** (Chicago: University of Chicago Press, 1980).

[2] C. I. Lewis and C. H. Langford developed S5 in their *Symbolic Logic* (New York: Dover, 1932).

[3] See M. J. Alban, "Independence of the Primitive Symbols of Lewis' Calculi of Propositions," *Journal of Symbolic Logic* 8 (1943): 25–26; S. Hallden, "On the Decision Problem of Lewis's Calculus S5," *Norsk Mathematisk Tidsskrift* 31 (1949): 89–94.

[4] See W. T. Parry, "Modalities in the Survey System of Strict Implication," *Journal of Symbolic Logic* 4 (1939): 131–54; B. Sobocinski, "Remarks About the Axiomatizations of Certain Modal Systems," *Notre Dame Journal of Formal Logic* 5 (1964): 71–80.

[5] George Lakoff, "Linguistics and Natural Logic," in Donald Davidson and Gilbert Harman (eds.), *Semantics of Natural Language* (Dordrecht: D. Reidel, 1972): 545–665, 549.

[6] See Arthur N. Prior, *Formal Logic* (Oxford: Oxford University Press, 1962): 312.

[7] See Alan Ross Anderson and Nuel D. Belnap, Jr., *Entailment: The Logic of Relevance and Necessity,* Volume 1 (Princeton: Princeton University Press, 1975).

[8] S4, like S5, was first formulated in C. I. Lewis and C. H. Langford's *Symbolic Logic*. G. H. von Wright developed M in *An Essay on Modal Logic* (Amsterdam: North-Holland, 1951). M is equivalent to a modal system, T, developed by R. Feys in "Les logiques nouvelles de modalités," *Revue Néoscolastique de Philosophie* 40 (1937): 517–33. The equivalence of these systems was proved by B. Sobocinski, "Note on a Modal System of Feys-von Wright," *The Journal of Computing Systems* 1 (1953): 171–78.

10

COUNTERFACTUALS

Concern about the logic of conditional sentences was a major motivation for the development of symbolic modal logic. The truth-functional conditional, whose meaning a simple truth table captures, is the weakest interpretation of the English conditional expression *if* that preserves the basic intuition that a conditional is false if its antecedent is true and its consequent is false. The strict conditional, which is not truth-functional, strengthens the interpretation of *if* by combining the truth-functional conditional with necessity.

Whether or not the truth-functional or the strict conditional succeeds in rendering the meanings of most English conditional sentences correctly, both certainly fail to account for a certain class of conditionals. Typically expressed in the subjunctive mood in English, these conditionals are called *counterfactual* conditionals, or, more simply, *counterfactuals,* because they generally signal the presupposition that the antecedent is false. These conditionals are all counterfactual.

(1) a. If the American Revolution had failed, the U.S. would be a member of the British Commonwealth.
   b. If I were to ask my boss for a raise, I'd be fired.
   c. If anyone were to challenge Mike, he wouldn't know how to respond.
   d. If Mozart had lived to the age of 60, he might have written twice as many operas.

Neither the truth-functional nor the strict conditional represents the logical structure of these sentences. According to the truth table for the truth-functional conditional, any conditional with a false antecedent is true. But, since counterfactuals usually have false antecedents, almost all counterfactuals would be true. We can see that counterfactuals are not automatically true when their antecedents are indeed counter-to-fact by examining these variants of the sentences in (1).

273

(2) a. If the American Revolution had failed, France would be a communist country ruled by the Pope.
   b. If I were to ask my boss for a raise, Romania would invade South Africa.
   c. If anyone were to challenge Mike, he would know how to respond.
   d. If Mozart had lived to the age of 60, he might have become queen of England.

Assuming the falsity of the antecedents of these conditionals, they should all be true—indeed, obviously true—if here *if* expresses a truth-functional conditional. So should the conditionals in (1). But this is obviously wrong. (1)a, b, and d are all much more plausible than (2)a, b, and d. Furthermore, (1)c and (2)c seem to contradict each other. If there is even a remote possibility that someone might challenge Mike, then 'If anyone were to challenge Mike, he would know how to respond' and 'If anyone were to challenge Mike, he wouldn't know how to respond' can't both be true. The counterfactual conditional, therefore, is not a variety of truth-functional conditional. It would be a mistake to symbolize any of the above as $\mathscr{A} \to \mathscr{B}$.

The counterfactual conditional isn't a strict conditional either. Several argument forms that are valid for strict as well as truth-functional conditionals do not hold for counterfactuals. One is *strengthening the antecedent,* the following argument form:

$$p \rightarrow 3\ q$$
$$\therefore (p\ \&\ r) \rightarrow 3\ q$$

These arguments show that an analogous argument form does not hold for counterfactuals:

(3) a. If I were to pinch you, you wouldn't die.
   ∴ If I were to pinch you and cut your head off, you wouldn't die.
   b. If Miami had beaten Denver, they'd be in the Super Bowl.
   ∴ If Miami had beaten Denver but had lost every other game this season, they'd be in the Super Bowl.
   c. If Beethoven hadn't written the Fifth Symphony, he'd still be remembered as a great composer.
   ∴ If Beethoven hadn't written the Fifth Symphony and, in fact, hadn't written any music at all, he'd still be remembered as a great composer.

These arguments are clearly invalid. So, it would also be a mistake to symbolize counterfactuals as having the form $\mathscr{A} \rightarrow 3\ \mathscr{B}$.

A second argument form leading to the same conclusion is *transitivity:*

$$p \rightarrow 3\ q$$
$$q \rightarrow 3\ r$$
$$\therefore p \rightarrow 3\ r$$

This argument is valid. But corresponding arguments containing counterfactuals are invalid:

(4) a. If the U.S. were to withdraw from the United Nations, Israel would, too.
   If Israel were to withdraw from the U.N., the U.N. budget would be little affected.
   ∴ If the U.S. were to withdraw from the United Nations, the U.N. budget would be little affected.

   b. If I. M. Pei hadn't designed the Kennedy Library, he would nevertheless be famous.
   If I. M. Pei had never been born, he wouldn't have designed the Kennedy Library.
   ∴ If I. M. Pei had never been born, he would nevertheless be famous.

   c. If Sadat hadn't been assassinated, he would probably be alive today.
   If Sadat had died when he was very young, he wouldn't have been assassinated.
   ∴ If Sadat had died when he was very young, he would probably be alive today.

These arguments, too, show that counterfactuals are not strict conditionals. Actually, the arguments in (3) already show that transitivity fails for counterfactuals; because $(p\ \&\ r) \dashv\!\!\!3\ p$ is valid, we could see strengthening the antecedent as a special case of transitivity.

We therefore cannot construe counterfactuals as either truth-functional or strict conditionals. But developing an account of them is important, because counterfactuals are crucial to reasoning concerning cause-and-effect and to decision making. Let's take a simple example of causal reasoning: Suppose that Reggie takes his hyperactive daughter Mona to the doctor, who says that Mona is hyperactive because she eats too much sugar. Reggie, inferring that Mona would be calmer if she were to eat less sugar, begins feeding his daughter apples rather than cookies for snacks. Notice that Reggie does this because he infers a counterfactual conditional—*Mona would be calmer if she were to eat less sugar*—from a causal statement. The counterfactual plays a critical role in linking a statement about causes and effects to actions.

Counterfactuals are similarly important in decision making. Reaching a decision involves considering various actions and thinking about what would happen if one were to perform them. Suppose that Rhonda is trying to decide whether to quit her job and start her own business. She may think, "If I were to quit, I might be short on money for a while; but, if I were to quit, and I could get a loan, then money wouldn't be a problem. If I were to start my own business, I would have a lot more independence but also a lot more responsibility and a lot more work. If I were to start my own business and succeed, then I would make a lot more money than I can by keeping my job. But, if I were to fail, I would lose a lot of money and be out of work besides." These counterfactuals don't determine what Rhonda should do. That depends on how likely she is to succeed and on how much she values independence, security, money, and so on. But the counterfactuals do play an essential role in her decision-making process. Their counterfactual character is important; Rhonda's reasoning begins, "If I were to quit, I might be short on money for a while; but, if I were to quit, and I could get a loan, then money wouldn't be a problem." If these conditionals were interpreted as material or strict, Rhonda's statement would be a contradiction. To reflect what goes into decision making, we must use conditionals for which strengthening the antecedent fails.

This chapter, then, outlines a theory of counterfactuals. Actually, it outlines three. An American philosopher, Robert Stalnaker, developed the first rigorous formal theory of counterfactuals in the late 1960s.[1] Another American philosopher, David Lewis, developed a related theory in the early 1970s.[2] We'll first construct C, a system that displays what these two approaches have in common. We'll then extend C to CS, which is Stalnaker's analysis, and CL, which is Lewis's.

Stalnaker intends his theory as an account of all English conditionals; Lewis restricts his attention to subjunctive, counterfactual conditionals. We focus here on that more restricted realm and leave as an open question whether the theory properly accounts for other uses of *if* in English. Most of the above examples, translated into the indicative, would establish analogous points about indicative conditionals.

10.1 THE MEANING OF COUNTERFACTUALS

We'll introduce a new connective, $\square\rightarrow$, to symbolize the counterfactual *if ... then*. We can read formulas of the form $\mathscr{A} \square\rightarrow \mathscr{B}$ as 'if it were the case that $\mathscr{A}$, then it would be the case that $\mathscr{B}$.' Whenever $\mathscr{A}$ and $\mathscr{B}$ are formulas, then $(\mathscr{A} \square\rightarrow \mathscr{B})$ is a formula. To develop a logic for a language including this connective, we need to have some understanding of what counterfactual conditionals mean, and, in particular, under what kinds of conditions they are true.

Recall that a strict conditional $\mathscr{A} \dashv \mathscr{B}$ is true just in case $\mathscr{B}$ is true in every possible world in which $\mathscr{A}$ is true. Interpreting *if* as a strict conditional thus leads us to read English conditionals as meaning that if the antecedent is true in a possible world, the consequent is true there as well. In any circumstance where the antecedent holds, the consequent must hold also.

This analysis cannot be appropriate for counterfactuals. We have already seen that it would lead us to accept argument forms that plainly fail to be valid. Such analysis would also lead us to count some sentences we intuitively perceive to be true as false.

(5) a. If you were to jump off the top of a tall building, you'd be severely injured.
   b. If Babe Ruth hadn't switched from pitching to the outfield, he wouldn't have hit 714 home runs.

Both sentences in (5) seem true. Yet we would have to call them false if we construed the *if*s as expressing strict conditionals. Surely there is a possible world where you jump off a tall building as part of a stunt and land safely in a net. Perhaps there is even a world where your guardian angel swoops down to catch you before you hit the sidewalk. Similarly, there is undoubtedly a possible world where Babe Ruth remained a pitcher but played outfield or some other position on days when he wasn't pitching, hitting as many home runs as he actually did. The point of these examples is that many possible worlds are unusual and even weird. We generally ignore these worlds when we evaluate a counterfactual conditional as true or false. But a strict conditional pays attention to *every* world.

We might adopt the strategy of limiting our thinking about counterfactuals to "normal," "typical," or "nonweird" worlds. So, perhaps we could say that $\mathscr{A} \,\Box\!\!\rightarrow \mathscr{B}$ is true if $\mathscr{B}$ is true in all the normal worlds in which $\mathscr{A}$ is true. On this interpretation of the conditional, however, strengthening the antecedent and transitivity inferences would still be valid, because this strategy treats normality and weirdness as absolute notions. If $\mathscr{C}$ is true in all the normal worlds in which $\mathscr{A}$ is true, then $\mathscr{C}$ must be true in all the normal worlds in which $\mathscr{A}$ and $\mathscr{B}$ are true. How we determine what is too far out to consider, however, depends on the conditional sentence we're trying to evaluate. Consider this sentence:

(6) If you were to jump off the top of a tall building, and your guardian angel were to swoop you up, you'd be severely injured.

(6) seems false. If we were to adopt the above strategy, however, we would have to say that (6) is true, because there are no normal worlds in which your guardian angel catches you. So, what we consider "normal" and "far out" depends, specifically, on the conditional's antecedent. Worlds in which you're severely injured are normal, relative to the worlds in which you jump off tall buildings. But such worlds are strange relative to worlds in which you jump and your guardian angel swoops you up. When we consider what would happen in a world in which you jump, we change nothing but what your jump forces us to change.

The worlds we need to consider, then, are those where we hold everything we can constant, changing only what we must in order to make the antecedent of the conditional true. Suppose that we arrange possible worlds around the actual world according to their similarity to it. Some possible worlds are very similar; the actual world, for example, is very similar to a world in which you rub your nose as you read this sentence. Unless you are a baseball coach or a spy transmitting information by hand signals, your nose rub is unlikely to affect very much. So, we can think of that world as very close to the actual world. Other worlds differ greatly from each other. Our world is quite different from one that would have resulted from a Nazi victory in World War II, and even more different from one in which the Cuban missile crisis escalated into an all-out nuclear exchange. These worlds, then, are relatively far from the actual world.

In evaluating a counterfactual, we want to leave unchanged as much as possible yet alter enough to make the antecedent of the conditional true. We therefore want to examine worlds where the antecedent holds. But we aren't interested in all those worlds. We want to look only at those that are closest to the actual world. These worlds are different enough to make the antecedent true but no more different than that condition requires.

We can formulate a semantics for counterfactuals, then, by focusing on a special set of possible worlds: those that make the antecedent true yet are closest to our world. So, relative to an interpretation of the formulas concerned

> $\mathscr{A} \;\square\!\!\rightarrow\; \mathscr{B}$ is true in a world, w, iff $\mathscr{B}$ is true in all the worlds that make $\mathscr{A}$ true and are closest to w.

Say that an $\mathscr{A}$-world is simply a world in which $\mathscr{A}$ is true. Using this terminology, we can say that $\mathscr{A} \;\square\!\!\rightarrow\; \mathscr{B}$ is true in w iff $\mathscr{B}$ is true in all the $\mathscr{A}$-worlds closest to w.[3]

$\mathscr{A} \;\square\!\!\rightarrow\; \mathscr{B}$, then, is weaker than $\mathscr{A} \;\dashv\; \mathscr{B}$. If $\mathscr{A} \;\dashv\; \mathscr{B}$ is true, then $\mathscr{B}$ is true in every $\mathscr{A}$-world. That fact implies that $\mathscr{B}$ is true in all the closest $\mathscr{A}$-worlds. So $\mathscr{A} \;\dashv\; \mathscr{B}$ implies $\mathscr{A} \;\square\!\!\rightarrow\; \mathscr{B}$. The reverse does not hold; there may be worlds where $\mathscr{A}$ is true and $\mathscr{B}$ false, even though $\mathscr{A} \;\square\!\!\rightarrow\; \mathscr{B}$ is true.

$\mathscr{A} \;\square\!\!\rightarrow\; \mathscr{B}$, however, is stronger than $\mathscr{A} \rightarrow \mathscr{B}$. If $\mathscr{A}$ is true in the actual world, then, since the actual world is at least as close to itself as any other world, we can reduce the counterfactual to mean that $\mathscr{B}$ is true in the actual world. So $\mathscr{A} \;\square\!\!\rightarrow\; \mathscr{B}$ implies that, if $\mathscr{A}$ is true in the actual world, so is $\mathscr{B}$. But $\mathscr{A} \rightarrow \mathscr{B}$ does not entail $\mathscr{A} \;\square\!\!\rightarrow\; \mathscr{B}$; the former speaks only of the actual world, while the latter says something about the closest $\mathscr{A}$-worlds. The counterfactual, then, lies between the truth-functional and strict conditionals in strength.

Problems

Language ML* is language ML plus the $\square\!\!\rightarrow$ connective. Symbolize the following in ML*.

1. If the moon smiled, she would resemble you (Sylvia Plath).

2. If playing today, [Frank Baker] would hit 40 home runs a year (Fred Lieb).

3. If Satan had never fallen, Hell had been made for thee! (Percy Bysshe Shelley).

4. I could not love thee, dear, so much/Loved I not honor more (Richard Lovelace).

▶ 5. If I had only known how he would end up, I'd have been nicer to the guy (Virginia Mayo, of Ronald Reagan).

6. If people knew how hard I have to work to gain my mastery, it wouldn't seem wonderful at all (Michelangelo).

7. If there were no bad people, there would be no good lawyers (Charles Dickens).

8. If we applied some of our economic policies to the Sahara, soon there would be no sand (Jan Pietrzak).

9. Do you want to know how the government can spend 9.2 times the amount of reported outlays? Well, it's magic, and if we in the private sector tried it, we would probably end up in jail (J. Peter Grace).

▶ 10. If the MX were studied carefully, it would emerge that we cannot have it in a properly survivable mode, nor do we need it (McGeorge Bundy).

11. If I could just get back up to broke, I'd quit (Farmer, quoted by T. J. Attebury).

12. Were I to await perfection, my book would never be finished (Tai T'ung).

13. What a dull world it would be if every imaginative maker of legends were stigmatized as a liar (Heywood Broun).

14. However, if niceness were a surefire guarantee of success, Gerald Ford would still be in the White House and Schrafft's would still be selling cucumber sandwiches and martinis to ladies who drink at lunch counters (Vincent Canby).

▶ 15. I should not really object to dying if it were not followed by death (Thomas Nagel).

16. If arms spending cannot safely be pared enough and aid to people cannot humanely be reduced enough to lower the deficit, the logical resort is to taxes (Marvin Stone).

17. If you pitched inside to [Larry Lajoie], he'd tear a hand off the third baseman, and if you pitched outside he'd knock down the second baseman (Ed Walsh).

18. We think too small. Like the frog at the bottom of the well. He thinks the sky is only as big as the top of the well. If he surfaced, he would have an entirely different view (Mao Tse-Tung).

19. If the projected budget deficits were smaller, interest rates would be lower because Treasury borrowing and inflationary expectations would ease. Yet if only interest rates were lower, the budget deficit *would* be smaller (Andrew Tobias).

▶ 20. Don't knock the weather; nine-tenths of the people couldn't start a conversation if it didn't change once in a while (Frank McKinney Hubbard).

21. But even if we [build a robot that will learn and that will show some of the things that are typically human such as insight, intuition, creativity, and inspiration], it would be so troublesome, so difficult, we would have to put so much into it, that it wouldn't be cost effective. It wouldn't be worth our while (Isaac Asimov).

10.2 DEDUCTION RULES FOR COUNTERFACTUALS

In this section we'll develop system C, which corresponds, roughly, to the theory of counterfactuals that Lewis and Stalnaker have in common.[4] We'll build C by starting with the modal logic S5. Our natural deduction system for truth-functional and modal sentential connectives extends to incorporate counterfactuals. The rules for $\Box\to$, in fact, are particularly simple. Only the reiteration rule presents any real complications. To exploit the information that $\mathscr{A} \Box\to \mathscr{B}$, we can rely on the fact that $\mathscr{A} \Box\to \mathscr{B}$ implies $\mathscr{A} \to \mathscr{B}$. A counterfactual conditional asserts that, if it were the case that $\mathscr{A}$, it would be the case that $\mathscr{B}$. So, if $\mathscr{A}$ is the case, we can conclude that $\mathscr{B}$ is the case.

Counterfactual Exploitation ($\Box\to$E)

n.	$\mathscr{A} \Box\to \mathscr{B}$	
m.	$\mathscr{A}$	
p.	$\mathscr{B}$	$\Box\to$E, n, m

We can use this rule to show that $p \Box\to q$, $q \Box\to r$, and p together imply r:

1.	$p \Box\to q$	A
2.	$q \Box\to r$	A
3.	p	A
4.	Show r	
5.	q	$\Box\to$E, 1, 3
6.	r	$\Box\to$E, 2, 5

The counterfactuals, for the purposes of this rule, act just like ordinary truth-functional conditionals.

To introduce counterfactuals into a deduction, we can treat them much like ordinary or strict conditionals. The central idea behind any form of conditional proof is that one can show a conditional to be true by assuming its antecedent and proving its consequent. We'll call this method for establishing counterfactuals *counterfactual proof*:

Counterfactual Proof

n.	Show $\mathscr{A} \:\square\!\!\rightarrow\: \mathscr{B}$	
	$\square\!\rightarrow$	
n + 1.	$\mathscr{A}$	A$\square\!\rightarrow$P
	$\vdots$	
n + p.	$\mathscr{B}$	

The symbol $\square\!\rightarrow$ marks the proof as counterfactual. It indicates that reiteration into the proof is restricted. $\mathscr{A}$, marked A$\square\!\rightarrow$P, is the assumption for counterfactual proof.

Without a reiteration rule, we can use counterfactual proof only in the most trivial cases. We can, for example, show that $p \:\square\!\!\rightarrow\: p$ is valid:

> 1. Show $p \:\square\!\!\rightarrow\: p$
> $\square\!\rightarrow$
> 2. p A$\square\!\rightarrow$P

and that $(p \:\&\: q) \:\square\!\!\rightarrow\: q$ is valid:

> 1. Show $(p \:\&\: q) \:\square\!\!\rightarrow\: q$
> $\square\!\rightarrow$
> 2. $p \:\&\: q$ A$\square\!\rightarrow$P
> 3. q &E, 2

But we can't prove $\mathscr{A} \:\square\!\!\rightarrow\: \mathscr{B}$ unless $\mathscr{A}$ implies $\mathscr{B}$.

The complications involved in the logic of counterfactuals thus arise with respect to reiteration into counterfactual proofs. A counterfactual proof, in effect, takes us to another possible world. We entertain the assumption that $\mathscr{A}$ is true to see what follows. We try to see what would be true in one of the closest $\mathscr{A}$-worlds. If we can conclude that $\mathscr{B}$ would be true in our arbitrary $\mathscr{A}$-world, then we can conclude that $\mathscr{A} \:\square\!\!\rightarrow\: \mathscr{B}$ is true in our world.

Within a counterfactual proof, then, we talk about what is true, not in our world, but in an arbitrary closest world in which the assumption for counterfactual proof is true. We can reiterate information into that proof only if it would still hold in any such world.

Formulas with modal connectives as main connectives certainly reiterate into counterfactual proofs. Any modally closed formula holds true in every possible world if it's true in any world at all. So, any such formula true in our world is true in every other world. In particular, it is true in the closest worlds where the counterfactual assumption holds. Formulas with $\square$, $\lozenge$, $\dashv$, and $\boxminus$ as main connectives consequently reiterate without difficulty.

Counterfactual Reiteration 1 ($\square{\rightarrow}$R1)

$$
\begin{array}{lll}
\text{n.} & \mathcal{A} & \\
& \square{\rightarrow} & \\
\text{n + p.} & \mathcal{A} & \square{\rightarrow}\text{R1, n}
\end{array}
$$

Here $\mathcal{A}$ is modally closed.

This rule thus justifies applications of reiteration of the following kinds:

$$
\begin{array}{lll}
\text{n.} & \square\mathcal{A} & \\
& \square{\rightarrow} & \\
\text{n + p.} & \square\mathcal{A} & \square{\rightarrow}\text{R1, n}
\end{array}
$$

$$
\begin{array}{lll}
\text{n.} & \lozenge\mathcal{A} & \\
& \square{\rightarrow} & \\
\text{n + p.} & \lozenge\mathcal{A} & \square{\rightarrow}\text{R1, n}
\end{array}
$$

$$
\begin{array}{lll}
\text{n.} & \mathcal{A} \dashv \mathcal{B} & \\
& \square{\rightarrow} & \\
\text{n + p.} & \mathcal{A} \dashv \mathcal{B} & \square{\rightarrow}\text{R1, n}
\end{array}
$$

$$
\begin{array}{lll}
\text{n.} & \mathcal{A} \boxminus \mathcal{B} & \\
& \square{\rightarrow} & \\
\text{n + p.} & \mathcal{A} \boxminus \mathcal{B} & \square{\rightarrow}\text{R1, n}
\end{array}
$$

The first reiteration rule allows us to see how counterfactual proof works in some more interesting cases. We can now show that strict conditionals imply counterfactual conditionals:

$$
\begin{array}{lll}
\text{1.} & p \dashv q & \text{A} \\
\text{2.} & \text{Show } p \;\square{\rightarrow}\; q & \\
& \square{\rightarrow} & \\
\text{3.} & \lceil p & \text{A}\square{\rightarrow}\text{P} \\
\text{4.} & \mid p \dashv q & \square{\rightarrow}\text{R1, 1} \\
\text{5.} & \lfloor q & \dashv\text{E, 3, 4}
\end{array}
$$

In fact, we can show that strict conditionals imply not only the truth but the necessary truth of counterfactuals:

$$
\begin{array}{lll}
\text{1.} & p \dashv\!\!\!\dashv q & \text{A} \\
\text{2.} & \text{Show } \Box(p \;\Box\!\!\rightarrow q) & \\
& \quad \Box & \\
\text{3.} & \quad\lceil \text{Show } p \;\Box\!\!\rightarrow q & \\
& \quad\;\; \Box\!\!\rightarrow & \\
\text{4.} & \quad\;\;\lceil p & \text{A}\Box\!\!\rightarrow\text{P} \\
\text{5.} & \quad\;\;\mid p \dashv\!\!\!\dashv q & \Box\text{R}, \Box\!\!\rightarrow\text{R1, 1} \\
\text{6.} & \quad\;\;\lfloor q & \dashv\!\!\!\dashv\text{E, 4, 5}
\end{array}
$$

Notice that $p \dashv\!\!\!\dashv q$ must reiterate into a modal proof and then into a counterfactual subproof. The formula must qualify to reiterate into both. Both reiteration rules thus appear in the justification for line 5.

The modal reiteration rule, together with counterfactual exploitation, allows us to demonstrate that, if a counterfactual conditional is necessarily true, the corresponding strict conditional holds.

$$
\begin{array}{lll}
\text{1.} & \Box(p \;\Box\!\!\rightarrow q) & \text{A} \\
\text{2.} & \text{Show } p \dashv\!\!\!\dashv q & \\
& \quad \Box & \\
\text{3.} & \quad\lceil p & \text{A}\Box\text{P} \\
\text{4.} & \quad\mid \Box(p \;\Box\!\!\rightarrow q) & \Box\text{R, 1} \\
\text{5.} & \quad\mid p \;\Box\!\!\rightarrow q & \Box\text{E, 4} \\
\text{6.} & \quad\lfloor q & \Box\!\!\rightarrow\text{E, 3, 5}
\end{array}
$$

So, $p \dashv\!\!\!\dashv q$ and $\Box(p \;\Box\!\!\rightarrow q)$ are equivalent.

A final example of the use of $\Box\!\!\rightarrow$R1 shows that a modalized version of transitivity is valid, as we might expect from the equivalence of $p \dashv\!\!\!\dashv q$ and $\Box(p \;\Box\!\!\rightarrow q)$.

$$
\begin{array}{lll}
\text{1.} & \Box(p \;\Box\!\!\rightarrow q) & \text{A} \\
\text{2.} & \Box(q \;\Box\!\!\rightarrow r) & \text{A} \\
\text{3.} & \text{Show } \Box(p \;\Box\!\!\rightarrow r) & \\
& \quad\lceil \Box & \\
\text{4.} & \quad\mid \text{Show } p \;\Box\!\!\rightarrow r & \\
& \quad\mid\;\; \Box\!\!\rightarrow & \\
\text{5.} & \quad\mid\;\;\lceil p & \text{A}\Box\!\!\rightarrow\text{P} \\
\text{6.} & \quad\mid\;\;\mid \Box(p \;\Box\!\!\rightarrow q) & \Box\text{R}, \Box\!\!\rightarrow\text{R1, 1} \\
\text{7.} & \quad\mid\;\;\mid \Box(q \;\Box\!\!\rightarrow r) & \Box\text{R}, \Box\!\!\rightarrow\text{R1, 2} \\
\text{8.} & \quad\mid\;\;\mid p \;\Box\!\!\rightarrow q & \Box\text{E, 6} \\
\text{9.} & \quad\mid\;\;\mid q \;\Box\!\!\rightarrow r & \Box\text{E, 7} \\
\text{10.} & \quad\mid\;\;\mid q & \Box\!\!\rightarrow\text{E, 5, 8} \\
\text{11.} & \quad\mid\;\;\lfloor r & \Box\!\!\rightarrow\text{E, 9, 10}
\end{array}
$$

Again, the initial assumptions had to reiterate into a counterfactual subproof within a modal proof and so had to fulfill the requirements of both modal and counterfactual reiteration rules.

A strict conditional, as we've seen, implies the corresponding counterfactual conditional. This fact is useful enough to be recorded as a derived rule.

Strict and Counterfactual Conditionals ($\dashv\ \square\rightarrow$)

$$\frac{\text{n.} \qquad \mathscr{A} \dashv \mathscr{B}}{\text{n + p.} \quad \mathscr{A} \ \square\rightarrow \mathscr{B} \qquad \dashv\square\rightarrow, \text{n}}$$

Our first reiteration rule is that any modally closed formula reiterates into a counterfactual proof. The rule says nothing about counterfactuals; despite the box in $\square\rightarrow$, the connective is not modal, and formulas of the form $\mathscr{A}\ \square\rightarrow \mathscr{B}$ are not modally closed.

Our second reiteration rule tells how to use the information given in counterfactual conditionals. To formulate it, we need a definition.

> **Definition.** A formula, $\mathscr{A}$, is *tantamount to* a formula, $\mathscr{B}$, at line n of a proof if (a) $\mathscr{A}$ is $\mathscr{B}$, (b) $\mathscr{A} \boxminus \mathscr{B}$ is free at n, or (c) $\mathscr{A} \ \square\rightarrow \mathscr{B}$ and $\mathscr{B}\ \square\rightarrow \mathscr{A}$ are both free at *n*. Also, *if $\mathscr{A}$ is tantamount to $\mathscr{B}$ at n, and $\mathscr{B}$ is tantamount to $\mathscr{C}$ at n, then $\mathscr{A}$ is tantamount to $\mathscr{C}$ at n.*

One formula is tantamount to another, in other words, if it is at least counterfactually equivalent to that formula. Every formula is tantamount to itself and to every logically equivalent formula. Formulas that have been established as strictly equivalent in a proof are tantamount to each other. And, if counterfactual conditionals in both directions are established between two formulas, the formulas are tantamount to each other.

This definition expresses a very simple semantic idea. At any given stage in constructing a proof, we are trying to reflect what's true in some possible world, say, w. A formula, $\mathscr{A}$, is tantamount to a formula, $\mathscr{B}$, at that stage just in case the $\mathscr{A}$-worlds closest to w are the $\mathscr{B}$-worlds closest to w. That is why we can expect formulas tantamount to each other to have the same counterfactual consequences.

The second reiteration rule specifies some circumstances under which we can use counterfactual conditionals to obtain information within a counterfactual proof:

Counterfactual Reiteration 2 ($\square\rightarrow$R2)

$$\begin{array}{ll} \text{n.} & \mathscr{A} \ \square\rightarrow \mathscr{B} \\ & \square\rightarrow \\ \text{n + p.} & \mathscr{B} \qquad\qquad \square\rightarrow\text{R2, n } (m_1, \ldots, m_n) \end{array}$$

Here (a) $\mathscr{A}$ is the assumption for the counterfactual proof ($A\square\rightarrow P$) into which the rule is being applied, or (b) $\mathscr{A}$ is tantamount to a conjunction of the free formulas in that proof.

This rule sounds complex but is usually easy to apply. It says that we can use a counterfactual, $\mathscr{A} \,\square\!\!\rightarrow \mathscr{B}$, to write $\mathscr{B}$ within a counterfactual proof if $\mathscr{A}$ is the assumption of that proof or is tantamount to the information so far supplied in that proof. The lines $m_1, \ldots, m_n$ written in parentheses in the justification show that it's legitimate to apply the rule. If we justify the application by appealing to the assumption for the counterfactual proof, we can cite its line number. If we appeal to the fact that the antecedent is tantamount to another formula, we can cite the line number of that formula. Or, if the formula is a conjunction, we can cite the line numbers of the conjuncts.

To demonstrate $\square\!\!\rightarrow$R2, let's show that $p \,\square\!\!\rightarrow q$ implies $p \,\square\!\!\rightarrow (p \,\&\, q)$.

1. $p \,\square\!\!\rightarrow q$ A
2. Show $p \,\square\!\!\rightarrow (p \,\&\, q)$
        $\square\!\!\rightarrow$
3. p A$\square\!\!\rightarrow$P
4. q $\square\!\!\rightarrow$R2, 1 (3)
5. $p \,\&\, q$ &I, 3, 4

On line 4, we use the counterfactual on line 1 to justify q. The antecedent of the counterfactual, p, is the assumption for counterfactual proof.

Another example shows a different aspect of the reiteration rule:

1. $p \,\square\!\!\rightarrow q$ A
2. $p \,\&\!\!\!\backslash\, r$ A
3. Show $r \,\square\!\!\rightarrow q$
        $\square\!\!\rightarrow$
4. r A$\square\!\!\rightarrow$P
5. q $\square\!\!\rightarrow$R2, 1 (2, 4)

Here p is not itself the assumption for counterfactual proof, but it is tantamount to that assumption. Through line 4, r is the only formula in the counterfactual proof. So, the conjunction of formulas is simply r. Since p is tantamount to r, by virtue of line 2, we are able to deduce q.

Our next example is an argument form extremely similar to the invalid transitivity principle. Nevertheless, it is valid.

1. $p \,\square\!\!\rightarrow q$ A
2. $(p \,\&\, q) \,\square\!\!\rightarrow r$ A
3. Show $p \,\square\!\!\rightarrow r$
        $\square\!\!\rightarrow$
4. p A$\square\!\!\rightarrow$p
5. q $\square\!\!\rightarrow$R2, 1 (4)
6. r $\square\!\!\rightarrow$R2, 2 (4, 5)

We need two applications of our second reiteration rule to show this argument form valid. In the first application, the antecedent of $p \,\square\!\!\rightarrow q$ is simply the A$\square\!\!\rightarrow$P, p. So, we are able to deduce q on line 5. But then the antecedent of $(p \ \& \ q) \ \square \rightarrow r$ is tantamount to—in fact, *is*—the conjunction of the formulas so far present in the counterfactual proof. So, we can write the consequent of $(p \ \& \ q) \ \square \rightarrow r$, namely, r, on line 6.

A final aspect of the second reiteration rule emerges in the following example. This argument form, too, is valid, despite its similarity to transitivity.

$$
\begin{array}{lll}
1. & p \ \square \rightarrow q & \text{A} \\
2. & q \ \square \rightarrow p & \text{A} \\
3. & q \ \square \rightarrow r & \text{A} \\
4. & \text{Show } p \ \square \rightarrow r & \\
& \qquad \square \rightarrow & \\
5. & \lceil p & \text{A} \square \rightarrow \text{P} \\
6. & \lfloor r & \square \rightarrow \text{R2, 3 (1, 2)}
\end{array}
$$

On line 5 we assume p. But p is tantamount to q, because of lines 1 and 2. Since line 3 tells us that, if q were the case, r would be the case, we can deduce r.

To see why this reiteration rule is justified, recall that a counterfactual proof with $\mathscr{A}$ as its assumption is concerned, in effect, with truth in an arbitrary closest $\mathscr{A}$-world. If $\mathscr{A} \ \square \rightarrow \mathscr{B}$ appears free, $\mathscr{B}$ is true in all the closest $\mathscr{A}$-worlds. So, we should be able to conclude that $\mathscr{B}$ is true in the arbitrary $\mathscr{A}$-world we selected in the counterfactual proof.

If $\mathscr{A}$ is not the assumption of the counterfactual proof, but is instead tantamount to the conjunction of the free formulas in the proof, then $\mathscr{A}$ has the same counterfactual consequences as that conjunction. Since $\mathscr{A} \ \square \rightarrow \mathscr{B}$ is free, $\mathscr{B}$ is a counterfactual consequence of $\mathscr{A}$. So, $\mathscr{B}$ must also be a counterfactual consequence of the conjunction.

In Lewis's system, reiteration is legitimate in two other circumstances as well. These circumstances arise relatively infrequently in analyzing natural language arguments. The linguistic facts, moreover, are unclear. So, we'll be content with two forms of reiteration in system C; we'll postpone Lewis's other reiteration rules until the next section, when we present CL.

We need one further rule, however, to complete a system faithful to the intentions that Stalnaker and Lewis have in common. Recall that the rule of $\diamondsuit$ exploitation in S5 allows us to infer $\mathscr{B}$ from $\diamondsuit \mathscr{A}$ and $\mathscr{A} \ \dashv \ \mathscr{B}$, provided that $\mathscr{B}$ is modally closed. The idea behind the rule is this: If $\diamondsuit \mathscr{A}$ is true in a world, w, then $\mathscr{A}$ is true in some possible world w1. If $\mathscr{A} \ \dashv \ \mathscr{B}$ is true in w, then $\mathscr{B}$ is true in every $\mathscr{A}$-world, so $\mathscr{B}$ is also true in w1. But, if $\mathscr{B}$ is modally closed, then it is true in every world if it's true at all. So, $\mathscr{B}$ must be true in w, the world we started with.

Now, if $\diamondsuit \mathscr{A}$ is true in w, then $\mathscr{A}$ is true in some world or worlds. Among the worlds where $\mathscr{A}$ is true, some are closer to w than others. Consider the worlds that make $\mathscr{A}$ true and that are closest to w. If $\mathscr{A} \ \square \rightarrow \mathscr{B}$ is true, then $\mathscr{B}$ is true in all the closest $\mathscr{A}$-worlds to w. So $\mathscr{B}$ is true in some worlds, namely, the closest $\mathscr{A}$-worlds. If $\mathscr{B}$ is modally closed, therefore, $\mathscr{B}$ is true in every world; in particular, $\mathscr{B}$ is true in w.

This reasoning shows that we can justify a new rule for exploiting possibility statements when our language contains counterfactuals. We'll therefore introduce a new rule, which we'll call possibility exploitation*.

Possibility Exploitation*($\Diamond$E*)

n. $\Diamond \mathscr{A}$
m. $\underline{\mathscr{A} \;\Box\!\!\to\; \mathscr{B}}$
p. $\mathscr{B}$ $\qquad\qquad$ $\Diamond$E*, n

Here $\mathscr{B}$ must be modally closed.

To see how this rule works, let's consider an argument form that reflects a very common form of reasoning. Suppose we want to show that a certain supposition is absurd. We might well argue that, if it were true, then something ridiculous would follow. That is, we might argue that, if $\mathscr{A}$ were true, then $\mathscr{B}$ would be true; but $\mathscr{B}$ is impossible, so $\mathscr{A}$ must be impossible as well.

1. $p \;\Box\!\!\to\; q$ $\qquad\qquad$ A
2. $\neg \Diamond q$ $\qquad\qquad$ A
3. Show $\neg \Diamond p$
4. $\quad$ $\Diamond p$ $\qquad\qquad$ AIP
5. $\quad$ Show $p \;\Box\!\!\to\; \Diamond q$
$\qquad\quad$ $\Box\!\!\to$
6. $\qquad$ p $\qquad\qquad$ A$\Box\!\!\to$P
7. $\qquad$ q $\qquad\qquad$ $\Box\!\!\to$R2, 1 (6)
8. $\qquad$ $\Diamond q$ $\qquad\qquad$ $\Diamond$I, 7
9. $\quad$ $\Diamond q$ $\qquad\qquad$ $\Diamond$E*, 4, 5
10. $\quad$ $\neg \Diamond q$ $\qquad\qquad$ R, 2

To show $\neg \Diamond p$, we begin an indirect proof, assuming $\Diamond p$. Now, it's tempting to use $\Diamond$E*. But q is not modally closed. $\Diamond q$ is modally closed, however, and contradicts $\neg \Diamond q$ on line 2. So we can use $\Diamond$E* to obtain $\Diamond q$, which completes our indirect proof.

System C, then, consists of S5 together with five rules: $\Box\!\!\to$E, counterfactual proof, $\Box\!\!\to$R1, $\Box\!\!\to$R2, and $\Diamond$E*.

Problems

Prove that each of the following formulas is valid in C.

1. $p \;\Box\!\!\to\; (p \lor q)$

2. $(p \,\&\, q) \;\Box\!\!\to\; \neg(\neg p \lor \neg q)$

3. $(p \to q) \;\Box\!\!\to\; (\neg p \lor q)$

4. $\Box q \to (p \;\Box\!\!\to\; q)$

▶ 5. $\neg \Diamond p \to (p \;\Box\!\!\to\; q)$

6. $(p \;\Box\!\!\to\; q) \to (p \;\Box\!\!\to\; (p \,\&\, q))$

7. $(p \:\square\!\!\rightarrow (p \& q)) \rightarrow (p \:\square\!\!\rightarrow q)$

8. $(p \:\square\!\!\rightarrow (p \:\square\!\!\rightarrow q)) \rightarrow (p \:\square\!\!\rightarrow q)$

9. $(p \:\square\!\!\rightarrow (p \rightarrow q)) \rightarrow (p \:\square\!\!\rightarrow q)$

▶ 10. $(p \& (p \:\square\!\!\rightarrow q)) \rightarrow q$

11. $((p \:\square\!\!\rightarrow q) \& \neg q) \rightarrow \neg p$ (This justifies a derivable rule, $\square\!\!\rightarrow$E*, parallel to $\rightarrow$E*.)

12. $((p \:\square\!\!\rightarrow q) \& \square\neg q) \rightarrow \square\neg p$

13. $((p \:\square\!\!\rightarrow q) \& \Diamond p) \rightarrow \Diamond q$

14. $(p \:\&\!\!\!3\: q) \rightarrow ((p \:\square\!\!\rightarrow q) \& (q \:\square\!\!\rightarrow p))$

15. $(p \:\&\!\!\!3\: q) \rightarrow \square((p \:\square\!\!\rightarrow q) \& (q \:\square\!\!\rightarrow p))$

16. $\square((p \:\square\!\!\rightarrow q) \& (q \:\square\!\!\rightarrow p)) \rightarrow (p \:\&\!\!\!3\: q)$

17. $((p \:\square\!\!\rightarrow q) \& (p \:\square\!\!\rightarrow r)) \rightarrow (p \:\square\!\!\rightarrow (q \& r))$

18. $(p \:\square\!\!\rightarrow (q \& r)) \dashv\!3 ((p \:\square\!\!\rightarrow q) \& (p \:\square\!\!\rightarrow r))$

19. $((p \:\square\!\!\rightarrow q) \& \Diamond p) \dashv\!3 \neg(p \:\square\!\!\rightarrow \neg q)$

▶ 20. $((p \:\square\!\!\rightarrow q) \& (p \:\square\!\!\rightarrow \neg q)) \rightarrow \neg\Diamond p$

Show that these argument forms are valid in C.

21. $p \:\square\!\!\rightarrow (p \& q); \therefore p \:\square\!\!\rightarrow q$

22. $p \:\square\!\!\rightarrow q; \therefore p \:\square\!\!\rightarrow (p \& q)$

23. $p \dashv\!3 q; q \:\square\!\!\rightarrow p; p \:\square\!\!\rightarrow r; \therefore q \:\square\!\!\rightarrow r$

24. $p \:\&\!\!\!3\: q; \therefore (p \:\square\!\!\rightarrow r) \leftrightarrow (q \:\square\!\!\rightarrow r)$

▶ 25. $p \:\&\!\!\!3\: q; \therefore (r \:\square\!\!\rightarrow p) \leftrightarrow (r \:\square\!\!\rightarrow q)$

26. $p \:\square\!\!\rightarrow q; q \:\square\!\!\rightarrow p; \therefore (p \:\square\!\!\rightarrow r) \leftrightarrow (q \:\square\!\!\rightarrow r)$

27. $p \:\square\!\!\rightarrow q; q \dashv\!3 r; (p \& r) \:\square\!\!\rightarrow s; \therefore p \:\square\!\!\rightarrow s$

28. $p \dashv\!3 q; q \:\square\!\!\rightarrow r; q \:\square\!\!\rightarrow p; \therefore p \:\square\!\!\rightarrow r$

29. $q \:\square\!\!\rightarrow p; p \dashv\!3 (q \& r); \therefore q \:\square\!\!\rightarrow r$

30. $p \:\square\!\!\rightarrow q; q \dashv\!3 r; \therefore \neg r \rightarrow \neg p$

31. $p \:\square\!\!\rightarrow (p \& q); (p \& q) \:\square\!\!\rightarrow r; \neg r; \therefore \neg p$

32. $\neg p \:\square\!\!\rightarrow \neg q; r; (\neg p \& \neg q) \:\square\!\!\rightarrow \neg r; \therefore p$

33. $p \:\square\!\!\rightarrow q; r \dashv\!3 s; (p \& q) \:\square\!\!\rightarrow r; \therefore p \:\square\!\!\rightarrow s$

34. $p \:\square\!\!\rightarrow q; \Diamond p; \therefore \Diamond(p \& q)$

▶ 35. $q \:\square\!\!\rightarrow p; p \dashv\!3 (q \& r); \therefore \Diamond q \rightarrow \Diamond r$

36. $p \: \square \!\! \rightarrow (p \: \& \: q); \: \neg \Diamond q; \: \therefore \: \neg \Diamond p$

37. $p \: \square \!\! \rightarrow q; \: \neg r \: \& \: p; \: \therefore \: \Diamond (q \: \& \: \neg r)$

38. $p \: \square \!\! \rightarrow q; \: q \: \square \!\! \rightarrow p; \: p \: \square \!\! \rightarrow r; \: \neg r; \: \therefore \: \neg q \: \& \: \neg p$

39. $p \: \square \!\! \rightarrow q; \: q \dashv 3 \: r; \: r \: \square \!\! \rightarrow p; \: p \: \square \!\! \rightarrow s; \: \therefore \: r \: \square \!\! \rightarrow s$

▶ **40.** $(p \: \& \: r) \: \square \!\! \rightarrow s; \: p \: \square \!\! \rightarrow (p \: \& \: q); \: (p \: \& \: q) \: \square \!\! \rightarrow r; \: \therefore \: p \: \square \!\! \rightarrow s$

41. $(p \: \& \: r) \: \square \!\! \rightarrow s; \: q \: \square \!\! \rightarrow p; \: q \: \square \!\! \rightarrow r; \: p \: \square \!\! \rightarrow q; \: \therefore \: p \: \square \!\! \rightarrow s*$

42. $r \: \square \!\! \rightarrow p; \: p \: \square \!\! \rightarrow (q \: \& \: p); \: p \: \square \!\! \rightarrow s; \: (q \: \& \: p) \: \square \!\! \rightarrow r; \: \therefore \: r \: \square \!\! \rightarrow s*$

43. $p \: \square \!\! \rightarrow q; \: (p \: \& \: q) \: \square \!\! \rightarrow r; \: r \: \square \!\! \rightarrow s; \: r \: \square \!\! \rightarrow p; \: \therefore \: p \: \square \!\! \rightarrow s*$

44. $p \: \square \!\! \rightarrow q; \: r \: \square \!\! \rightarrow s; \: p \: \square \!\! \rightarrow r; \: q \: \square \!\! \rightarrow p; \: r \: \square \!\! \rightarrow q; \: \therefore \: q \: \square \!\! \rightarrow s*$

▶ **45.** $p \: \square \!\! \rightarrow q; \: q \: \square \!\! \rightarrow s; \: r \: \square \!\! \rightarrow q; \: p \: \square \!\! \rightarrow r; \: q \: \square \!\! \rightarrow p; \: \therefore \: r \: \square \!\! \rightarrow s*$

46. $p \dashv 3 \: q; \: p \: \square \!\! \rightarrow r; \: \Diamond q; \: q \: \square \!\! \rightarrow p; \: \therefore \: \Diamond r*$

47. $\square \neg r; \: q \: \square \!\! \rightarrow r; \: q \: \square \!\! \rightarrow \neg p; \: \neg p \dashv 3 \: q; \: \therefore \: \square p*$

48. $p \: \square \!\! \rightarrow (p \: \& \: q); \: \square s; \: (p \: \& \: q) \: \square \!\! \rightarrow \neg r; \: (p \: \& \: \neg r) \: \square \!\! \rightarrow \neg s; \: \therefore \: \neg \Diamond p*$

49. $\Diamond p; \: p \: \square \!\! \rightarrow q; \: r \dashv 3 \: s; \: (p \: \& \: q) \: \square \!\! \rightarrow r; \: \therefore \: \Diamond s*$

Show that each of the following arguments is valid.

50. If John attended the meeting, he'd be able to convince at least a few people to vote against the development proposal. If we were able to get most of the neighborhood to sign a petition, then the developers would change their proposal. Either we'll get most of the neighborhood to sign a petition, or John will attend the meeting. So, John will be able to convince at least a few people to vote against the development proposal, unless the developers change their proposal.

51. If God existed, He would exist necessarily. And it's necessarily true that, if God existed, He would be omnipotent. Therefore, if He existed, God would be necessarily omnipotent.

52. If God were all-powerful, then He could make a rock so heavy that He could not lift it. But if God were omnipotent, then He could lift any rock that He could make. Therefore, God can't be all-powerful.

53. If Geraldine were to sign the contract, she'd lose money. If Geraldine were to lose money, we'd lose money too. Geraldine will sign the contract. So, we'll lose money.

54. If the United States didn't agree to arms limitation talks, tensions with the Soviets would remain high. But even if the U.S. were to agree to such talks, tensions between the U.S.S.R. and the U.S. would remain high. Thus, tensions between the United States and the Soviets will remain high.

► **55.** It's true that, if 4 were odd, then 2 would be odd. But it's also true that, if 4 were odd, 2 would still be even. Since nothing can be both odd and even, it's necessarily false that 4 is odd.

56. If Napoleon had not attacked Russia, he would have been able to maintain his grip on Europe for much longer. If Napoleon had refrained from attacking Russia, and had held onto his European empire for many more years, then the British Empire's dominance would have been much diminished. So, if Napoleon had not attacked Russia, the dominance of the British Empire would have been much diminished.

57. If Richard were to challenge Donna for the chairmanship, then Richard would challenge and be promptly defeated. If Richard launched his challenge to Donna, and were quickly defeated, then he'd be despondent and embarrassed. Hence, if Richard were to challenge Donna for the chairmanship, he'd be embarrassed.

58. If the president retaliated against Libya with military force, the public would be ambivalent. But if Libya were to direct terrorist attacks toward Americans on American soil, then, if the president were to retaliate militarily, the American public would not be ambivalent at all. So, Libya will not engage in terrorism against Americans on U.S. soil unless the president cannot retaliate militarily.

59. It's not true that, if Nathan were to publish several more papers, he'd get tenure. So, it's possible for Nathan to publish several additional papers and still not get tenure.

► **60.** If Kelly were to refuse the operation, she'd never regain full use of her arm. But if Kelly refused the operation and didn't regain full use of the arm, she'd regret not having the operation. So, either Kelly will agree to the operation, or she'll regret not doing so.

61. If Johnson left the firm, then Jones would leave, too. Furthermore, if Jones were to leave the firm, Johnson would leave. So, Robinson would leave if Johnson left only if Robinson would leave if Jones left.

62. If the federal deficit were smaller, then interest rates would be lower. But if they were lower, then the deficit would be smaller. Moreover, if interest rates were lower, then, necessarily, the inflation rate would also be lower. But if the inflation rate were lower, interest rates would be lower. So, if the deficit were smaller, the rate of inflation would be lower, and, if the inflation rate were lower, the deficit would be smaller.

63. Lily can talk her way out of some tough situations, but she can't perform miracles. So, even if Lily were to talk her way out of a tough situation, she wouldn't perform any miracles.

64. If Japan had not attacked Pearl Harbor, the American public would not have wanted to enter the War. Japan could have refrained from the

Pearl Harbor attack; in fact, if Roosevelt had not already been eager to give assistance to the Allied forces, Japan would not have attacked. If there had been no Pearl Harbor attack, and the public had not wanted to enter the War, then FDR would not have been eager to assist the Allied forces. And, if he had not been eager to help, the U.S. would have entered the conflict much later. So the U.S. might have entered World War II much later.

65. If Lincoln had not been assassinated, Reconstruction would not have left such a strong mark on the character of the South. It could have happened that Lincoln wasn't assassinated. So, Reconstruction didn't have to mark the Southern character so strongly.

66. If you were to profit from our mutual arrangement, then so would I. If I were to profit, then so would you. If you were to profit, of course, you'd be able to pay back some of the money you owe. If I profit from the arrangement, and you pay back some of the money you owe, I'll be able to help you avoid a nasty battle with the guys from Cleveland. Hence, I won't be able to help you only if you can't manage to profit from our mutual arrangement.

67. If the party were to maintain its current economic policy, there would be a flight of capital to other countries. But the party wouldn't change its economic policy if it tightened its control over the economy. If the party maintained its policy, and capital fled to other countries, then it would tighten its grip on the economy. If the party were to maintain its current policy, however, the nation would have to pay large amounts of foreign debt in hard currency, and this it cannot do. So, it's impossible for the party to tighten its grip on the economy.

10.3 System CS

The rules we've seen so far are widely accepted. They give us a system of logic for counterfactuals that capture, in essence, the commonalities in the approaches of Lewis and Stalnaker. In this section, we consider further rules that describe the full system of Stalnaker.

To obtain Stalnaker's theory, CS, we need a fairly simple rule to the effect that, to negate a counterfactual, we can simply negate its consequent, provided that its antecedent is possible.

Stalnaker's Rule (S)

n.	$\neg(\mathcal{A} \mathbin{\square\!\!\rightarrow} \mathcal{B})$	
m.	$\mathcal{A} \mathbin{\square\!\!\rightarrow} \neg\mathcal{B}$	S, n

Note that this rule goes in only one direction. $\neg(\mathcal{A} \ \Box \rightarrow \mathcal{B})$ does not follow from $\mathcal{A} \ \Box \rightarrow \neg\mathcal{B}$, though it does follow from $(\mathcal{A} \ \Box \rightarrow \neg\mathcal{B})$ & $\Diamond\mathcal{A}$.

Stalnaker's rule is not justified by what we've said about the meanings of counterfactuals. But it is justified if we assume that there is never more than one closest $\mathcal{A}$-world. Symbolically, adopting Stalnaker's rule has the effect of limiting to at most one member the set of the closest $\mathcal{A}$-worlds to a world, w, no matter what w and $\mathcal{A}$ are. So, the rule has the effect of making a counterfactual $\mathcal{A} \ \Box \rightarrow \mathcal{B}$ true just in case $\mathcal{B}$ is true in *the* closest $\mathcal{A}$-world.

We can summarize the disagreement between Lewis and Stalnaker as a disagreement about how to negate a counterfactual. In each of the following examples, b is the negation of a, according to Lewis, while c is the negation of a, according to Stalnaker.

(7) a. If this match were struck, it would light.
   b. If this match were struck, it might not light.
   c. If this match were struck, it wouldn't light.

(8) a. If Russia had lost at Stalingrad, Hitler would have shifted most of his manpower to the Western Front.
   b. If Russia had lost at Stalingrad, Hitler might not have shifted most of his manpower to the Western Front.
   c. If Russia had lost at Stalingrad, Hitler would not have shifted most of his manpower to the Western Front.

(9) a. If the Cubs had beaten the Padres, they would have beaten the Tigers.
   b. If the Cubs had beaten the Padres, they might not have beaten the Tigers.
   c. If the Cubs had beaten the Padres, they wouldn't have beaten the Tigers.

Stalnaker's rule has a number of important consequences. The first important consequence of Stalnaker's rule is a principle sometimes called *counterfactual excluded middle*: $((\mathcal{A} \ \Box \rightarrow \mathcal{B}) \vee (\mathcal{A} \ \Box \rightarrow \neg\mathcal{B}))$.

1. ~~Show~~ $((p \ \Box \rightarrow q) \vee (p \ \Box \rightarrow \neg q))$		
2. $\neg((p \ \Box \rightarrow q) \vee (p \ \Box \rightarrow \neg q))$	AIP	
3. $\neg(p \ \Box \rightarrow q)$ & $\neg(p \ \Box \rightarrow \neg q)$	$\neg \vee$, 2	
4. $\neg(p \ \Box \rightarrow q)$	&E, 3	
5. $\neg(p \ \Box \rightarrow \neg q)$	&E, 3	
6. $p \ \Box \rightarrow \neg q$	S, 4	

The second is that, in CS, we must define *might*-counterfactuals differently from Lewis. Just as the connective $\Box \rightarrow$ symbolizes *if . . . would* conditionals in English, another connective, $\diamond \rightarrow$, symbolizes the English *if . . . might*. We can read $\mathscr{A} \diamond \rightarrow \mathscr{B}$ as 'if it were the case that $\mathscr{A}$, then it might be the case that $\mathscr{B}$.' Whenever $\mathscr{A}$ and $\mathscr{B}$ are formulas, $(\mathscr{A} \diamond \rightarrow \mathscr{B})$ will be a formula, too. We can thus use $\diamond \rightarrow$ to symbolize any of these sentences:

(10) a. If Ken had done just a few things differently, it might not have turned out so badly.
   b. Alice might not be so wealthy if she had taken George's investment advice.
   c. Had the U.S. lost the battle of Midway, Japan might have invaded California.
   d. If Nicholas were to write the proposal, it might be approved.

We can symbolize each as something of the form $\mathscr{A} \diamond \rightarrow \mathscr{B}$.

As we'll see in the next section, Lewis defines $\mathscr{A} \diamond \rightarrow \mathscr{B}$ as $\neg(\mathscr{A} \square \rightarrow \neg\mathscr{B})$. This definition is unacceptable in CS; we can show, using the definition together with Stalnaker's rule, that, if $\diamond\mathscr{A}$, then $\mathscr{A} \diamond \rightarrow \mathscr{B}$ and $\mathscr{A} \square \rightarrow \mathscr{B}$ are equivalent. These proofs demonstrate this fact for the specific formulas $p \diamond \rightarrow q$ and $p \square \rightarrow q$.

1.	$\diamond p$	A
2.	$p \square \rightarrow q$	A
3.	Show $p \diamond \rightarrow q$	
4.	Show $\neg(p \square \rightarrow \neg q)$	
5.	$p \square \rightarrow \neg q$	AIP
6.	Show $p \square \rightarrow \square(q \,\&\, \neg q)$	
	$\square \rightarrow$	
7.	p	A$\square \rightarrow$P
8.	q	$\square \rightarrow$R2, 2 (7)
9.	$\neg q$	$\square \rightarrow$R2, 5 (7)
10.	$\square(q \,\&\, \neg q)$	!, 8, 9
11.	$\square(q \,\&\, \neg q)$	$\diamond$E*, 1, 6
12.	$q \,\&\, \neg q$	$\square$E, 11
13.	q	&E, 12
14.	$\neg q$	&E, 12

1.	$p \diamond \rightarrow q$	A
2.	Show $p \square \rightarrow q$	
3.	$\neg(p \square \rightarrow \neg q)$	Df$\diamond \rightarrow$, 1
4.	$p \square \rightarrow \neg\neg q$	S, 3
5.	$p \square \rightarrow q$	$\neg\neg$, 4

We need the hypothesis that $\mathscr{A}$ is possible to deduce $\mathscr{A} \diamond \rightarrow \mathscr{B}$ from $\mathscr{A} \square \rightarrow \mathscr{B}$, although we can deduce $\mathscr{A} \square \rightarrow \mathscr{B}$ from $\mathscr{A} \diamond \rightarrow \mathscr{B}$ alone. Lewis's definition of *might* in terms of *would* thus makes little sense in CS; it makes *might* and *would* almost equivalent.

In CS, we'll define *might* in terms of *would* somewhat differently:

CS Definition of *Might*-Counterfactual (CSDf◇→)

n. $\dfrac{\mathscr{A} \diamondsuit\to \mathscr{B}}{}$

m. $\diamondsuit(\mathscr{A} \,\square\to \mathscr{B})$ CSDf◇→, n

This rule inverts. To see how to use it, let's show that $p \diamondsuit\to (q \,\&\, r)$ implies $p \diamondsuit\to r$.

1.	$p \diamondsuit\to (q \,\&\, r)$	A
2.	Show $p \diamondsuit\to r$	
3.	$\diamondsuit(p \,\square\to (q \,\&\, r))$	CSDf◇→, 1
4.	Show $(p \,\square\to (q \,\&\, r)) \,\square\to \diamondsuit(p \,\square\to r)$	
	$\square\to$	
5.	$p \,\square\to (q \,\&\, r)$	A$\square$→P
6.	Show $p \,\square\to r$	
	$\square\to$	
7.	p	A$\square$→P
8.	$q \,\&\, r$	$\square$→R2, 5 (7)
9.	r	&E, 8
10.	$\diamondsuit(p \,\square\to r)$	◇I, 6
11.	$\diamondsuit(p \,\square\to r)$	◇E*, 3, 4
12.	$p \diamondsuit\to r$	CSDf◇→, 11

Finally, in CS a counterfactual with a true antecedent has whatever truth value its consequent has. Normally, we presuppose that the antecedents of counterfactuals are false. But, if the antecedent happens to be true, we don't tend to reject the counterfactual as meaningless, truth-valueless, or otherwise deviant. In fact, counterfactuals with true antecedents and consequents seem to be true, while those with true antecedents and false consequents are false.

That a counterfactual with a true antecedent and a false consequent is false follows from very general facts about conditionals. If such a counterfactual were true, then $\square$→E—which allows us to move from $\mathscr{A}$ and $\mathscr{A} \,\square\to \mathscr{B}$ to $\mathscr{B}$—would not be a sound rule. But the claim that any counterfactual with a true antecedent and a true consequent is true requires additional support. There is linguistic evidence for it. Consider these questions and answers:

(11) a. Q: If you were to accept the job in Riverside, would you sell your house?

     A1: Yes; I have accepted the Riverside job, and I am selling my house.

     A2: No; I have accepted the Riverside job, but I'm not selling the house.

b. Q: If Caesar had crossed the Rubicon, would he have become consul?
   A: Yes; in fact, Caesar did cross the Rubicon, and he did become consul.
c. Q: If the Russians had withdrawn from World War I, would the Russian Revolution have failed?
   A: No; the Russians did withdraw, but the revolution succeeded.

(11)b suggests that the counterfactual 'If Caesar had crossed the Rubicon, he would have become consul' is true, since Caesar did cross the Rubicon and did become consul. And (11)c indicates that the counterfactual 'If the Russians had withdrawn from World War I, the Russian Revolution would have failed' is false, because the Russians did withdraw, but the revolution nevertheless succeeded. Thus, (11)a shows that, if you've already accepted the Riverside job, the truth value of the counterfactual 'If you were to accept the Riverside job, you'd sell your house' depends solely on whether you sell your house.

Stalnaker's rule, at first glance, seems to say nothing about such counterfactuals. But, as the following proof shows, $p \,\square\!\!\rightarrow q$ must be true if both p and q are true.

$$
\begin{array}{lll}
1. & p & \text{A} \\
2. & q & \text{A} \\
3. & \text{Show } p \,\square\!\!\rightarrow q & \\
4. & \quad \neg(p \,\square\!\!\rightarrow q) & \text{AIP} \\
5. & \quad p \,\square\!\!\rightarrow \neg q & \text{S, 4} \\
6. & \quad \neg q & \square\!\!\rightarrow\text{E, 1, 5} \\
7. & \quad q & \text{R, 2}
\end{array}
$$

Adding Stalnaker's rule and CSDf$\diamond\!\!\rightarrow$ to system C thus yields system CS.

Problems

Show that each of the following is valid in CS.

1. $(p \,\square\!\!\rightarrow q) \rightarrow (p \,\diamond\!\!\rightarrow q)$

2. $(p \,\square\!\!\rightarrow (q \vee r)) \rightarrow ((p \,\square\!\!\rightarrow q) \vee (p \,\square\!\!\rightarrow r))$

3. $(p \,\&\, q) \rightarrow (p \,\square\!\!\rightarrow q)$

4. $(\diamond p \,\&\, \square\neg q) \rightarrow (p \,\square\!\!\rightarrow (\neg p \vee \neg q))$

▶ **5.** $p \rightarrow ((p \,\square\!\!\rightarrow q) \leftrightarrow q)$

6. $((p \,\&\, \neg r) \,\&\, (q \dashv3 r)) \rightarrow (p \,\square\!\!\rightarrow \neg q)$

7. $(p \,\square\!\!\rightarrow (p \rightarrow q)) \rightarrow (\diamond p \rightarrow (p \,\square\!\!\rightarrow q))$

8. $(p \,\square\!\!\rightarrow (q \rightarrow r)) \rightarrow ((p \,\square\!\!\rightarrow \neg q) \vee (p \,\square\!\!\rightarrow r))$

9. $p \,\&\, \neg r$
  $(p \,\&\, q) \,\square\!\!\rightarrow r$
  $\therefore p \,\square\!\!\rightarrow \neg q$

▶ **10.** $p \rightarrow\!\!\!3\ q$
  $p \vee q$
  $\neg(p \,\square\!\!\rightarrow r)$
  $\therefore (q \,\square\!\!\rightarrow \neg p) \vee (q \,\square\!\!\rightarrow \neg r)$

11. $(p \,\&\, q) \,\square\!\!\rightarrow r$
  $(p \,\&\, \neg q) \,\square\!\!\rightarrow s$
  $\therefore (r \vee s) \vee \neg p$

12. $(p \,\&\, q) \,\square\!\!\rightarrow r$
  $q \,\square\!\!\rightarrow p$
  $\neg q \,\square\!\!\rightarrow s$
  $\therefore s \vee r$

13. $\neg q \,\square\!\!\rightarrow s$
  $q \,\square\!\!\rightarrow r$
  $\therefore \neg r \rightarrow s$

14. $p \,\square\!\!\rightarrow s$
  $s \rightarrow\!\!\!3\ p$
  $s \,\square\!\!\rightarrow (q \vee r)$
  $\therefore (p \,\square\!\!\rightarrow q) \vee (p \,\square\!\!\rightarrow r)$

▶ **15.** $p \,\square\!\!\rightarrow (q \vee r)$
  $(p \,\&\, q) \,\square\!\!\rightarrow s$
  $(p \,\&\, r) \,\square\!\!\rightarrow s$
  $\therefore p \,\square\!\!\rightarrow s*$

16. $p \,\square\!\!\rightarrow (\neg q \rightarrow r)$
  $(p \,\&\, q) \,\square\!\!\rightarrow s$
  $(p \,\&\, r) \,\square\!\!\rightarrow t$
  $\therefore p \,\square\!\!\rightarrow (s \vee t)*$

17. $p \,\square\!\!\rightarrow q$
  $q \,\square\!\!\rightarrow r$
  $r \rightarrow\!\!\!3\ t$
  $p \,\&\!\!3\ s$
  $s \,\square\!\!\rightarrow \neg t$
  $\therefore \neg \lozenge s \vee (q \,\square\!\!\rightarrow \neg p)*$

18. $p \,\square\!\!\rightarrow (q \vee r)$
  $(p \,\&\, q) \,\square\!\!\rightarrow s$
  $(p \,\&\, r) \,\square\!\!\rightarrow t$
  $s \,\square\!\!\rightarrow p$
  $s \,\square\!\!\rightarrow t$
  $\therefore p \,\square\!\!\rightarrow t*$

10.4 System CL

To capture Lewis's theory of counterfactuals, we need to add to C two more variants of reiteration. Both involve *might*-counterfactuals. Before presenting those rules, therefore, let's examine Lewis's definition of *might*.

Let $\lozenge \!\rightarrow$ represent *if . . . might* just as $\square\!\!\rightarrow$ represents *if . . . would*. How can we define $\lozenge \!\rightarrow$ in terms of $\square\!\!\rightarrow$? Think about how to negate *might* conditionals. Suppose it's not true that, if Nicholas were to write the proposal, it might be approved. Then, it seems, if Nicholas were to write the proposal, it wouldn't be approved. To say that it's false that Japan might have invaded California if the U.S. had lost at Midway seems to amount to saying that, if the U.S. had been defeated at the battle of Midway, Japan still wouldn't have invaded California. In general, then, sentences of the form 'if it were the case that A, it might be the case that B' appear to be equivalent to negations of sentences of the form 'if it were the case that A, it wouldn't have been the case that B.' So we can define $\lozenge \!\rightarrow$:

CL Definition of *Might*-Counterfactual (CLDf◇→)

$$
\begin{array}{ll}
\text{n.} & \mathscr{A} \diamond\!\!\rightarrow \mathscr{B} \\
\hline
\text{m.} & \neg(\mathscr{A} \,\square\!\!\rightarrow\, \neg\mathscr{B}) \qquad \text{CLDf}\diamond\!\!\rightarrow, \text{n}
\end{array}
$$

This rule is invertible; the implication holds in both directions. According to this definition, the sentences in each of these pairs are contradictory:

(12) a. This match would light if it were struck.
   b. This match might not light if it were struck.

(13) a. If Mitch were to have an affair, Harriet would divorce him.
   b. If Mitch were to have an affair, Harriet might not divorce him.

(14) a. Had Hitler attacked at Dunkirk, he would have been able to take all of Europe.
   b. Had Hitler attacked at Dunkirk, he still might not have been able to take all of Europe.

The definition of $\diamond\!\!\rightarrow$ in terms of $\square\!\!\rightarrow$ determines the truth conditions of formulas of the form $\mathscr{A} \diamond\!\!\rightarrow \mathscr{B}$, given an interpretation of $\mathscr{A}$ and $\mathscr{B}$:

$\mathscr{A} \diamond\!\!\rightarrow \mathscr{B}$ is true in a world, w, iff $\mathscr{B}$ is true in at least one of the worlds that make $\mathscr{A}$ true and are closest to w.

In other words, $\mathscr{A} \diamond\!\!\rightarrow \mathscr{B}$ is true just in case $\mathscr{B}$ is true in one of the closest $\mathscr{A}$-worlds. So, assume that this match might not light if it were struck. If we change the world only as much as we must to make it true that the match is struck, then among the possible outcomes is a world in which the match does not light.

To use the definition in a proof, let's show that $p \diamond\!\!\rightarrow (q \,\&\, r)$ implies $p \diamond\!\!\rightarrow r$.

$$
\begin{array}{lll}
\text{1.} & p \diamond\!\!\rightarrow (q \,\&\, r) & \text{A} \\
\text{2.} & \text{Show } p \diamond\!\!\rightarrow r & \\
\text{3.} & \quad \text{Show } \neg(p \,\square\!\!\rightarrow\, \neg r) & \\
\text{4.} & \quad\quad p \,\square\!\!\rightarrow\, \neg r & \text{AIP} \\
\text{5.} & \quad\quad \neg(p \,\square\!\!\rightarrow\, \neg(q \,\&\, r)) & \text{CLDf}\diamond\!\!\rightarrow, 1 \\
\text{6.} & \quad\quad \text{Show } p \,\square\!\!\rightarrow\, \neg(q \,\&\, r) & \\
 & \quad\quad\quad \square\!\!\rightarrow & \\
\text{7.} & \quad\quad\quad p & \text{A}\square\!\!\rightarrow\text{P} \\
\text{8.} & \quad\quad\quad \neg r & \square\!\!\rightarrow\text{R2, 4 (7)} \\
\text{9.} & \quad\quad\quad \neg q \vee \neg r & \vee\text{I, 8} \\
\text{10.} & \quad\quad\quad \neg(q \,\&\, r) & \neg\&, 9 \\
\text{11.} & \quad p \diamond\!\!\rightarrow r & \text{CLDf}\diamond\!\!\rightarrow, 3
\end{array}
$$

CL contains four reiteration rules. We've already seen two—□→R1 and □→R2—in section 10.2. The third reiteration rule justified by Lewis's approach to counterfactuals pertains to counterfactuals with true antecedents. As we saw in the last section, a counterfactual with a true antecedent seems to have the truth value of its consequent. In Lewis's theory, this is true, apparently, whether the counterfactual contains *would* or *might*. We can express this information in a truth table:

$\mathscr{A}$	$\mathscr{B}$	$\mathscr{A}\,\square\!\rightarrow\mathscr{B}$	$\mathscr{A}\,\diamond\!\rightarrow\mathscr{B}$
T	T	T	T
T	F	F	F
F	T	?	?
F	F	?	?

If counterfactuals with true antecedents have the truth values of their consequents, then we can justify a third version of reiteration.

Counterfactual Reiteration 3 (□→R3)

n.	$\mathscr{A}$	
m.	$\mathscr{B}$	
p.	Show $\mathscr{A}\,\square\!\rightarrow\mathscr{C}$	
q.	$\underline{\mathscr{A}}$	A□→P
q + r.	$\mathscr{B}$	□→R3, m (n, q)

According to this rule, any formula at all reiterates into a counterfactual proof if the assumption for that proof appears free above. That is, if we know that the antecedent of the counterfactual is true, then all we need to establish the counterfactual is the truth of its consequent. We don't need to move to another world, for the actual world is the closest $\mathscr{A}$-world if $\mathscr{A}$ is actually true.

To illustrate, let's show that p and q imply $p \,\square\!\rightarrow q$.

1.	p	A
2.	q	A
3.	Show $p \,\square\!\rightarrow q$	
	□→	
4.	$\lceil p$	A□→P
5.	$\lfloor q$	□→R3, 2 (1, 4)

Because the assumption for the counterfactual proof, p, is free on line 1, we can reiterate anything we like into the proof. In particular, we can reiterate q to get the consequent of the counterfactual $p \,\square\!\rightarrow q$.

The fourth rule allows us to apply counterfactuals. But it is more complicated to state and to justify than $\Box\rightarrow$R2.

Counterfactual Reiteration 4 ($\Box\rightarrow$R4)

n. $\quad\mathscr{A}\,\Box\rightarrow\mathscr{B}$

$\quad\quad\Box\rightarrow$

n + p. $\mathscr{B}\quad\quad\quad\quad\Box\rightarrow$R4, n ($m_1, \ldots, m_n$)

Here (a) $\mathscr{A}$ must be among the free formulas in the counterfactual proof into which the rule is being applied, and (b) $\mathscr{A}\,\Diamond\rightarrow\mathscr{C}$ must be free, where $\mathscr{C}$ is the A$\Box\rightarrow$P in that proof.

Here $m_1, \ldots, m_n$ are the lines on which the free formulas in the proof appear, together with the line on which $\mathscr{A}\,\Diamond\rightarrow\mathscr{C}$ appears. This rule countenances inferences of the general form

$$\mathscr{A}\,\Box\rightarrow\mathscr{B}$$
$$\text{Show }\mathscr{D}_1\,\Box\rightarrow\mathscr{E}$$
$$\mathscr{D}_1$$
$$\mathscr{D}_2$$
$$\vdots$$
$$\mathscr{D}_n$$
$$\mathscr{B}$$

if $\mathscr{A}$ is some $\mathscr{D}_i$ among $\mathscr{D}_1, \ldots, \mathscr{D}_n$ and the formula $\mathscr{A}\,\Diamond\rightarrow\mathscr{D}_1$ is free.

A simple example will make the rule clearer. We've already seen that we can prove $p\,\Box\rightarrow r$ from $p\,\Box\rightarrow q$, $q\,\Box\rightarrow p$, and $q\,\Box\rightarrow r$. The third reiteration rule also allows us to deduce $p\,\Box\rightarrow r$ from $p\,\Box\rightarrow q$, $q\,\Diamond\rightarrow p$, and $q\,\Box\rightarrow r$.

1. $p\,\Box\rightarrow q$ $\quad\quad$ A
2. $q\,\Diamond\rightarrow p$ $\quad\quad$ A
3. $q\,\Box\rightarrow r$ $\quad\quad$ A
4. Show $p\,\Box\rightarrow r$

$\quad\quad\Box\rightarrow$

5. $\lceil p$ $\quad\quad\quad\quad$ A$\Box\rightarrow$P
6. $\mid q$ $\quad\quad\quad\quad$ $\Box\rightarrow$R2, 1 (5)
7. $\lfloor r$ $\quad\quad\quad\quad$ $\Box\rightarrow$R4, 3 (2, 5, 6)

We begin by using counterfactual proof to derive $p\,\Box\rightarrow r$. So, we assume p. Appearing above is $p\,\Box\rightarrow q$, so we can deduce q by $\Box\rightarrow$R2; p is the assumption for counterfactual proof. Finally, we use $\Box\rightarrow$R4 to apply $q\,\Box\rightarrow r$.

We can do this because we've deduced q in the counterfactual proof, and we have $q \diamond\!\!\rightarrow p$ free on line 2.

Why does $\square\!\!\rightarrow$R4 work? Given our understanding of the meanings of counterfactuals, in other words, why is this rule sound? If we have $\mathcal{A} \square\!\!\rightarrow \mathcal{B}$, for example, we know that $\mathcal{B}$ is a counterfactual consequence of $\mathcal{A}$. If $\mathcal{C}$ is the assumption for counterfactual proof, we need to be able to conclude that $\mathcal{B}$ is also a counterfactual consequence of $\mathcal{C}$. This conclusion will be true if all counterfactual consequences of $\mathcal{A}$ are also counterfactual consequences of $\mathcal{C}$. It will be true, in other words, if any formula true in the closest $\mathcal{A}$-worlds is also true in the closest $\mathcal{C}$-worlds. This holds if the collection of the closest $\mathcal{A}$-worlds includes all the closest $\mathcal{C}$-worlds. Since $\mathcal{A}$ is inside the counterfactual proof, $\mathcal{A}$ is a counterfactual consequence of $\mathcal{C}$; $\mathcal{C} \square\!\!\rightarrow \mathcal{A}$ is true. So the closest $\mathcal{C}$-worlds are $\mathcal{A}$-worlds. If we also know that $\mathcal{A} \diamond\!\!\rightarrow \mathcal{C}$, then we know that one of the closest $\mathcal{A}$-worlds is a $\mathcal{C}$-world. Because the closest $\mathcal{C}$-worlds are $\mathcal{A}$-worlds, there can't be any $\mathcal{C}$-worlds closer than that world. And this justifies deriving $\mathcal{B}$ within the counterfactual proof.

Adding CLDf$\diamond\!\!\rightarrow$, $\square\!\!\rightarrow$R3, and $\square\!\!\rightarrow$R4 to C gives us Lewis's theory of counterfactual conditionals, CL.

Problems

Show that each of the following is valid in CL.

    1. $(p \,\&\, q) \rightarrow (p \square\!\!\rightarrow q)$

    2. $(p \,\&\, \neg q) \rightarrow \neg(p \square\!\!\rightarrow q)$

    3. $(p \,\&\, \neg q) \square\!\!\rightarrow (p \square\!\!\rightarrow \neg q)$

    4. $p \rightarrow ((p \square\!\!\rightarrow q) \vee (p \square\!\!\rightarrow \neg q))$

▶   **5.** $p \rightarrow ((p \square\!\!\rightarrow q) \leftrightarrow q)$

    6. $p \rightarrow ((p \square\!\!\rightarrow q) \leftrightarrow (p \diamond\!\!\rightarrow q))$

    7. $\square p \rightarrow ((p \dashv 3\, q) \leftrightarrow \square(p \diamond\!\!\rightarrow q))$

    8. $\diamond p \rightarrow (\diamond(p \square\!\!\rightarrow q) \rightarrow \diamond q)$

    9. $(q \,\&\, \neg \diamond p) \rightarrow (\neg p \diamond\!\!\rightarrow q)$

▶ **10.** $((p \square\!\!\rightarrow q) \,\&\, \neg q) \rightarrow (\neg q \square\!\!\rightarrow \neg p)$

11. q	**12.** q
$\neg p \square\!\!\rightarrow \neg q$	$\neg p \square\!\!\rightarrow \neg q$
$\therefore q \square\!\!\rightarrow p$	$p \dashv 3\, r$
	$\therefore q \square\!\!\rightarrow r$
13. p	**14.** $p \square\!\!\rightarrow q$
$q \dashv 3\, r$	$\neg q \,\&\, \neg r$
$\neg q \square\!\!\rightarrow \neg p$	$\therefore \diamond \neg p$
$\therefore p \square\!\!\rightarrow r$	

15. q
$\neg p \;\square\!\!\rightarrow \neg q$
$\therefore (p \;\square\!\!\rightarrow q) \& (q \;\square\!\!\rightarrow p)$

16. $q \& \neg r$
$\neg r \;\diamondsuit\!\!\rightarrow p$
$\therefore p \;\diamondsuit\!\!\rightarrow q$

17. $q \;\diamondsuit\!\!\rightarrow \neg p$
q
$\therefore \neg p \;\diamondsuit\!\!\rightarrow q$

18. $p \& \neg r$
$q \;\dashv\; r$
$\therefore \neg q \;\diamondsuit\!\!\rightarrow p$

19. q
$p \;\dashv\; r$
$q \;\diamondsuit\!\!\rightarrow \neg r$
$\therefore \neg p \;\diamondsuit\!\!\rightarrow q$

▶ **20.** $p \;\diamondsuit\!\!\rightarrow q$
$p \;\square\!\!\rightarrow r$
$\neg r$
$\therefore \neg r \;\square\!\!\rightarrow \neg(p \& q)$

21. $p \;\square\!\!\rightarrow q$
$\therefore p \;\square\!\!\rightarrow (p \;\square\!\!\rightarrow q)$

22. $r \& (p \;\diamondsuit\!\!\rightarrow \neg q)$
$p \;\square\!\!\rightarrow \neg r$
$\therefore p \rightarrow q$

23. $p \;\square\!\!\rightarrow r$
$p \;\diamondsuit\!\!\rightarrow q$
p
$\therefore q \rightarrow r$

24. $p \;\diamondsuit\!\!\rightarrow q$
$r \;\dashv\; s$
$p \;\square\!\!\rightarrow r$
$\therefore (p \& q) \;\square\!\!\rightarrow s$

▶ **25.** $r \;\dashv\; s$
$(p \& q) \;\diamondsuit\!\!\rightarrow \neg s$
$p \;\diamondsuit\!\!\rightarrow q$
$\therefore p \;\diamondsuit\!\!\rightarrow \neg r$

26. $(p \& s) \;\square\!\!\rightarrow r$
$(p \& s) \;\diamondsuit\!\!\rightarrow q$
$\neg r$
$\therefore \neg r \;\square\!\!\rightarrow (\neg p \vee (\neg s \vee \neg q))$

27. $(p \vee q) \;\square\!\!\rightarrow r$
$(p \vee q) \;\diamondsuit\!\!\rightarrow p$
$\therefore p \;\square\!\!\rightarrow r$

28. $(p \vee q) \;\square\!\!\rightarrow r$
$(p \vee q) \;\diamondsuit\!\!\rightarrow p$
$(p \vee q) \;\diamondsuit\!\!\rightarrow q$
$\therefore (p \;\square\!\!\rightarrow r) \& (q \;\square\!\!\rightarrow r)$

29. $p \;\diamondsuit\!\!\rightarrow q$
$q \;\square\!\!\rightarrow p$
$q \;\diamondsuit\!\!\rightarrow \neg r$
$\therefore p \;\diamondsuit\!\!\rightarrow \neg r$

30. $q \;\diamondsuit\!\!\rightarrow p$
$p \;\diamondsuit\!\!\rightarrow \neg r$
$q \;\square\!\!\rightarrow r$
$\therefore p \;\diamondsuit\!\!\rightarrow \neg q$

31. $p \;\diamondsuit\!\!\rightarrow q$
$p \;\square\!\!\rightarrow s$
$(p \& q) \;\diamondsuit\!\!\rightarrow r$
$\therefore (p \& r) \;\square\!\!\rightarrow s$

32. $p \;\diamondsuit\!\!\rightarrow q$
$p \;\square\!\!\rightarrow (q \rightarrow (r \rightarrow s))$
$(p \& q) \;\diamondsuit\!\!\rightarrow r$
$\therefore ((p \& q) \& r) \;\square\!\!\rightarrow s$

33. $p \;\diamondsuit\!\!\rightarrow q$
$(p \& r) \;\diamondsuit\!\!\rightarrow s$
$(p \& q) \;\diamondsuit\!\!\rightarrow r$
$\therefore p \;\diamondsuit\!\!\rightarrow (r \& s)$

34. $p \;\diamondsuit\!\!\rightarrow q$
$(p \& q) \;\diamondsuit\!\!\rightarrow r$
$p \;\square\!\!\rightarrow (r \rightarrow s)$
$\therefore (p \& r) \;\square\!\!\rightarrow s$

▶ **35.** $p \& \neg r$
$(p \& q) \;\square\!\!\rightarrow r$
$\therefore p \;\square\!\!\rightarrow \neg q$

36. $\diamondsuit(p \& r)$
$(p \& q) \;\square\!\!\dashv\; \neg r$
$\therefore \diamondsuit(p \;\square\!\!\rightarrow \neg q)$

37. $q \rightarrow\!\!\!3\ r$
$\neg r\ \&\ p$
$\therefore\ p\ \square\!\!\rightarrow \neg q$

38. $p\ \square\!\!\rightarrow q$
$p\ \square\!\!\rightarrow r$
$\neg r\ \&\ q$
$\neg p\ \square\!\!\rightarrow \neg q$
$\therefore\ \neg p\ \&\ \neg q\ \&\ \neg s$

39. $r \rightarrow\!\!\!3\ s$
$(p\ \&\ q)\ \square\!\!\rightarrow r$
p
$\neg q\ \square\!\!\rightarrow \neg p$
$\therefore\ p\ \square\!\!\rightarrow s$

40. $(p\ \&\ q)\ \square\!\!\rightarrow r$
$\neg r\ \&\ p$
$\neg q\ \square\!\!\rightarrow \neg p$
$\therefore\ s$

41. $(p\ \&\ q)\ \square\!\!\rightarrow r$
$\neg q\ \square\!\!\rightarrow \neg p$
$\neg(p \rightarrow r)$
$\therefore\ \square\neg s \rightarrow \square\neg p$

42. $p\ \square\!\!\rightarrow (q \rightarrow r)$
$(p\ \&\ q)\ \diamond\!\!\rightarrow \neg r$
$\therefore\ p\ \square\!\!\rightarrow \neg q$

43. $p\ \diamond\!\!\rightarrow q$
$(p\ \&\ q)\ \square\!\!\rightarrow r$
$\therefore\ p\ \diamond\!\!\rightarrow (q\ \&\ r)$

44. $p\ \diamond\!\!\rightarrow q$
$(p\ \&\ q)\ \diamond\!\!\rightarrow r$
$\therefore\ p\ \diamond\!\!\rightarrow r$

45. $p \rightarrow\!\!\!3\ q$
$q\ \diamond\!\!\rightarrow \neg r$
$p\ \square\!\!\rightarrow r$
$\therefore\ q\ \diamond\!\!\rightarrow \neg p$

46. $p\ \square\!\!\rightarrow q$
$p\ \diamond\!\!\rightarrow s$
$q \rightarrow\!\!\!3\ r$
$\therefore\ (p\ \&\ r)\ \diamond\!\!\rightarrow s$

47. $p \rightarrow\!\!\!3\ q$
$q\ \square\!\!\rightarrow p$
$q\ \diamond\!\!\rightarrow \neg r$
$\therefore\ p\ \diamond\!\!\rightarrow \neg r$

48. $p\ \boxminus\ q$
$\therefore\ (p\ \diamond\!\!\rightarrow r) \leftrightarrow (q\ \diamond\!\!\rightarrow r)$

49. $p\ \boxminus\ q$
$\therefore\ (r\ \diamond\!\!\rightarrow p) \leftrightarrow (r\ \diamond\!\!\rightarrow q)$

▶ **50.** $p\ \square\!\!\rightarrow q$
$q\ \square\!\!\rightarrow p$
$\therefore\ (p\ \diamond\!\!\rightarrow r) \leftrightarrow (q\ \diamond\!\!\rightarrow r)$

51. $p\ \diamond\!\!\rightarrow q$
$\therefore\ p\ \diamond\!\!\rightarrow (p \rightarrow q)$

52. $p \rightarrow\!\!\!3\ (q\ \&\ r)$
$q\ \diamond\!\!\rightarrow \neg r$
$\therefore\ q\ \diamond\!\!\rightarrow \neg p$

53. $p\ \diamond\!\!\rightarrow q$
$\therefore\ \diamond(p\ \&\ q)$

54. $\diamond(p\ \diamond\!\!\rightarrow q)$
$\therefore\ \diamond(p\ \&\ q)$

55. $p \rightarrow\!\!\!3\ \neg q$
$\diamond p$
$\therefore\ p\ \diamond\!\!\rightarrow \neg q$

56. $\diamond p$
$\square\neg q$
$\therefore\ p\ \diamond\!\!\rightarrow (\neg p \vee \neg q)$

57. $r\ \&\ p$
$q \rightarrow\!\!\!3\ \neg r$
$\therefore\ p\ \diamond\!\!\rightarrow \neg q$

58. $q \rightarrow\!\!\!3\ r$
$p\ \diamond\!\!\rightarrow \neg s$
$(p\ \&\ r)\ \square\!\!\rightarrow s$
$\therefore\ p\ \diamond\!\!\rightarrow \neg q$

59. $p \dashv q$
$p \diamond\!\!\rightarrow r$
$\therefore (q \diamond\!\!\rightarrow \neg p) \vee (q \diamond\!\!\rightarrow r)$

60. $p \dashv q$
$q \diamond\!\!\rightarrow p$
$p \;\square\!\!\rightarrow \neg r$
$\therefore q \diamond\!\!\rightarrow \neg r*$

61. $q \;\square\!\!\rightarrow \neg r$
$p \dashv q$
$p \diamond\!\!\rightarrow r$
$\therefore q \diamond\!\!\rightarrow \neg p*$

62. $r \diamond\!\!\rightarrow \neg s$
$\neg q \dashv r$
$r \;\square\!\!\rightarrow p$
$p \;\square\!\!\rightarrow s$
$\therefore p \diamond\!\!\rightarrow q*$

63. $p \;\square\!\!\rightarrow q$
$p \diamond\!\!\rightarrow \neg s$
$q \;\square\!\!\rightarrow r$
$(p \;\&\; r) \;\square\!\!\rightarrow s$
$\therefore q \diamond\!\!\rightarrow \neg p*$

64. $p \;\square\!\!\rightarrow (p \;\&\; q)$
$p \diamond\!\!\rightarrow s$
$(p \;\&\; q) \;\square\!\!\rightarrow r$
$\therefore (p \;\&\; r) \diamond\!\!\rightarrow s*$

65. $(p \;\&\; q) \;\square\!\!\rightarrow r$
$p \diamond\!\!\rightarrow \neg s$
$r \dashv s$
$\therefore p \diamond\!\!\rightarrow \neg q*$

66. $\diamond q$
$p \dashv q$
$\neg\diamond r$
$\therefore (p \diamond\!\!\rightarrow \neg r) \vee (q \diamond\!\!\rightarrow \neg p)*$

67. $p \;\&\; \neg r$
$(p \;\&\; q) \;\square\!\!\rightarrow r$
$\therefore p \diamond\!\!\rightarrow \neg q*$

68. $(p \;\&\; q) \;\square \dashv \neg r$
$\diamond(p \;\&\; r)$
$\therefore \diamond(p \diamond\!\!\rightarrow \neg q)*$

69. $q \;\square\!\!\rightarrow p$
$r \;\square\!\!\rightarrow q$
$r \diamond\!\!\rightarrow \neg s$
$q \;\square\!\!\rightarrow s$
$p \;\square\!\!\rightarrow r$
$\therefore p \diamond\!\!\rightarrow \neg q*$

▶ **70.** $\neg r \;\square\!\!\rightarrow \neg s$
$p \diamond\!\!\rightarrow s$
$\neg r \;\square\!\!\rightarrow p$
$(p \;\&\; q) \;\square\!\!\rightarrow \neg r$
$\therefore p \diamond\!\!\rightarrow \neg q*$

71. q
$\neg p \;\square\!\!\rightarrow \neg q$
$p \dashv r$
$\square s \dashv \square\neg r$
$\therefore \diamond\neg s*$

72. $p \diamond\!\!\rightarrow \neg s$
p
$(p \;\&\; q) \;\square\!\!\rightarrow r$
$r \dashv s$
$\therefore \neg q \diamond\!\!\rightarrow p*$

73. $q \diamond\!\!\rightarrow p$
$p \;\square\!\!\rightarrow q$
$q \;\square\!\!\rightarrow (r \rightarrow s)$
$p \diamond\!\!\rightarrow r$
$\therefore (p \;\&\; r) \;\square\!\!\rightarrow s*$

74. $p \;\square\!\!\rightarrow r$
$r \diamond\!\!\rightarrow (p \;\&\; q)$
$r \;\square\!\!\rightarrow s$
$p \diamond\!\!\rightarrow q$
$\therefore (p \;\&\; q) \;\square\!\!\rightarrow s*$

75. $q \diamond\!\!\rightarrow p$
$q \;\square\!\!\rightarrow r$
$p \diamond\!\!\rightarrow \neg r$
$s \dashv t$
$p \;\square\!\!\rightarrow s$
$\therefore (p \;\&\; t) \diamond\!\!\rightarrow \neg q*$

76. $p \diamond\!\!\rightarrow q$
$(p \;\&\; q) \diamond\!\!\rightarrow r$
$(p \;\&\; r) \diamond\!\!\rightarrow s$
$(p \;\&\; s) \diamond\!\!\rightarrow t$
$\therefore p \diamond\!\!\rightarrow t*$

77. If the U.S. had entered the League of Nations, it might not have collapsed so quickly. If the U.S. had entered the League of Nations, then the league would have had more international respect. Of course, if it had commanded more international respect, the U.S. would have entered. If the league had enjoyed more international respect, it would have been able to stop Italy from conquering Ethiopia. Thus, if the U.S. had joined the League of Nations, and the league had been able to stop Italy from conquering Ethiopia, then it might not have collapsed so quickly.

78. Interest rates will rise unless Congress enacts a tax increase. But rates would rise even if Congress were to raise taxes. If interest rates were to rise, the unemployment level would rise. The unemployment level will not rise. It follows that, if interest rates were to rise, Congress might not enact a tax increase.

79. If philosophers were to turn away from linguistic philosophy, their thinking would become more relevant to American culture as a whole. If philosophers did abandon linguistic methods and make their thinking more relevant to the culture in its entirety, they would find more of an audience outside philosophy itself. But, if philosophers turned away from linguistic methods, their work might have no more effect outside philosophy than it does now. Thus, philosophers could find more of an audience outside philosophy without having any more effect outside philosophy than they do now.

80. If the earth had been visited recently by extraterrestrial beings, then the government would have kept the information silent although the visits had occurred. If some reported UFO sightings were authentic, then the earth would have been visited recently by extraterrestrials. But if those sightings were authentic, then our current understanding of our place in the universe might be seriously mistaken. Therefore, either our understanding of our place in the universe might be seriously mistaken, or, even if there had been recent extraterrestrial visits to the earth and the government had kept information about them silent, it might be the case that no reported UFO sightings are authentic.

81. If Jill had to pay more than $400 to repair her car, she'd be better off getting a new one. But if Jill had to pay more than $400 to get her car fixed, somebody would be cheating her. If someone were to cheat Jill, she'd be better off with a new car. But, even if Jill were better off with a new car, she might not buy one. So, unless it's true that, if someone were to cheat her, Jill might not buy a new car, it's possible that Jill would be better off with a new car even though she might not have to pay more than $400 to repair her old one.

82. Lynn would get a promotion if Ken did (because of affirmative action rules). And Ken would get a promotion if Kim did (because he's a friend

of the boss). If Ken were to get a promotion, he might not accept it. If he were able to bring in the Bergman account, however, Ken would be promoted, and he'd accept. So, if Lynn were to get a promotion, Ken might not be able to bring in the Bergman account.

83. The central administration wouldn't increase the department's budget if it were to fail to improve its performance. But the department wouldn't perform better if the administration didn't increase its budget allocation. If the department were to fail to improve, then its chairman would be replaced. If the central administration were to refuse to increase the department's budget, it would begin to lose some good people. If the department's chairman were replaced, however, it might not lose any good people. So, the central administration might increase the department's budget if the chairman were replaced.

It's possible to define modalities within the theory of counterfactuals. Lewis suggests several different possible definitions:

$$\Box \mathscr{A} \leftrightarrow$$
   a. $((p \rightarrow p) \Box\!\!\rightarrow \mathscr{A})$
   b. $(\neg \mathscr{A} \Box\!\!\rightarrow (p \,\&\, \neg p))$
   c. $(\neg \mathscr{A} \Box\!\!\rightarrow \mathscr{A})$

$$\Diamond \mathscr{A} \leftrightarrow$$
   d. $(\mathscr{A} \Diamond\!\!\rightarrow (p \rightarrow p))$
   e. $(\mathscr{A} \Diamond\!\!\rightarrow \mathscr{A})$
   f. $\neg(\mathscr{A} \Box\!\!\rightarrow (p \,\&\, \neg p))$

Throughout the following problems, assume that $\Box \mathscr{A} \leftrightarrow \neg \Diamond \neg \mathscr{A}$.

84. Does a imply b? Does a imply c?

85. Does b imply c? Does b imply a?

86. Does c imply b? Does c imply a?

87. Does d imply e? Does d imply f?

88. Does e imply f? Does e imply d?

89. Does f imply e? Does f imply d?

90. Does a imply d? e? f?

91. Does b imply d? e? f?

92. Does c imply d? e? f?

Which of a–f imply the following?

93. $\Box p \rightarrow p^*$

94. $\Box(p \rightarrow p)^*$

95. $(\Box p \,\&\, \Box(p \rightarrow q)) \rightarrow \Box q^*$

96. $\Box p \rightarrow \Box\Box p^*$

97. $\Diamond p \rightarrow \Box\Diamond p^*$

98. $\Diamond\Box p \rightarrow \Box p^*$

99. $\Diamond\Diamond p \rightarrow \Diamond p^*$

100. $\Box p \rightarrow \Diamond p^*$

101. $\boxdot p \rightarrow \Box p*$ **102.** $\Box p \rightarrow \boxdot p*$

103. $\diamondsuit p \rightarrow \diamondsuit p*$ **104.** $\diamondsuit p \rightarrow \diamondsuit p*$

Lewis defines $\mathscr{A} < \mathscr{B}$ (read '$\mathscr{A}$ is more possible than $\mathscr{B}$' or '$\mathscr{A}$ is less farfetched than $\mathscr{B}$') as $\diamondsuit \mathscr{A} \,\&\, ((\mathscr{A} \vee \mathscr{B}) \,\Box\!\rightarrow (\mathscr{A} \,\&\, \neg \mathscr{B}))$. $\mathscr{A} \leq \mathscr{B}$, in turn, is defined as $\neg(\mathscr{B} < \mathscr{A})$. Prove each of the following, where possible.

105. $p \leq p*$

106. $(p < q) \rightarrow (p \leq q)*$

107. $((p < q) \,\&\, (q < r)) \rightarrow (p < r)*$

108. $((p \leq q) \,\&\, (q \leq r)) \rightarrow (p \leq r)*$

109. $(p \,\Box\!\rightarrow q) \leftrightarrow (\diamondsuit p \rightarrow ((p \,\&\, q) < (p \,\&\, \neg q)))*$

110. $(p \,\diamondsuit\!\rightarrow q) \leftrightarrow (\diamondsuit p \,\&\, ((p \,\&\, q) \leq (p \,\&\, \neg q)))*$

The American logician Donald Nute has proposed a logic of counterfactuals in which $\mathscr{A} \,\Box\!\rightarrow \mathscr{B}$ counts as true just in case $\mathscr{B}$ is true in all sufficiently close $\mathscr{A}$-worlds, where this may include more than just the closest $\mathscr{A}$-worlds. Nute advocates this account, in part, to make the following rule sound:

Nute's Rule

n.	$\underline{(\mathscr{A} \vee \mathscr{B}) \,\Box\!\rightarrow \mathscr{C}}$	
m.	$(\mathscr{A} \,\Box\!\rightarrow \mathscr{C}) \,\&\, (\mathscr{B} \,\Box\!\rightarrow \mathscr{C})$	N, n

Although Nute's rule has a good deal of intuitive plausibility, it is legitimate in neither Lewis's nor Stalnaker's theory. Nute's understanding of the meanings of counterfactuals leads him to reject $\Box\!\rightarrow$R4 and $\Box\!\rightarrow$R3, as well as a good part of $\Box\!\rightarrow$R2. Nute can replace $\Box\!\rightarrow$R2 with a rule

$\Box\!\rightarrow$RN

n.	$\mathscr{A} \,\Box\!\rightarrow \mathscr{B}$
	$\Box\!\rightarrow$
n + p.	$\mathscr{B}$

where (a) $\mathscr{A}$ is the A$\Box\!\rightarrow$P of the counterfactual proof, or where $\mathscr{C}$ is the conjunction of the free formulas in that proof, (b) $\mathscr{A}$ is $\mathscr{C}$, or (c) $\mathscr{A} \,\varepsilon\!\!\;\vdash\, \mathscr{C}$ is free. This rule is the same as $\Box\!\rightarrow$R2, but without the clause allowing counterfactual equivalence to establish $\mathscr{A}$ and $\mathscr{C}$ as tantamount.

111. Show, using □→R4, that adding Nute's rule to Lewis's or Stalnaker's full system would force us to count this unacceptable inference as valid:

$$p \,\square\!\rightarrow r$$
$$p < q \quad \text{(for the definition of this, see above)}$$
$$\therefore q \,\square\!\rightarrow r*$$

112. Show, using □→R2, that adding Nute's rule even to our basic system would force us to count this unacceptable inference as valid:

$$p \,\square\!\rightarrow r$$
$$p < q \quad \text{(for the definition of this, see above)}$$
$$\therefore q \,\square\!\rightarrow r*$$

Show that each of the following is valid in Nute's system.

113. $p \,\square\!\rightarrow q$
$r \dashv 3 s$
$(p \,\&\, q) \,\square\!\rightarrow r$
$\therefore (p \lor r) \,\square\!\rightarrow s*$

114. $q \dashv 3 p$
$p \dashv 3 (q \,\&\, r)$
$\therefore (p \lor q) \,\square\!\rightarrow r*$

115. $(p \lor s) \,\square\!\rightarrow q$
$(p \,\&\, q) \,\square\!\rightarrow r$
$\therefore p \,\square\!\rightarrow r*$

116. $q \,\square\!\rightarrow s$
$((p \,\&\, q) \lor (q \,\&\, s)) \,\square\!\rightarrow r$
$p \,\square\!\rightarrow q$
$\therefore (p \lor q) \,\square\!\rightarrow r*$

117. Should we include Nute's rule in a system of counterfactual logic? Explain.**

Notes

[1] See R. Stalnaker, "A Theory of Conditionals," in W. L. Harper, R. Stalnaker, and G. Pearce (eds.), *Ifs* (Dordrecht: D. Reidel, 1981): 41–56.

[2] See D. Lewis, *Counterfactuals* (Cambridge: Harvard University Press, 1973); and D. Lewis, "Counterfactuals and Comparative Possibility," in W. L. Harper, R. Stalnaker, and G. Pearce (eds.), *Ifs* (Dordrecht: D. Reidel, 1981): 57–86.

[3] This is essentially Lewis's truth condition; throughout this chapter we'll accept the "limit assumption." See D. Lewis, *Counterfactuals* (Cambridge: Harvard University Press, 1973): 19–21; and D. Lewis, "Counterfactuals and Comparative Possibility" in W. L. Harper, R. Stalnaker, and G. Pearce (eds.), *Ifs* (Dordrecht: D. Reidel, 1981): 63–64. CL will thus correspond to Lewis's Analysis 2. For arguments against the limit assumption, see R. Stalnaker, "A Defense of Conditional Excluded Middle," and W. L. Harper, "A Sketch of Some Recent Developments in the Theory of Conditionals," in W. L. Harper, R. Stalnaker, and G. Pearce (eds.), *Ifs* (Dordrecht: D. Reidel, 1981): 87–106, 3–40. See also Stalnaker's *Inquiry* (Cambridge: MIT Press, 1984): 140–42.

[4] The systems of this chapter stem from a deduction system Richmond Thomason devised for Stalnaker's approach in "A Fitch-Style Formulation of Conditional Logic," *Logique et Analyse* 52 (1970): 397–412.

11

OBLIGATION

The logical systems we've developed so far neglect a kind of argument that is both very common and very important. Consider, for example, these arguments:

(1) If we use military force, we'll endanger the lives of the hostages. So, if putting the hostages' lives in danger is unacceptable, we shouldn't use military force.

(2) Jones is responsible for the crash only if he ought to have avoided it. Jones could have avoided the crash only if he could have stopped more quickly. But he couldn't stop more quickly. Therefore, Jones isn't responsible.

Intuitively, these arguments seem valid. Yet translating them into modal logic yields obviously invalid argument forms.

(3) a. $p \dashv q$
  $\therefore \neg r \dashv \neg s$
 b. $r \dashv p$
  $\diamond q \dashv \diamond s$
  $\neg \diamond s$
  $\therefore \neg r$

The problem is simple: We can't translate *should, ought,* or *unacceptable* into our symbolic language.

This chapter, therefore, will add to our system of sentential modal logic two new logical connectives. One, O, represents English expressions such as *should* and *ought to.* The other, P, represents expressions such as *it's permissible that, it's acceptable that, it's OK,* and *may.* These connectives allow us to render in symbolic terms arguments falling into the realm of what Aristotle called *practical reasoning:* They pertain to what a person may do or ought to do, given certain goals, abilities, and circumstances. They allow us to treat *moral* arguments, pertaining to what a person may do or ought to do, considered from a moral point of view. Systems for analyzing moral and practical reasoning are called *deontic logics,* from the Greek word δεοντως, meaning *duly* or *as it should be,* which itself comes from the verb meaning *to bind.*[1] The logic of obligation, permission, and similar concepts remains a topic of very active investigation and even controversy today.

11.1 DEONTIC CONNECTIVES

The central notion of deontic logic is obligation. There are various kinds of obligation: moral, practical, legal, religious, prudential, and so on. Some of the following sentences express certain kinds of obligations more readily than others. All of them can express the idea that it ought to be the case that John drop out of school, though they are not quite equivalent:

(4) a. John *should* drop out of school.
  b. John *ought* to drop out of school.
  c. John *has an obligation* to drop out of school.
  d. *It's obligatory that* John drop out of school.
  e. *It's right* for John to drop out of school.
  f. John *has a duty* to drop out of school.
  g. John *is obliged* to drop out of school.
  h. It's John's *responsibility* to drop out of school.
  i. *It's incumbent on* John to drop out of school.
  j. John *must* drop out of school.
  k. *It's necessary* for John to drop out of school.
  l. John *is required* to drop out of school.
  m. John *has* to drop out of school.

Note that some of these sentences—such as c, d, f, g, h, i, and l—strongly suggest that John has some quite specific obligation or responsibility to a particular person, or due to a particular situation. The others seem far more general. In the next chapter, we discuss this contrast in some detail. Notice also that j, k, and m are ambiguous; they could be expressing either deontic necessity—obligation—or a purely modal sense of necessity. As we noted in Chapter 9, many English expressions can express several different kinds of modality. Words such as *must, have to,* and *necessary* can function modally— signifying that it's impossible to do otherwise—or deontically, signifying that it's forbidden, or immoral, or unwise to do otherwise. Similarly, *may* and *can, might,* and *could* are capable of expressing either deontic or modal concepts. We often rely on context in natural language to distinguish deontic notions from other sorts of modality. As the children's game "Mother, may I?" tries to teach, *may* is generally deontic, while *can* and *could* are generally modal. But this is at best a rough guide.

The obligation connective O is singular. We can read it, for convenience, as *it is obligatory that, it ought to be the case that,* or *it should be the case that.* Clearly, obligation is not truth-functional. If we were to try to fill in a truth table capturing its meaning, we could not establish a value for any row, let alone every row.

$\mathscr{A}$	O$\mathscr{A}$
T	?
F	?

Given that the state has recently raised taxes, we may infer neither that it should have raised taxes nor that it shouldn't have. Similarly, from the fact that Liu doesn't own a Mercedes-Benz we can infer neither that Liu should own one nor that Liu shouldn't. Presumably, some things are as they should be, while others aren't. Whether something is or isn't so tells us nothing about whether it *should* be so. The Scottish philosopher David Hume (1711–1776), putting the point more generally, stressed that we cannot infer *ought* from *is.*

Closely related to obligation is a second singulary connective, P, which we can read as *it is permissible that.* This concept appears in natural language in many different guises. Each of the following, for example, can express the idea that it's permissible for Susan to elope. Again, however, they aren't quite equivalent.

(5) a. *It's permissible* for Susan to elope.
  b. *It's acceptable* for Susan to elope.
  c. *It's OK* for Susan to elope.
  d. Susan *is allowed* to elope.
  e. Susan *may* elope.
  f. Susan *can* elope.
  g. Susan *could* elope.
  h. *It's possible* for Susan to elope.

Although a–d are all explicitly deontic, e–h admit either a deontic or a modal interpretation. Whether these should translate as P*p* or as $\Diamond p$ must be decided by context.

Just as possibility may be defined in terms of necessity, permission may be defined in terms of obligation. If it's permissible for Susan to elope, she has no obligation not to. Similarly, if it's OK for Fred to bring his brother to the party, then it's not true that he has to leave his brother at home. In general, something is permissible if its negation isn't obligatory:

DEFINITION. P$\mathscr{A}$ iff $\neg$O$\neg\mathscr{A}$.

Consequently, permission is also nontruth-functional. From the fact that something is or isn't so, we can conclude nothing about whether it is permissible.

These connectives allow us to translate English expressions such as *it is forbidden that* or *it shouldn't be the case that.* Each of these sentences, for example, captures the essential idea behind the commandment 'Thou shalt not covet thy neighbor's ox,' though, again, they're not equivalent:

(6) a. You *are forbidden* to covet your neighbor's ox.
    b. *It is wrong* for you to covet your neighbor's ox.
    c. *It's immoral* to covet your neighbor's ox.
    d. Coveting your neighbor's ox *is forbidden.*
    e. Coveting your neighbor's ox *is prohibited.*
    f. Coveting your neighbor's ox *is wrong.*
    g. Coveting your neighbor's ox *is immoral.*
    h. You *may not* covet your neighbor's ox.
    i. You *aren't allowed* to covet your neighbor's ox.
    j. Coveting your neighbor's ox *is not allowed.*
    k. Coveting your neighbor's ox *is unacceptable.*
    l. You *shouldn't* covet your neighbor's ox.
    m. *It's your duty not* to covet your neighbor's ox.

Notice that b, c, g, and m seem to convey moral obligations explicitly; a, d, i, and j seem to convey legal obligation, or some other obligation deriving from rules and regulations. Forbiddance, too, is a nontruth-functional notion. The last several sentences above suggest ways of defining forbiddance in terms of permission as well as obligation. If it's forbidden for you to cross the line, then crossing the line is not permitted; you have an obligation not to cross it. '$\mathscr{A}$ is forbidden' is equivalent to both $\neg P\mathscr{A}$ and $O\neg\mathscr{A}$.

Armed with our two connectives, we can translate into symbolic notation a great many moral and practical arguments. In doing so, as the above examples suggest, we must beware of certain indeterminacies. Certain words—*can, may, must,* etc.—can bear either deontic or modal significance. But just as modal terms can express several different senses of necessity or possibility, deontic terms themselves vary in significance. At times they carry moral import, as in 'You should not commit murder.' At times, they express merely prudence, that is, saying that something is or isn't in a person's best interests. Consider 'You should ask Evelyn out next weekend' or 'You should apply for that job.' These sentences may advise you to follow a certain course of action because of the benefits it may bring you, not because of its moral quality. Sometimes deontic expressions have legal force: 'You should not pass on the right' and 'You should notify the landlord thirty days in advance' specify the legal status of certain actions or states of affairs without indicating their moral character or the benefits or harms they might bring. Finally, deontic expressions may convey attitudes about practical matters. The questions 'What should I do?' and 'What may I do?' (equivalently, 'What are my options?') may be requests for some practical guidance and advice.

In this regard, deontic expressions are similar to modal ones. Which interpretation we should give a deontic expression seems to be a pragmatic matter, determined by context. Normally, for our logical purposes, the interpretation itself does not matter, so long as it remains constant throughout the argument. If we say, 'John shouldn't drive to Toledo,' and Joe concludes 'It's not OK for John to drive to Toledo,' the validity of the inference doesn't seem to depend on whether we all have in mind moral, legal, prudential, or practical obligations, so long as we all have the same kind of obligation in mind. Provided that the source of obligation remains the same throughout an argument, the specific identity of that source makes little difference to validity. If we are speaking in prudential terms, however, meaning that it would be to John's disadvantage to drive to Toledo, and Joe concludes that John is under some kind of legal obligation to avoid driving to Toledo, then the inference would surely be illegitimate. We can't conclude that something is immoral from the fact that it's illegal, or vice versa. Nor can we infer that something is in a person's best interests from the fact that it is their moral duty. Similar points could be made about the other kinds of obligation. Each is independent of the rest. 'Double parking is illegal here,' 'Double parking here is immoral,' 'Double parking here isn't in your best interests' (or 'is stupid'), and 'All things considered, you shouldn't double park here' are miles apart in meaning. Thus, in any given argument, we must be careful to symbolize one and only one sense of obligation as O. For arguments containing several different senses of obligation, it would be possible to introduce several different obligation connectives: perhaps O_m for moral obligation, O_l for legal obligation, O_p for prudential obligation, and so on. We return to this theme near the end of this chapter.

Another problem in symbolizing sentences concerns conditional obligation sentences. Many statements of obligation, permission, and the like are conditional:

(7) a. If you promised you'd be there, you should go.
   b. You may take the car only if you fill it up with gas.
   c. If you can't handle responsibility, then you shouldn't try to act like a grown-up.

In Chapter 9, we began using the fishhook to symbolize conditionals. In Chapter 10, we introduced another connective to represent conditionals. The appropriateness of these in deontic logic is unclear. Notice that, in each sentence in (7), the deontic expression appears in one portion of the conditional. The placement in English is a good guide to placing the deontic connective in a symbolic formula. So we could symbolize these sentences as

(8) a. $p \dashv3 Oq$
     $p \rightarrow Oq$
     $p \,\Box\!\!\rightarrow Oq$
   b. $Ps \dashv3 q$
     $Ps \rightarrow q$
     $Ps \,\Box\!\!\rightarrow q$
   c. $\neg\Diamond r \dashv3 O\neg s$
     $\neg\Diamond r \rightarrow O\neg s$
     $\neg\Diamond r \,\Box\!\!\rightarrow O\neg s$

Later, we discuss in some detail whether the arrow, the fishhook, or a counterfactual conditional is most appropriate for these sorts of sentences.

The formulas above are in a new symbolic language built by adding deontic connectives to our modal language. Adding the deontic connectives allows us to define a deontic language, DL, that expands the modal language of Chapter 9. O and P are new singulary connectives: If $\mathscr{A}$ is a formula, then O$\mathscr{A}$ and P$\mathscr{A}$ are formulas.

The semantics of this language is somewhat complicated to explain. All classical and modal sentential connectives have their previously defined meanings. Sentence letters, as in modal logic, take truth values in possible worlds. The truth value of a formula in a possible world depends on the truth values of its atomic components in various possible worlds, in a way determined by the formula's connectives.

To provide a semantics for obligation, we need to consider the truth conditions for formulas of the form O$\mathscr{A}$. We need to ask under what circumstances 'It is obligatory that $\mathscr{A}$' is true in a possible world. Our answer entails the notion of an *ideal* world. We may imagine each possible world as looking to certain other worlds for its moral or practical values. These worlds need not be perfect, but they do need to be ideal in some respects. In particular, the worlds that our world holds up as ideal should be ones in which all the obligations that hold in our world are fulfilled. We can imagine ourselves as thinking of how things ought to be and measuring the current state of affairs, or our own behavior, by that yardstick. But not all worlds need have the same ideals. We can perhaps imagine that some worlds are better than ours, in that none of our obligations go unfulfilled there. Perhaps the moral standards of such a world, however, are so high that new obligations arise in that world. In contrast, there may be worlds so destitute that they hold up our world as an ideal.

We can think, then, of each possible world w as associated with a nonempty set of possible worlds that serve as its ideals. The ideal worlds will differ in some respects, but they are the same in fulfilling all the obligations in w. Suppose that murder, for example, is forbidden in our world. Then, in those worlds we maintain as ideals, there are no murders. If we assume that sock colors, however, are indifferent, the ideal worlds will differ in who wears what color socks when. In some ideal worlds, everyone may wear black socks. In another, everyone may wear white. In yet another, everyone may wear just what they do in this world.

This example suggests that something is obligatory in a world just in case it holds in every world that is ideal relative to that world. Something should be the case if it is the case in every ideal world. So, we can define the truth conditions of obligation formulas in this way:

On a given interpretation, O$\mathscr{A}$ is true in world w iff $\mathscr{A}$ is true in all w's ideal worlds.

Once we've defined truth conditions for obligation, we've implicitly done the same for permission, since it's definable in terms of obligation.

> On a given interpretation, P$\mathscr{A}$ is true in world w iff $\mathscr{A}$ is true in at least one of w's ideals.

That is, something is obligatory in our world just in case it holds in every world that's ideal relative to ours. It's permissible in our world just in case it holds in some of our ideal worlds.

Problems

Symbolize the following sentences in DL.

1. John should keep his promise.

2. It's acceptable for Fred to resign quietly.

3. It's wrong for Smith to put the blame on Jackson.

4. It's OK for George to marry Charlotte.

▶ 5. Bertha was wrong to leave unannounced.

6. If you want to, you should go.

7. You may go if you want to.

8. If Mary doesn't want you to come, then you shouldn't.

9. Nancy should take the assignment only if she thinks she can complete it on time.

▶ 10. It's permissible for Verne to send Harriet a copy, unless he agreed to keep the agreement secret.

11. If Donna finds it to her advantage, and if nobody objects, then she should sign with the other company.

12. It makes no difference whether Ken finishes this week.

13. If it's OK for my sister to go, it's OK for me to go.

14. If it's OK for my sister to go, then it should be OK for me to go, too.

▶ 15. If it's wrong for Carl to forget, but OK for Penny, then the rules are unfair.

16. If Xenia should be allowed to graduate, it's wrong for either the chairman or the dean to stop her.

17. You should forget about the past and begin seeing John again if and only if you shouldn't have broken up with him in the first place.

18. Henry shouldn't have sounded so confident, unless he knows something we don't.

19. If Richard buys the house, and can't sell his condominium, he'll be in financial trouble; he ought to start trying to sell the condominium first.

▶ **20.** If you do this only if you ought to do it, then you won't do a good job.

21. Molly should leave, and Julie should keep quiet, unless they both want to upset Grace.

22. Joyce should try to finish the thesis this year, unless it makes no difference whether she finds a job.

23. It does not do to leave a live dragon out of your calculations, if you live near him (J. R. R. Tolkien).

24. He was a good old man, and it was right that he should have his fling (Edwin Arlington Robinson).

▶ **25.** You ought to play it mean. They ought to hate you on the field (Whitlow Wyatt).

26. One should be just as careful in choosing one's pleasure as in avoiding calamities (Chinese proverb).

27. One should absorb the color of life, but should never remember its details. Details are always vulgar (Oscar Wilde).

28. Defeat should never be a source of discouragement, but rather a fresh stimulus (Robert South).

A number of philosophers have thought that the logic of imperatives is essentially the same as the logic of obligation. If this is true, then we could perhaps symbolize sentences such as *Watch your back* as O*p*. Adopting this strategy of symbolization, translate the following sentences into our symbolic language.

29. Be true to your own highest conviction (William Ellery Channing).

▶ **30.** Execute every act of life as though it were thy last (Marcus Aurelius).

31. Do not remove a fly from your friend's forehead with a hatchet (Chinese proverb).

32. Measure not dispatch by the time of sitting, but by the advancement of business (Francis Bacon).

33. Make it your habit not to be critical about small things (Edward Everett Hale).

34. Never invest in anything that eats or needs repairing (Billy Rose).

▶ **35.** Rarely promise, but, if lawful, constantly perform (William Penn).

36. Make the most of the best and the least of the worst (Robert Louis Stevenson).

37. Take time to deliberate; but when the time for action arrives, stop thinking and go in (Andrew Jackson).

38. If it be possible, as much as lieth in you, live peaceably with all men (Romans 12:18).

39. Build broken walls and make every stone count (John Brown).

▶ 40. If you can't win, make the winner look great (Whitt N. Schultz).

41. Never speak evil of anyone if you do not know it for a certainty. If you know it for a certainty, ask yourself, "Why should I tell it?" (Johann Kaspar Lavater).

42. Don't ridicule your employees. If they were perfect, you'd be working for them (M. S. Forbes).

43. Don't find fault, find a remedy (Henry Ford).

44. So live that, when you die, even the undertaker will feel sorry for you (Mark Twain).

▶ 45. Be kind, for everyone you meet is fighting a hard battle (Philo).

46. Go thy way, eat thy bread with joy, and drink thy wine with a merry heart (Ecclesiastes 9:7).

47. Set yourself earnestly to see what you were made to do, and then set yourself earnestly to do it (Phillips Brooks).

48. Be fond of the man who jests at his scars, if you like, but never believe he is being on the level with you (Pamela Hansford Johnson).

49. Give your decisions, never your reasons; your decisions may be right, your reasons are sure to be wrong (William Mansfield).

▶ 50. Do, or do not! There is no try! (Donald F. Glut).

11.2 DEONTIC TABLEAUX

The method of semantic tableaux extends readily to cover our deontic connectives. We'll build our deontic system, D, on the modal logic S5. All the tableau rules of classical and modal sentential logic apply in deontic logic. In addition, however, we need five rules specific to moral and practical reasoning. As usual, we need rules for obligation and permission formulas on the left and others for obligation and permission formulas on the right. Additionally, we need an analogue of the survival rule in modal logic.

Let's begin by considering a formula of the form O$\mathscr{A}$ on the right side of a tableau branch. Under what circumstances can O$\mathscr{A}$ be false? We've said that O$\mathscr{A}$ is true in a world just in case $\mathscr{A}$ is true in each of that world's ideals. So, if O$\mathscr{A}$ is false in a world, $\mathscr{A}$ must be false in one of that world's ideals. The world in which $\mathscr{A}$ is false, however, isn't necessarily the same world as that in which O$\mathscr{A}$ is false. Just as in modal logic, then, we need to shift worlds. But our world shift is more specific, in a sense, than world shifts in modal logic. Here we are shifting, not just to any world, but to a world that is ideal relative to the original world. We've used a straight horizontal line to mark world shifts in modal logic. To mark deontic shifts, from a world to one of its ideals, we can use a horizontal line with circles on each end, called an *ideal world shift line*. The rule for obligation on the right thus takes this form:

Obligation Right (OR)

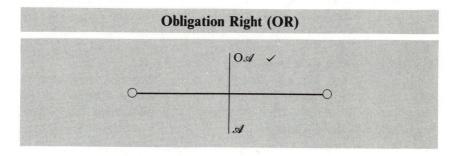

For example, suppose that it's false that you ought to join the religious cult that all your friends have been joining. That means, in semantic terms, that there's an ideal world in which you don't join. The ideal world may be different from this world; perhaps you decide to go ahead and join in the actual world. There may or may not be ideal worlds in which you join. But there must be some ideal world in which you don't; otherwise, you'd have an obligation to join in the real world. This rule looks very much like the rule for necessity on the right, □R. The similarity is not accidental; necessity and obligation are very similar concepts. But there are important differences. In comparing □R and OR, one such difference emerges. If □$\mathscr{A}$ is false, then $\mathscr{A}$ must be false in some possible world. But $\mathscr{A}$ could be false in any world; we can draw no conclusions about how that world relates to the actual world. If O$\mathscr{A}$ is false, then again $\mathscr{A}$ must be false in some possible world. In this case, however, we can say more: $\mathscr{A}$ must be false in a world that is ideal relative to our world. Thus, obligation is a weaker notion than necessity. Knowing that an obligation statement is false gives us more information than knowing that the corresponding necessity statement is false. Knowing that an obligation statement is true accordingly gives us less information than knowing that the corresponding necessity statement is true. □$\mathscr{A}$ asserts, in effect, that $\mathscr{A}$ is true in every possible world. But O$\mathscr{A}$ makes a much more limited claim: that $\mathscr{A}$ is true in every possible world that counts as ideal relative to our world. This diagram should make clear that O$\mathscr{A}$ is weaker than □$\mathscr{A}$.

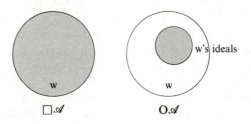

$$□\mathscr{A} \qquad\qquad O\mathscr{A}$$

(Shading indicates worlds in which $\mathscr{A}$ must be true for the formula in question to be true.)

Like the □R rule, OR gives rise to a world shift. Given an obligation formula on the right, we may shift worlds, drop the initial obligation connective, and place the resulting formula on the right side of the branch. This rule does not allow us to cross an already existing ideal world shift bar. As in modal logic, crossing a bar requires a survival rule.

Ideal world shift bars also arise through the application of PL, our rule for permission on the left. If P$\mathscr{A}$ is true, then $\mathscr{A}$ is true in some ideal world. Consequently, if P$\mathscr{A}$ appears on the left, we can draw an ideal shift line and write $\mathscr{A}$ on the left below the line.

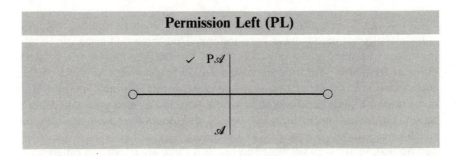

Permission Left (PL)

Which formulas should survive an ideal world shift? As in modal logic, a world shift brings about a striking change of perspective. When we shift from a world, w, to another, w1, we have no reason to expect that the ordinary, contingent matters of fact in w will still hold in w1. Formulas without modal or deontic connectives as main connectives are thus bound to perish. Necessity and possibility, however, look the same from every possible world, assuming S5's hypothesis that all worlds pertain to modal assertions made within any world. This principle remains true whether the worlds involved are ideal or not. So, modally closed formulas survive ideal world shifts as readily as they survive more typical world shifts, regardless of the side of the tableau they occupy.

The real question, then, concerns deontic formulas. Do formulas of the form O𝒜 survive ideal world shifts? Suppose that O𝒜 appears on the left. That means that O𝒜 is true in our original world w, so 𝒜 must be true in each of w's ideals. When we shift from w to its ideal w1, then, we can conclude that 𝒜 is true in w1. We can't, of course, conclude anything about O𝒜 itself. So, obligation formulas on the left survive, but at a cost: They lose their main connectives crossing the bar.

Obligation formulas on the right, in contrast, do not survive at all. Suppose that we know that O𝒜 is false in world w. This tells us that 𝒜 must be false in at least one of w's ideal worlds. But this tells us nothing about whether 𝒜 is true or false in w1 in particular. 𝒜 may well be true in some of the ideal worlds and false in others. Obligation formulas on the right, then, can give rise to world shifts by way of the rule OR but can't survive existing world shift bars on the tableau. Permission formulas, correlatively, can survive only on the right, losing their permission connectives in the process.

Ideal Survival (IS)

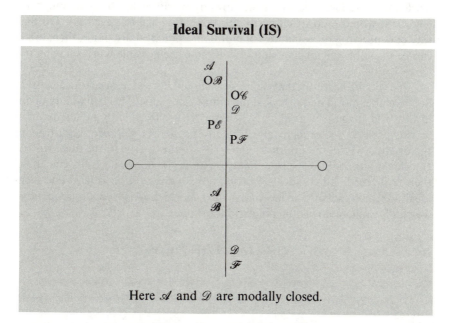

Here 𝒜 and 𝒟 are modally closed.

In summary, the only formulas surviving ideal world shifts are (a) modal formulas; (b) obligation formulas on the left, which lose their main connective, O, in the process; and (c) permission formulas on the right, which lose their main connective P in the process. Obligation formulas on the right, permission formulas on the left, and formulas with truth-functional connectives as main connectives perish.

These three rules suffice for the evaluation of many moral and practical arguments. Indeed, they come very close to capturing every such argument. These rules would have sufficed if we had not stipulated that every world should have some ideal worlds. We specified that the set of worlds ideal relative to w should be nonempty for each world w. This seems like a rather insignificant, even trivial requirement. But it has a startling, powerful consequence: that there can be no insoluble moral or practical dilemmas.

So far, our rules say nothing about such dilemmas. Suppose we place both $O\mathscr{A}$ and $O\neg\mathscr{A}$ on the left of a tableau branch, signifying that, in some world w, both $\mathscr{A}$ and $\neg\mathscr{A}$ are obligatory. This situation is a true dilemma. The tableau so far remains open; in fact, we so far have no way of applying any rules to obligation formulas on the left unless we already have available some ideal world shift line. So the tableau is simply

$$O\mathscr{A} \quad \Big|$$
$$O\neg\mathscr{A} \quad \Big|$$

Semantically, this tableau means that both $\mathscr{A}$ and $\neg\mathscr{A}$ are true in all worlds that are ideal relative to w. Now, no world makes both $\mathscr{A}$ and $\neg\mathscr{A}$ true, so it would follow that no worlds could be ideal relative to w. On this way of conceiving the semantics of moral and practical discourse, then, there can be insoluble dilemmas if and only if there are possible worlds having no ideal worlds.

We've said, however, that each world should have at least one ideal world. So, insoluble moral and practical dilemmas can't arise. The availability of an ideal world suggests a strategy: Whenever we are faced with a dilemma in a world, move to one of its ideal worlds. If $O\mathscr{A}$ and $O\neg\mathscr{A}$ were both true in the original world, then both $\mathscr{A}$ and $\neg\mathscr{A}$ would be true in the ideal world. But this is impossible. The rule that this implies is simple. If $O\mathscr{A}$ is true in w, then $\mathscr{A}$ is true in all w's ideals. Furthermore, there must be at least one such ideal, so there is a world, ideal relative to w, in which $\mathscr{A}$ is true. Consequently, we can formulate a rule for obligation formulas on the left:

Obligation Left (OL)

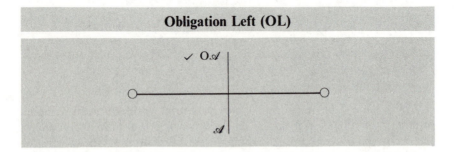

This rule, like OR, gives rise to a world shift. Tableaux treat obligation formulas on the left very differently from those on the right. $O\mathscr{A}$, appearing on the left, survives an ideal world shift as $\mathscr{A}$ and can generate its own world shift. $O\mathscr{A}$ on the right, however, can generate its own world shift but can't survive already existing world shifts.

The same reasoning leads to a rule for permission on the right:

Permission Right (PR)

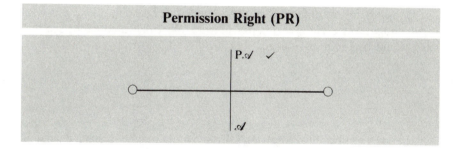

Like OL, this rule gives rise to an ideal world shift.

To see how OL blocks the possibility of a dilemma, consider the above tableau. We can now apply OL to the formula $O\mathscr{A}$, shifting to an ideal world:

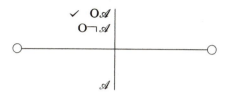

At this point, we can see that $O\neg\mathscr{A}$ survives the ideal world shift but loses its main connective in crossing the bar. So we can apply IS, the ideal survival rule, to reach a closed tableau:

The nature of OL and PR allows us to formulate some simple principles of strategy in constructing tableaux. Obligation formulas on the left can survive shifts generated by such formulas on the right, but those on the right can't survive shifts generated by formulas on the left. Similarly, permission formulas on the right can survive those generated by permission formulas on the left, but permission formulas on the left can't survive ideal world shifts. Consequently:

1. Attempt to apply OR and PL first.

2. Apply IS to see what survives.

3. Apply OL or PR only when no obligation formulas remain live on the right side of the branch.

Consider, for example, the simple argument: 'You ought to see a lawyer; so, you ought to see a lawyer or a psychiatrist.' We can evaluate it by symbolizing it and beginning a tableau:

$$Op \ \big| \ O(p \lor q)$$

If we were to begin by applying OL, we would quickly reach a dead end. We would shift worlds, writing *p* below the bar. But the obligation formula on the right would perish, yielding an open tableau:

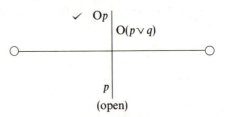

(open)

In deontic logic, as in modal logic, an argument form is valid if and only if it has a closed tableau. Reaching an open tableau thus does not determine that the argument form is invalid. The argument form is invalid only if every way of proceeding yields an open tableau.

Following the strategy rules we've outlined, we begin by applying OR to the formula on the right. This produces a world shift:

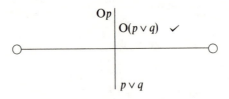

Now, *Op* on the left survives but loses its connective. So, by applying IS and then ∨R, we can close the tableau:

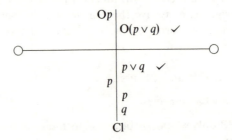

Just as we can think of S5 as advancing several theses about modality, so can we think of our system of deontic logic as advancing several principles about obligation. The first principle deals with conjunction.

Principle 1: ($\mathscr{A}$ & $\mathscr{B}$) is obligatory if and only if $\mathscr{A}$ and $\mathscr{B}$ are each obligatory.

As a second example, then, consider this tableau, which establishes the half of this principle that British philosopher Bernard Williams has called *the agglomeration principle:*

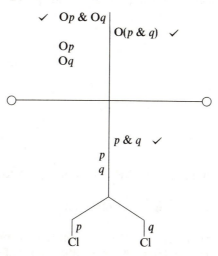

By beginning with &L, we derive two deontic formulas on the left, which then survive the ideal world shift arising from applying OR to the conclusion.

The agglomeration principle is one half of our biconditional relating obligation and conjunction. This tableau shows that the other half is valid as well:

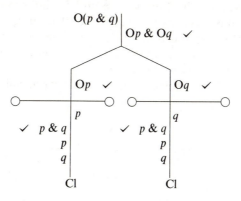

If we had started with the premise on the left, the conjunction on the right would have perished. By working toward a point where we could begin with

OR, we were able to construct two closely analogous tableaux. Note that the premise survives the ideal world shift of each branch.

A fourth example shows that another principle is valid as well. The German philosopher Immanuel Kant (1724–1804) first formulated a thesis that has become famous as the "*ought*-implies-*can*" principle:

Principle 2: Whatever is obligatory is possible.

Equivalently, we could say that nobody can be obligated to do the impossible. The rule OL is crucial to showing that this principle counts as valid in our system:

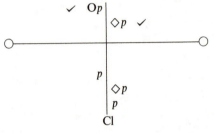

Our system of moral and practical reasoning advances a third principle:

Principle 3: What is necessary is obligatory.

Equivalently, we could say that the impossible is forbidden. Everyone is obliged to obey the laws of logic. This principle is an artifact of our approach to the semantics of moral and practical notions. Though there is some evidence from natural language in its favor, it shouldn't be taken too seriously. $O\mathscr{A}$ is true if $\mathscr{A}$ is true in all ideal worlds. But necessary truths are true in all worlds, and so they are true in all ideal worlds too. So, necessity implies obligation. Of course, the obligations that result have little content: They oblige us to do only what is necessary anyway. Notice that this thesis is not the same as the *ought*-implies-*can* principle, which is equivalent to the thesis that whatever is necessary is permissible. Principle 2 is thus weaker than the claim that whatever is necessary is obligatory.

Let's take another example to show that this principle is valid.

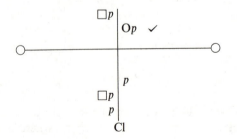

The modal formula crosses the bar with ease. Once again, beginning with OR leads to a closed tableau.

The fourth principle is the final in this series.

Principle 4: The necessary conditions of what is obligatory are themselves obligatory.[2]

Suppose that you have to do B in order to do A. Then, if you ought to do A, you ought to do B, too. You ought, in other words, to do whatever it takes to fulfill your obligations. Symbolizing this notion seems to require the use of modal connectives; we can translate it as follows: If $\mathcal{B}$ is a necessary condition of $\mathcal{A}$, and $O\mathcal{A}$, then $O\mathcal{B}$. If we treat conditionals modally, this translation has an instance $((p \dashv 3 q) \mathbin{\&} Op) \dashv 3 Oq$.

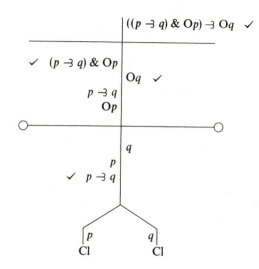

We must begin by applying $\dashv 3 R$ to the formula on the right, the validity of which we're testing. Applying $\dashv 3 R$ results in a standard modal world shift line. After applying &L, we apply OR, giving rise to an ideal world shift. The modal formula on the left survives; Op survives, but as p. The tableau closes, demonstrating the validity of the principle. Notice that the presence of modalities was crucial. If we had not interpreted "necessary condition" modally, the conditional would have perished, and the tableau would have remained open.

Although modal formulas survive ideal world shifts, deontic formulas don't survive ordinary modal world shifts. Adding deontic connectives to the language thus requires no revision in the modal survival rule. Formulas survive modal world shifts just in case they are modally closed. Formulas of the form $O\mathcal{A}$ aren't, so they don't survive modal world shifts.

Problems

Evaluate the validity of these arguments in D using semantic tableaux. If the argument contains statements of conditional obligation, try evaluating the argument (a) translating them with → and (b) translating them with ─3.

1. If we use military force, we'll endanger the lives of the hostages. So, if putting the hostages' lives in danger is unacceptable, we shouldn't use military force.

2. Jones is responsible only if he ought to have avoided the crash. Jones could have avoided the crash only if he could have stopped more quickly. But he couldn't stop more quickly. Therefore, Jones isn't responsible.

3. If you vote for Carol, the city's rate of development will be maintained or accelerated. Since the rate of the city's development should be slowed, you shouldn't vote for Carol.

4. If you vote for Carol, the city's rate of development will be maintained or accelerated. Since the rate of the city's development should be maintained or accelerated, you should vote for Carol.

▶ 5. If we elect Bob, we'll withdraw from the nuclear project. We should withdraw from the project. So, we should elect Bob.

6. If we elect Bob, we'll withdraw from the nuclear project. We shouldn't withdraw from the project. So, we shouldn't elect Bob.

7. If you borrowed the money from John, you should pay him $1,000 by the end of the year. You borrowed the money from him, so you have an obligation to pay him $1,000 by December 31.

8. If you are in graduate school, you should be studying hard. You should be in graduate school, but you aren't. Even so, in consequence, you ought to be studying hard.

9. If we increase incentives, productivity will improve, although it might not improve immediately. We ought to improve productivity. Therefore, we should increase incentives.

▶ 10. If we increase incentives, productivity will improve. So, we ought to increase incentives only if we ought to improve productivity.

11. I promised to take Alice to dinner, and I promised to attend the meeting. If I promised to go to the meeting, I have an obligation to attend; if I promised Alice I would take her to dinner, I should. But I can't take her to dinner and go to the meeting. So, I'm in big trouble!

12. It's permissible to criticize my actions on moral grounds only if I had an obligation to act differently. But I couldn't act any differently. So, it's wrong to criticize my actions on moral grounds.

13. If I said I would be there, I ought to be there. But being there is incompatible with stopping to save the life of an injured accident victim. Thus, if I gave my word to be there, I shouldn't save the life of an injured accident victim.

14. It's permissible, but by no means obligatory, for you to go to college. If you do go, you must devote several years of your life to study. So, it's OK but not obligatory for you to devote several years of your life to study.

15. If you bring about a true revolution, you'll destroy the lives of many innocent people. It's wrong to destroy the lives of many innocent people. So, it's wrong to bring about a true revolution.

16. You can bring about revolutionary reforms only if you destroy the lives of many innocent people. It's not wrong to bring about revolutionary reforms, but you have no duty to do so. Therefore, it's not wrong for you to destroy the lives of innocent people.

17. If you bring about revolution, many innocent lives will be destroyed. You ought to bring about revolution. So, it's permissible to destroy many innocent lives.

18. The marshal can maintain law and order only if he convicts these innocent men and sentences them to death. He has an obligation to maintain law and order. So, he ought to sentence these innocent men to die.

19. The sacrifice of much time and money is a necessary condition for us to complete this project. It's OK for us to complete the project, and so it's acceptable for us to sacrifice much time and money.

▶ 20. If you're like me, you can't resist temptation. So, if you're like me, you should give in to temptation.

21. It's wrong that it's possible for some people to avoid paying the taxes they owe to the government. But it would be possible to stop some people from avoiding their tax burden only if we cast aside our respect for individual liberties and rights to privacy. Therefore, it's necessarily true that it's permissible for us to cast aside our respect for individual liberties and rights to privacy.

22. If Jane is willing to have an abortion, then Jane is willing to kill something that might be a human being. If Jane can't think of any alternatives, she'll be willing to go through with the abortion. So, if Jane doesn't think of any alternatives, it's OK for her to kill something that might be a human being.

23. It's wrong to have an abortion only if a fetus is a human being with a right to life. It's possible that a fetus is a human being with a right to life. Therefore, abortion is wrong.

24. It's permissible to violate the right to life of the fetus only if it's obligatory to violate the fetus's right to life. But the mother will survive in this case if and only if we violate the fetus's right to life. Therefore, it is permissible that we violate the fetus's right to life only if we have an obligation to save the mother's life.

▶ 25. If we've acted too late, the enemy will attack by daybreak. If they attack by then, however, we won't be able to avoid an all-out conflict. So, if we've acted too late, we have no obligation to avoid an all-out conflict.

26. It's not morally wrong for you to treat yourself well by indulging yourself in luxuries. But you have a moral obligation to provide substantial aid to those less fortunate than you. So, providing substantial aid to those less fortunate than you is compatible with treating yourself well by indulging yourself in luxuries.

27. If I blew my part of the assignment, my friends are going to be awfully upset with me. It will be possible for me to rely on their help on the next assignment only if they aren't upset with me. So, I ought to rely on their help next time only if I didn't blow my part of this assignment.

28. If you take out the Smiths' daughter, you'll make your mother happy. Of course, you have no obligation to make your own mother happy. It's impossible, then, that you have any obligation to take out the Smiths' daughter.

29. This act conforms to the principle of utility just in case its tendency to augment the happiness of the community is greater than its tendency to diminish it. If this action conforms to the principle of utility, it ought to be done. So, this act ought to be done if its tendency to augment the community's happiness is greater than its tendency to diminish it.

▶ 30. This act conforms to the principle of utility just in case its tendency to augment the happiness of the community is greater than its tendency to diminish it. If this action conforms to the principle of utility, it is permissible. So, this act is permissible if its tendency to augment the community's happiness is greater than its tendency to diminish it.

31. If you go to Yale, you'll become a very successful lawyer. If you promised Yale you would go, you should. You have no obligation to become a very successful lawyer. Therefore, you didn't promise Yale you'd go there unless you're a fool.

These somewhat bizarre arguments are particular instances of various interpretations of the Golden Rule. Some of the interpretations are plausible; some aren't. Which arguments are valid in D? (**What do the results say about the proper interpretation of the Golden Rule?)

32. I may whip you, if I would be willing to have you whip me. Being a masochist, I'd be willing to have you whip me. Therefore, it's permissible for me to whip you.

33. You ought to put the mentally ill to death if you would wish others to put you to death, if you were mentally ill. If you were mentally ill, you'd wish others to put you to death. Thus, you ought to kill the mentally ill.

34. You should not sustain the life of a terminally ill patient undergoing great suffering if you wouldn't be willing to have others sustain your life in such a situation. You would be willing to have others sustain your life in such a case. So, you are obliged to sustain the life of suffering, terminally ill patients.

35. You may torture others only if you would be willing to be tortured yourself. You would be willing to be tortured. Thus, you may torture others.

36. You have an obligation to remain faithful to your wife only if you would wish your wife to be faithful to you. In fact, you wish your wife were unfaithful. It's all right, then, for you to be unfaithful to her.

37. It's wrong to steal from your employer only if you wouldn't be willing to have him steal from you. You are not willing to have your employer rob you. So, you shouldn't steal from him.

Iterations of deontic connectives are difficult to interpret. In D, are any of the formulas on the left equivalent to their mates on the right?

38. OOp Op **39.** PPp Pp ▶ **40.** OPp Pp

41. OPp Op **42.** POp Op **43.** POp Pp

44. POp OPp

Combinations of deontic and modal connectives also produce some formulas that are hard to interpret. In D, are any of the formulas on the left equivalent to their mates on the right?

45. $\Box$Op Op **46.** $\Diamond$Op Op **47.** O$\Box p$ Op

48. O$\Diamond p$ Op **49.** $\Box$Pp Pp ▶ **50.** $\Diamond$Pp Pp

51. P$\Box p$ Pp **52.** P$\Diamond p$ Pp **53.** $\Box$Op O$\Box p$

54. O$\Diamond p$ $\Diamond$Op ▶ **55.** P$\Box p$ $\Box$Pp **56.** P$\Diamond p$ $\Diamond$Pp

Which of the following formulas are valid in D?

57. O$p \lor$ O$\neg p$ **58.** P$p \lor$ P$\neg p$

59. $\neg$P$p \lor \neg$P$\neg p$ ▶ **60.** $\neg$O$p \lor \neg$O$\neg p$

61. (O$p \lor$ O$\neg p$) $\rightarrow \neg$(Pp & P$\neg p$)   **62.** $\neg$(Pp & P$\neg p$) $\rightarrow$ (O$p \lor$ O$\neg p$)

63. P(p & q) $\rightarrow$ (Pp & Pq)   **64.** (Pp & Pq) $\rightarrow$ P(p & q)

65. P($p \lor q$) $\rightarrow$ (P$p \lor$ Pq)   **66.** (P$p \lor$ Pq) $\rightarrow$ P($p \lor q$)

67. P($p \rightarrow q$) $\rightarrow$ (P$p \rightarrow$ Pq)   **68.** (P$p \rightarrow$ Pq) $\rightarrow$ P($p \rightarrow q$)

69. (O$p \rightarrow$ Oq) $\rightarrow$ O($p \rightarrow q$)   ▶ **70.** O($p \rightarrow q$) $\rightarrow$ (O$p \rightarrow$ Oq)

71. $O(p \lor q) \to (Op \lor Oq)$

72. $(Op \lor Oq) \to O(p \lor q)$

73. $P(p \lor \neg p)$

74. $O(p \lor \neg p)$

▸ 75. $\neg P(p \ \& \ \neg p)$

76. $\neg O(p \ \& \ \neg p)$

77. $P(p \leftrightarrow q) \to (Pp \leftrightarrow Pq)$

78. $(Pp \leftrightarrow Pq) \to P(p \leftrightarrow q)$

79. $O(p \leftrightarrow q) \to (Op \leftrightarrow Oq)$

▸ 80. $(Op \leftrightarrow Oq) \to O(p \leftrightarrow q)$

81. $Op \to Pp$

82. $Pp \to Op$

83. $Op \to \Diamond p$

84. $Pp \to \Diamond p$

85. $\Box p \to Pp$

86. $(p \dashv3 q) \to (Op \dashv3 Oq)$

87. $(Op \dashv3 Oq) \to (p \dashv3 q)$

One of the few plausible candidates for an unconditional obligation statement being a logical truth is $O(Op \to p)$.

88. Show that $O(Op \to p)$ is not valid in D.

89. Revise our tableau rules to make this formula valid.*

90. Do you think that natural-language sentences having the form above are valid?**

11.3 DEONTIC DEDUCTION

Our system of modal deduction from the last chapter extends easily to cover moral and practical reasoning. All the rules of truth-functional sentential logic and of S5 apply. In addition, we'll add two new rules, admit a new method of proof, and extend the reiteration rule to incorporate that proof method.

The first rule is simple: It allows us to introduce or exploit the permission connective by translating formulas containing it into those containing obligation. The rule relies on the definition of permission in terms of obligation.

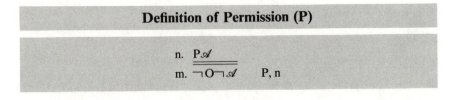

Definition of Permission (P)

n. $P\mathscr{A}$
m. $\neg O \neg \mathscr{A}$ P, n

This rule inverts; it serves to introduce P, as well as exploit it.

The new proof method, which forms the foundation of our system of deontic deduction, is *deontic proof*. It's similar in many ways to modal proof. How can we establish the truth of a statement of obligation? Thinking in semantic terms, we can say that O$\mathscr{A}$ is true in a world if $\mathscr{A}$ is true in all its ideal worlds. So, we can show that O$\mathscr{A}$ is true by showing that $\mathscr{A}$ is true in an arbitrary ideal world. This strategy suggests a method of proof:

Deontic Proof
n. Show O$\mathscr{A}$ $\bigcirc$ $\vdots$ $\mathscr{A}$

Of course, since we're trying to show that $\mathscr{A}$ is true in any ideal world, we must be very careful about what information we bring into the subproof. As a result, we must restrict reiteration into a deontic proof. The circle written after the *Show* line above, like the box written there in a modal proof, indicates that reiteration of formulas across the symbol is not automatic. Not every formula may cross the circle into the deontic proof.

A modal proof demonstrates the truth of $\square\mathscr{A}$ by showing that $\mathscr{A}$ is true in any possible world. The only information we can bring into the proof is that which holds in every possible world. Consequently, modally closed formulas, which are true in every world if they are true in any, reiterate into modal proofs. Other formulas don't. Their truth in one world does nothing to guarantee their truth in other worlds.

We can now clarify how we should restrict reiteration into deontic proofs. We are trying to show that $\mathscr{A}$ is true in an arbitrary possible world that is ideal relative to the original world. We should allow into the deduction, then, only information that holds in every such ideal world. Any modally closed formula true in our original world is true in all possible worlds, and, so, in all ideal worlds. Modally closed formulas, then, should reiterate into deontic proofs without difficulty.

Formulas with obligation as a main connective are slightly more complex. If something of the form O$\mathscr{B}$ is true in our original world, then $\mathscr{B}$ must be true in each of that world's ideals. So O$\mathscr{B}$ can reiterate into the deontic proof, but it loses its main connective in the process. That is, given O$\mathscr{B}$ on a line in a proof, we may write $\mathscr{B}$ in a deontic subproof immediately subordinate to that proof. The qualification "immediately subordinate to that proof" is important here. Since a formula of obligation reiterated into a deontic proof loses its main connective, reiterating the formula into a distant subproof all at once may not yield the same result as reiterating it one step at a time. Since crossing a circle results in the loss of an O connective, a formula loses an O for each circle it crosses.

Formulas that are not modally closed and don't have deontic main connectives don't reiterate at all into deontic proofs. Their truth or falsehood in a world implies nothing about their truth or falsehood in any of its ideal worlds.

Reiteration into Deontic Proofs (OR)

n. $\mathscr{A}$
○
q. $\mathscr{A}$ OR, n

Here $\mathscr{A}$ must be modally closed.

m. O$\mathscr{B}$
○
r. $\mathscr{B}$ OR, m

The deontic proof method, together with the extended reiteration rule, suffice for most deontic proofs. Let's take a very simple case to demonstrate the principle that whatever is necessary is obligatory.

1. $\Box p$ A
2. Show Op
   ○
3. $\lceil \Box p$ OR, 1
4. $\lfloor p$ $\Box$E, 3

To show our conclusion, whose main connective is obligation, we use deontic proof. The premise reiterates into the proof, since it's modally closed. At that point, applying necessity exploitation gives us p. In the context of a deontic proof, this suffices to show Op.

As another example, consider this proof of the distributivity of obligation over conjunction:

1. O(p & q)   A
2. Show Op & Oq
3. $\lceil$ Show Op
    | ○
4. | $\lceil p$ & q OR, 1
5. | $\lfloor p$ &E, 4
6. | Show Oq
    | ○
7. | $\lceil p$ & q OR, 1
8. | $\lfloor q$ &E, 7
9. $\lfloor$ Op & Oq &I, 3, 6

The conclusion of this proof is a conjunction, and so we proceed to prove each conjunct separately. To show that Op is true, we use deontic proof. $O(p$ & $q)$ reiterates into this proof but loses its initial connective, becoming simply p & q. This gives us p, which completes the deontic subproof. We have proved Op here by proving that p is true in any world ideal relative to our original one. We then proceed to do exactly the same for Oq.

This proof demonstrates half of the deontic thesis that a conjunction is obligatory just in case each conjunct is obligatory. The other half is the agglomeration principle.

1.	Op	A
2.	Oq	A
3.	Show $O(p$ & $q)$	
	○	
4.	p	R, 1
5.	q	R, 2
6.	p & q	&I, 4, 5

Another good illustration of a deontic proof is the demonstration that a medieval principle is valid in our system. The principle is that the necessary conditions of what is obligatory are themselves obligatory: You should do whatever it takes to fulfill your obligations. This proof also provides an interesting mixture of modal and deontic proofs.

1.	Show $((p \dashv\mathbf{3} q)$ & $Op) \dashv\mathbf{3} Oq$	
	□	
2.	$(p \dashv\mathbf{3} q)$ & Op	A□P
3.	Op	&E, 2
4.	$p \dashv\mathbf{3} q$	&E, 2
5.	Show Oq	
	○	
6.	$p \dashv\mathbf{3} q$	OR, 4
7.	p	OR, 3
8.	q	$\dashv\mathbf{3}$E, 6, 7

This proof begins like a typical proof in modal logic. To show a formula with a fishhook as main connective, we begin a modal proof, which begins with an assumption. The consequent has obligation as its main connective, so we begin a deontic proof. The modally closed $p \dashv\mathbf{3} q$ reiterates without difficulty, and Op reiterates as p. At that point, we can obtain q by applying $\dashv\mathbf{3}$ E.

Deontic proof, together with extended reiteration, thus constitutes a powerful deduction technique. But they don't suffice to prove every valid deontic formula. We've required that each possible world have a nonempty set of ideal worlds. Each world, then, has some ideals; in no world is moral or practical discourse completely empty. We are never left, in our system, in the circumstances described by the novelist V. S. Naipaul, where "it isn't that there isn't any right or wrong, it's that there isn't any right!" In D, there can be no insoluble moral or practical dilemmas. The question "What should I do?" makes sense, no matter how bad the situation. The answer is never that all possible courses of action are forbidden. So far, however, our proof system doesn't incorporate the requirement that each world have some ideals.

To capture this requirement, we can reflect on the thesis that all moral and practical problems have solutions. For this reason, no moral conflicts are irreconcilable: Never are O$\mathscr{A}$ and O$\neg\mathscr{A}$ both true. Assertions of conflicting obligation should thus be contradictory. Given our thesis about conjunctions, of course, we know that for any formula, $\mathscr{A}$, O$\mathscr{A}$ & O$\neg\mathscr{A}$ and O($\mathscr{A}$ & $\neg\mathscr{A}$) are equivalent. So, in our system, O($\mathscr{A}$ & $\neg\mathscr{A}$) is contradictory: Contradictions are never obligatory. O($\mathscr{A}$ & $\neg\mathscr{A}$), then, has the same logical status as $\mathscr{A}$ & $\neg\mathscr{A}$ itself.

We'll thus introduce a rule of obligation consistency:

Obligation Consistency (OC)

$$\text{n.} \quad \overline{\neg O(\mathscr{A} \ \& \ \neg\mathscr{A})} \qquad \text{OC}$$

Whenever we like in a proof, for any formula $\mathscr{A}$, we can write $\neg O(\mathscr{A} \ \& \ \neg\mathscr{A})$. This rule allows us to derive a rule corresponding to Kant's *ought*-implies-*can* principle:

Obligation Exploitation (OE)

$$\begin{array}{ll} \text{n.} & \dfrac{O\mathscr{A}}{} \\ \text{n + p.} & \Diamond\mathscr{A} \qquad \text{OE, n} \end{array}$$

This rule allows us to replace O's that function as main connectives with diamonds. Both these rules have the effect, then, of making O$\mathscr{A}$ true only if $\mathscr{A}$ is satisfiable. Deriving obligation exploitation from obligation consistency is not difficult. We can show that, if nobody's obligated to do something contradictory, then no one can be obligated to do the impossible.

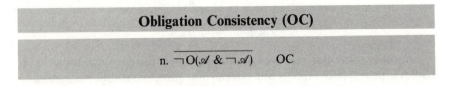

```
1. Op                      A
2. Show ◇p
3.  ┌ ¬◇p                  AIP
4.  │ Show O(p & ¬p)
    │  ◇
5.  │  ┌ p                 OR, 1
6.  │  │ ¬◇p               OR, 3
7.  │  └ ◇p                ◇I, 5
8.  └ ¬O(p & ¬p)           OC
```

Problems

Construct deductions to show that the following argument forms are valid in D.

1. $p ⥽ q$; Pp; ∴ Pq

2. Op; ∴ Pp

3. $p ⥽ q$; $¬Pq$; ∴ $O¬p$

4. $p ⥽ q$; $¬Oq$; ∴ $P¬p$

▶ 5. $¬Pp$; ∴ $P¬p$

6. $O¬p$; ∴ $¬Op$

7. $□p$; ∴ Pp

8. Pp; ∴ $◇p$

9. $□¬p$; ∴ $¬Op$

▶ 10. $□¬p$; ∴ $¬Pp$

11. $p ⥽ Oq$; p; ∴ Oq

12. $p ⥽ Oq$; $◇p$; ∴ $◇Oq$

13. $p ⥽ Oq$; p; ∴ Pq

14. $p ⥽ Oq$; $¬◇q$; ∴ $¬p$

▶ 15. $p ⥽ Oq$; $¬◇q$; ∴ $¬◇p$

16. $p ⥽ Oq$; $P¬q$; ∴ $¬p$

17. $p ⥽ Oq$; $O¬q$; ∴ $¬p$

18. $◇(p ⥽ q)$; Pp; ∴ Pq

19. $◇(p ⥽ q)$; Op; ∴ Oq

▶ 20. $p ⥽ Oq$; $◇r ⥽ O¬q$; ∴ $p ⥽ □¬r$

21. $p ⥽ Oq$; $¬◇(q \& r)$; ∴ $p ⥽ O¬r$

22. $p ⥽ r$; $r ⥽ □¬q$; ∴ $p ⥽ O¬q$

▶ 23. $p → r$; $r → □¬q$; ∴ $p → ¬Pq$

24. $O¬p ⅋ q$; $◇q$; ∴ $◇¬Pp$

25. $p ⅋ q$; ∴ $Op ⅋ Oq$

26. $p ⅋ q$; ∴ $Pp ⅋ Pq$

27. $Op ⅋ q$; $O¬q$; ∴ $O¬Op$

28. $p ⥽ q$; $◇r ⥽ ¬q$; ∴ $Or ⥽ ¬p$

29. $Pp ⥽ Oq$; $¬◇q$; ∴ $O¬p$

30. $p ⥽ q$; $r ⥽ Op$; $¬Oq$; ∴ $¬r$

▶ 31. $p ⥽ Oq$; $◇q ⥽ Or$; $Pr ⥽ ◇s$; ∴ $□¬s ⥽ □¬p$

Alan Ross Anderson proposed reducing deontic logic to modal logic by (a) translating O𝒜 as $¬𝒜 ⥽ s$, where s is a sentence letter representing "the sanction"—that is, something terrible—and (b) adopting the special axiom $◇¬s$, which represents the statement that it's possible to avoid the sanction. Show that each of the following is valid in S5 given Anderson's reduction.

32. $(Op \& Oq) ↔ O(p \& q)$

33. $□p → Op$

34. $Op → ◇p$

35. $((p ⥽ q) \& Op) → Oq$

A variant of Anderson's technique reduces deontic logic to modal logic by (a) translating O𝒜 as $m ⥽ 𝒜$, where m is a sentence letter symbolizing 'the moral law is fulfilled,' and (b) adopting the special axiom $\Diamond m$, representing the statement that it's possible to fulfill the moral law. Show that, given this reduction, each of the following is valid in S5.

▶ **36.** $(Op \mathbin{\&} Oq) \leftrightarrow O(p \mathbin{\&} q)$ **37.** $\Box p \to Op$

38. $Op \to \Diamond p$ **39.** $((p \mathbin{\dashv} q) \mathbin{\&} Op) \to Oq$

11.4 MORAL AND PRACTICAL REASONING

Logical systems are mathematically precise models of reasoning in certain linguistic realms. In constructing such models, logicians must make assumptions about the meanings of the expressions they are trying to study. Logicians thus attempt to create systems that count valid just those arguments in natural language that speakers of that language, on reflection, believe are valid. The analysis of moral and practical reasoning still abounds with controversy. This section of the chapter, therefore, is an attempt to indicate what about our system is controversial and to point a direction toward a more complex but more comprehensive and adequate logic of moral and practical reasoning.

In this chapter, we break down into four the principles that underlie our system. We can state these principles informally as follows:

> $\mathscr{A} \mathbin{\&} \mathscr{B}$ should be the case iff $\mathscr{A}$ and $\mathscr{B}$ each should be the case.
> Nobody is obliged to do the impossible.
> You should do what you have to.
> You should do whatever it takes to fulfill your obligations.

Although the fourth principle is the most complicated to state, it stands more securely than the others. First formulated six centuries ago, it forms the basis for almost every system of deontic logic. We can express the basic idea using each of the fundamental deontic concepts of obligation, permission, and forbiddance. If you are obligated to do something, you are thereby obligated to do what's necessary to bring it about. If, for example, you should go to the meeting tonight, and, to go to the meeting, you must miss dinner, you should miss dinner. We can put this in terms of permission: The necessary conditions of what is permissible are themselves permissible. If it's OK for you to leave work early, and, to do that, you must skip the staff meeting, then it must be OK for you to skip the staff meeting. If it weren't, then it wouldn't be OK for you to leave early.

Equivalently, we can say, in terms of forbiddance: The sufficient conditions of what is forbidden are forbidden. If you shouldn't kill your neighbor, and pulling the trigger will be sufficient to kill him, then it follows that you shouldn't pull the trigger.

In spite of the strong support for this principle, it has been questioned. Some have used two "paradoxes" of deontic logic to draw some disturbing consequences from the principle. Called the robber and victim paradoxes, they rely on the same idea. Both attack the forbiddance version of our thesis:

The Robber Paradox. The robber can repent a crime only if he has committed a crime. It's wrong for him to commit a crime. Thus, it's wrong for him to repent a crime.

The Victim Paradox. The victim of the robbery can legitimately report it to the police only if a robbery has occurred. Robberies shouldn't occur. So, the victim should not legitimately report a robbery to the police.

Both these arguments seem to have the form $p \rightarrow q; O\neg q; \therefore O\neg p$. But this is just the forbiddance version of the principle.

It's easy to see why these arguments are valid, given our approach to the semantics of moral and practical discourse. A thief can repent only in those worlds in which the thief has committed a robbery. But, in the morally ideal worlds, a thief commits no robbery. So, in ideal worlds, the thief doesn't repent.

Part of the problem here is surely that our system of deontic logic ignores questions of time and tense. The robber can repent *now* only if he has committed a crime *earlier*. So, he can repent at a point in time only if he has committed a crime at an earlier time. Suppose that the robber does steal Aunt Martha's bracelet on Tuesday. Before that time, our ideal worlds should be ones in which he doesn't commit the robbery. Afterward, however, the ideal worlds should be those in which he repents the crime and returns the bracelet. There is no point in maintaining ideal worlds in which the actual events of the past haven't happened. To formulate a more adequate system of deontic logic, then, we should associate a set of ideal worlds with a possible world at each moment. As the world changes, the set of ideal worlds should change as well. At each stage, furthermore, the ideal worlds should reflect exactly the events up to that time in the world in question. The ideal and real worlds differ, if at all, with respect to the future.

We can't even sketch what the resulting logic would look like without discussing tense logic, the study of reasoning involving time. Tense logic is far beyond the scope of this chapter or, indeed, this book. But we can see how such logic would help us solve the robber and victim paradoxes. The arguments would be invalid. We could have a possible world in which robbery is always wrong but sometimes occurs. In this world, on Tuesday, our thief has stolen Aunt Martha's bracelet. We can conclude that he shouldn't have committed the crime, and so, that he shouldn't have been in a position where he ought to repent. But we can't conclude that he shouldn't repent: In fact, all the worlds that are ideal relative to that world at that time share its past, so, in all those worlds, the crime has already occurred. And, in some or even all of those worlds, the thief may repent. The robber and victim paradoxes, therefore, point the way toward important developments of our system of deontic logic.

The thesis that you should do what you have to do reflects a fundamental feature of our approach to the meanings of moral and practical concepts. So long as we wish to think of obligations in terms of what would be true in various possible worlds, we are committed to this thesis, since necessary truths are true in all worlds. Saying that everything necessary is obligatory sounds strange. But the thesis is equivalent to the assertion that whatever is permissible is possible, that is, that you *may* do only what you *can* do. Perhaps one can have permission to do the impossible, but that permission surely isn't worth very much.

In any case, even if natural language usage doesn't justify the principle, we can define obligation in natural language in terms of obligation in D: We can say that $\mathscr{A}$ is obligatory (in English) if and only if $\mathscr{A}$ is obligatory (in D) and $\mathscr{A}$ is not necessary. So, we lose nothing by adopting the principle that whatever is necessary is obligatory and gain simplicity and convenience in return.

Our principle concerning the relation of obligation and conjunction is uncontroversial in one direction. If you should do both A and B, then you should surely do each of A and B. If Pat, for example, should go to the meeting and protest the committee's action, then she should go to the meeting, and she should protest the committee's action. If Frank ought to pay the bill and shut up, it follows that Frank should pay the bill and that Frank should shut up. The other direction of the biconditional, however, raises a problem. In fact, the same problem afflicts our other principle, that only what is possible can be obligatory: that *ought* implies *can*.

The problem concerns moral dilemmas. We might say that a person faces an insoluble moral dilemma if that person has obligations that contradict each other; if, in other words, for some sentence $\mathscr{A}$ with that person as subject, $O\mathscr{A}$ and $O\neg\mathscr{A}$. Our system asserts that no dilemmas are insoluble. For any formula $\mathscr{A}$, $O\mathscr{A}$ contradicts $O\neg\mathscr{A}$. This assertion rests both on the *ought*-implies-*can* principle and on the agglomeration principle. Suppose that $O\mathscr{A}$ and $O\neg\mathscr{A}$ are both true. By our Kantian principle alone, we can conclude only that $\mathscr{A}$ and $\neg\mathscr{A}$ are each possible: $\diamondsuit\mathscr{A}$ and $\diamondsuit\neg\mathscr{A}$. But this possibility isn't contradictory; in fact, it's typical. However, the agglomeration principle is that $O\mathscr{A}$ and $O\mathscr{B}$ together imply $O(\mathscr{A} \& \mathscr{B})$. Here, that thesis allows us to deduce $O(\mathscr{A} \& \neg\mathscr{A})$. And now, by applying the *ought*-implies-*can* principle, we arrive at $\diamondsuit(\mathscr{A} \& \neg\mathscr{A})$, which is contradictory. The combination of the agglomeration and *ought*-implies-*can* principles is thus responsible for denying the possibility of real moral or practical dilemmas.

Are there insoluble dilemmas? Are there irreconcilable conflicts of obligation? Many philosophers think that there are. Indeed, many hold that such dilemmas are fundamental features of the moral life. Perhaps the first and, even now, most commonly cited example of a moral dilemma in philosophical literature occurs in Plato's *Republic*. Suppose that you have borrowed a knife from a friend, promising to return it when he needed it. One afternoon he knocks on your door and demands the knife. Seeing that he is in a very agitated state of mind, you learn that his neighbor has just insulted him, and he wants the knife to seek revenge. What should you do? If you return the knife, you're contributing to an assault or even a murder that you might be able to prevent. If you don't, you're breaking your promise. So, you seem to face a dilemma. The promise obligates you to return the knife, but other general moral principles obligate you not to return it. If the knife dilemma constitutes a real example of moral conflict, then we must reject either the agglomeration principle or the thesis that only what is possible can be obligatory.

Interestingly, most philosophers find Plato's dilemma soluble. Plato himself thought you shouldn't return the knife; if your friend were rational, he would want you not to return it to him when he was in a homicidal rage. Your friendship thus directs you not to return the knife, despite your promise. Kant, thinking about the same circumstance, believed you should return the knife. You should be willing to break a promise only if you're willing to have everyone act in the same way. To break even one promise, in Kant's view, is to attack the institution of promising itself. Though they reach different conclusions, Plato and Kant each find one course of action in this seeming dilemma preferable to the other. They believe that one obligation takes precedence over the other.

Such circumstances have led philosophers to distinguish two kinds of obligations. The British philosopher W. D. Ross distinguished *actual* from *prima facie* obligations.[3] Actual obligations are things that absolutely ought to be done. They are obligations, all things considered; they are what, in the final analysis, ought to be done. Prima facie obligations, in contrast, ought to be done, all other things being equal (or, in the Latin phrase, *ceteris paribus*). They are obligatory at first glance, but they can be rebutted. This distinction allows us to express the conflict between Plato and Kant by saying that both hold that you have a prima facie duty to return the knife and a prime facie duty to withhold it. That is, both agree that there is a conflict of prima facie obligations. But they differ about the nature of your actual obligations. Plato thinks that your obligation to withhold the knife is actual, and rebuts your obligation to keep your promise. Kant contends that the obligation to return the knife is actual.

In both cases, the conflict of obligations means that we must evaluate competing principles, goals, or values. Considerations of promise-keeping must be weighed against the harm that may result to your maniacal friend and his neighbor. Prima facie obligations, according to Ross, become actual obligations if no other moral considerations come into play. But prima facie obligations can be conflicting. Both $O\mathscr{A}$ and $O\neg\mathscr{A}$ can be true at the same time, if we understand the obligation involved as prima facie. In such a case, however, only one of the obligations is actual. We must decide which moral factors take precedence.

If the options represented by $\mathscr{A}$ and $\neg\mathscr{A}$ are truly equal, in the sense that the moral scale is perfectly balanced between them, then we might say that we have an insoluble dilemma. But it seems more natural to say that, in such a case, neither $\mathscr{A}$ nor $\neg\mathscr{A}$ are obligatory. The moral forces, we might say, are equal but opposite: They cancel each other. Although, then, we may have prima facie obligations to both $\mathscr{A}$ and $\neg\mathscr{A}$, these obligations balance each other. Reflecting on values doesn't lead us to rank either obligation over the other. So, neither $\mathscr{A}$ nor $\neg\mathscr{A}$ is actually obligatory in such a case.

If we think that the problems of competing moral considerations always, in principle, have solutions, then we will be led to think that actual obligations cannot conflict. Our commitment to the impossibility of dilemmas may thus seem correct for actual obligation, but incorrect for prima facie obligation. For this reason, our system leads us to assume that there are no *insoluble* moral conflicts. If we interpret our rules and methods of proof concerning the obligation connective as pertaining to actual obligation alone, we can admit the possibility of prima facie conflicts while upholding our analysis of some kinds of moral and practical reasoning. Of course, this possibility suggests that our system is incomplete; we need to add some way of formalizing the notion of prima facie obligation. And it appears that, to formalize prima facie obligation, we'll have to surrender the analogue of the agglomeration principle or the analogue of *ought* implies *can,* or both.

There is, however, a simpler solution. Conflicts of obligation arise when different directives come from different sources. The conflict may be between two moral principles, between two principles of etiquette, between a moral principle and a prudential consideration, or between a practical concern and a matter of propriety. The logical structure of these conflicts seems to be the same. Factor f asserts $\mathscr{A}$ as obligatory, while factor f' asserts $\neg\mathscr{A}$ as obligatory. Faced with such a situation, we can ask the practical question, "What should we do?" Which of the two factors should we rank as more important? Which imperative should predominate? In Ross's sense, we are trying to determine our actual obligation. To stress that this is a practical question, however, and not necessarily a moral one, we can call this our *practical obligation.*

This way of representing obligation points the way toward a further development of our system, which has the virtue of using the same logical principles for both prima facie and actual obligation. Or, to put things more generally, we can use the same principles for practical obligations as for obligations from any particular source. We can place indices on the obligation connective to indicate the source of the obligation. Thus we can let $O_f\mathscr{A}$ mean that $\mathscr{A}$ is obligatory due to factor or imperative f; $O_{f'}\mathscr{A}$, that $\mathscr{A}$ is obligatory due to factor or imperative f'; and, in particular, $O_\pi\mathscr{A}$, that $\mathscr{A}$ is practically obligatory. (We use the Greek letter π because it is the first letter in the Greek word $\pi\rho\alpha\xi\iota\varsigma$, from which our word *practical* derives, and because we are already using p as a sentence letter.) We can then let all our rules and proof methods stand, amended so that we replace O with O_f.

Semantically, this substitution means that we should assign each possible world, not one set of ideal worlds, but a set of ideal worlds for each distinct imperative or factor that can serve as a source of obligation. The principles of agglomeration and *ought* implies *can* would hold for each imperative. But, of course, though $O_f \mathscr{A}$ and $O_f \mathscr{B}$ would entail $O_f(\mathscr{A} \& \mathscr{B})$, we could deduce nothing from $O_f \mathscr{A}$ and $O_{f'} \neg \mathscr{A}$, so conflicts between obligations would be possible.

Once again, developing this logic fully is beyond the scope of this chapter. But the system of moral and practical reasoning we've developed here forms the heart of this more complex system.

To conclude, let's return to the question of conditional obligation. It relates in some ways to the issue of moral conflict. 'If you promised to take Sarah to the movies, you should' can be read as asserting a prima facie or an actual obligation. In either case, the obligation is conditional; if you didn't promise, there's no obligation, unless it arises from some other consideration. We've hesitated about choosing any particular translation of the conditional in these sentences. If we interpret the conditional truth-functionally, then the sentence above should be true if you didn't promise. Yet that doesn't seem to suffice. Consider this example:

(9) You didn't promise to commit an ax murder tonight.
∴ If you promised to commit an ax murder tonight, you should.

So, we might try a strict conditional. But that would make a form for this argument come out valid:

(10) I can't be in New York tomorrow.
∴ If I'm in New York tomorrow, I should blow up the U.N.

Furthermore, both the strict and truth-functional conditionals make instances of strengthening the antecedent valid:

(11) If you promised to take Sarah to the movies tonight, you should.
∴ If you promised to take Sarah to the movies tonight, but taking her would kill her, then you should take her to the movies tonight.

Perhaps, then, we should translate conditional obligation sentences using the counterfactual connective $\square \rightarrow$.

In fact, conditional prima facie obligation sentences do seem to be counterfactual. When we say that if you promised to do something you should do it, we seem to be saying that, in the circumstances in which you promised that are otherwise most similar to the actual world, you have an obligation to fulfill your promise. If something odd happens—if the person you've promised to take to the movies runs off to get married, joins the Foreign Legion, leaves on a polar expedition, or tries to kill you—then the obligation to take that person to the movies is canceled.

Conditional statements of actual obligation, however, are more puzzling. Certain general moral rules, such as the Golden Rule—treat others only as you would want them to treat you—should probably be taken as strict conditionals, if they are taken as expressing actual obligation.

Problems

Which of the following formulas are valid in D?

1. $P(p \,\square\!\!\to q) \to (Pp \,\square\!\!\to Pq)$

2. $(Pp \,\square\!\!\to Pq) \to P(p \,\square\!\!\to q)$

3. $(Op \,\square\!\!\to Oq) \to O(p \,\square\!\!\to q)$

4. $O(p \,\square\!\!\to q) \to (Op \,\square\!\!\to Oq)$

▸ **5.** $(p \dashv\!\!\!3\, q) \to (Op \,\square\!\!\to Oq)$

6. $(Op \dashv\!\!\!3\, Oq) \to (p \,\square\!\!\to q)$

7. $O(Op \,\square\!\!\to p)$.

Which of the following argument forms are valid in D?

8. $p \,\square\!\!\to q; Pp; \therefore Pq$

9. $p \,\square\!\!\to q; \neg Pq; \therefore O\neg p$

▸ **10.** $p \,\square\!\!\to q; \neg Oq; \therefore P\neg p$

11. $p \,\square\!\!\to Oq; p; \therefore Oq$

12. $p \,\square\!\!\to Oq; \Diamond p; \therefore \Diamond Oq$

13. $p \,\square\!\!\to Oq; p; \therefore Pq$

14. $p \,\square\!\!\to Oq; \neg\Diamond q; \therefore \neg p$

▸ **15.** $p \,\square\!\!\to Oq; \neg\Diamond q; \therefore \neg\Diamond p$

16. $p \,\square\!\!\to Oq; P\neg q; \therefore \neg p$

17. $p \,\square\!\!\to Oq; O\neg q; \therefore \neg p$

18. $\Diamond(p \,\square\!\!\to q); Pp; \therefore Pq$

19. $\Diamond(p \,\square\!\!\to q); Op; \therefore Oq$

▸ **20.** $p \,\Diamond\!\!\to q; Pp; \therefore Pq$

21. $p \,\Diamond\!\!\to q; \therefore Pp \,\Diamond\!\!\to Pq$

22. $p \,\Diamond\!\!\to q; Op; \therefore Oq$

23. $p \,\Diamond\!\!\to q; Op; \therefore \Diamond Oq$

24. $p \,\Diamond\!\!\to q; \therefore Op \,\Diamond\!\!\to Oq$

▸ **25.** $p \,\square\!\!\to Oq; \Diamond r \,\square\!\!\to O\neg q; \therefore p \,\square\!\!\to \square\neg r$

26. $p \,\square\!\!\to Oq; \neg\Diamond(q \,\&\, r); \therefore p \,\square\!\!\to O\neg r$

27. $p \,\square\!\!\to r; r \,\square\!\!\to \square\neg q; \therefore p \,\square\!\!\to O\neg r$

28. $O\neg p \,\square\!\!\to q; q \,\square\!\!\to O\neg p; \Diamond q; \therefore \Diamond\neg Pp$

29. $p \,\square\!\!\to q; q \,\square\!\!\to p; \therefore (Op \,\square\!\!\to Oq) \,\&\, (Oq \,\square\!\!\to Op)$

▸ **30.** $p \,\square\!\!\to q; q \,\square\!\!\to p; \therefore (Pp \,\square\!\!\to Pq) \,\&\, (Pq \,\square\!\!\to Pp)$

31. $Op \,\square\!\!\to q; O\neg q; \therefore O\neg Op$

32. $p \,\square\!\!\to q; \Diamond r \,\square\!\!\to \neg q; \therefore Or \,\square\!\!\to \neg p$

33. $Pp \,\square\!\!\to Oq; \neg\Diamond q; \therefore O\neg p$

34. $p \,\square\!\!\to q; r \,\square\!\!\to Op; \neg Oq; \therefore \neg\Diamond r$

35. $p \,\square\!\!\to Oq; \Diamond q \,\square\!\!\to Or; Pr \,\square\!\!\to \Diamond s; \therefore \square\neg s \,\square\!\!\to \square\neg p$

Using counterfactual conditionals for conditional obligation sentences, and wherever else they are appropriate, evaluate the validity of these arguments.

36. If we use military force, we'll endanger the lives of the hostages. So, if putting the hostages' lives in danger is unacceptable, we shouldn't use military force.

37. Jones is responsible only if he ought to have avoided the crash. Jones could have avoided the crash only if he could have stopped more quickly. But he couldn't stop more quickly. Therefore, Jones isn't responsible.

38. If you were to vote for Carol, the city's rate of development would be maintained or accelerated. Since the rate of the city's development should be slowed, you shouldn't vote for Carol.

39. If you were to vote for Carol, the city's rate of development would be maintained or accelerated. Since the rate of the city's development should be maintained or accelerated, you should vote for Carol.

40. If we were to elect Bob, we'd withdraw from the nuclear project. We should withdraw from the project. So, we should elect Bob.

41. If we were to elect Bob, we'd withdraw from the nuclear project. We shouldn't withdraw from the project. So, we shouldn't elect Bob.

42. If you borrowed the money from John, you should pay him $1,000 by the end of the year. You borrowed the money from him, so you have an obligation to pay him $1,000 by December 31.

43. If you are in graduate school, you should study hard. You should be in graduate school, but you aren't. Even so, in consequence, you have an obligation to study hard.

44. If we were to increase incentives, productivity would improve, although it might not improve immediately. We ought to improve productivity. Therefore, we should increase incentives.

45. If we increased incentives, productivity would improve. So, we ought to increase incentives only if we ought to improve productivity.

46. I promised to take Alice to dinner, and I promised to attend the meeting. If I promised to go to the meeting, I have an obligation to attend; if I promised Alice I would take her to dinner, I should. But I can't take her to dinner and go to the meeting. So, I'm in big trouble!

47. If I said I would be there, I ought to be there. But being there is incompatible with stopping to save the life of an injured accident victim. Thus, if I gave my word to be there, I shouldn't save the life of an injured accident victim.

48. It's morally indifferent whether you go to college. But if you were to go, you'd have to devote several years of your life to study. So, it's morally indifferent whether you devote several years of your life to study.

49. If you brought about a true revolution, you'd destroy the lives of many innocent people. It's wrong to destroy the lives of many innocent people. So, it's wrong to bring about a true revolution.

50. You would bring about revolutionary reforms only if you destroyed the lives of many innocent people. It would be morally indifferent whether you bring about revolutionary reforms, therefore, only if it were indifferent whether you destroyed the lives of innocent people.

51. If you brought about revolution, many innocent lives would be destroyed. You ought to bring about revolution. So, it's permissible to destroy many innocent lives.

52. The marshal could maintain law and order only if he were to convict these innocent men and sentence them to death. He has an obligation to maintain law and order, so he ought to sentence these innocent men to die.

53. If you're like me, you can't resist temptation. So, if you're like me you should give in to temptation.

54. It's wrong that it's possible for some people to avoid paying the taxes they owe to the government. But it would be possible to stop some people from avoiding their tax burden only if we cast aside our respect for individual liberties and rights to privacy. Therefore, it's necessarily true that it's permissible for us to cast aside our respect for individual liberties and rights to privacy.

55. If Jane were willing to have an abortion, then Jane would be willing to kill something that might be a human being. If Jane couldn't think of any alternatives, she'd be willing to go through with the abortion. So, if Jane didn't think of any alternatives, it would be OK for her to kill something that might be a human being.

56. It would be permissible to violate the right to life of this fetus only if it were obligatory to violate the fetus's right to life. But the mother would survive if and only if we were to violate the fetus's right to life. Therefore, it would be permissible that we violate the fetus's right to life only if we had an obligation to save the mother's life.

57. If we'd acted too late, the enemy would attack by daybreak. If they were to attack by then, however, we wouldn't be able to avoid an all-out conflict. So, if we'd acted too late, we'd have no obligation to avoid an all-out conflict.

58. This act would conform to the principle of utility just in case its tendency to augment the happiness of the community would be greater than its tendency to diminish it. If this action conforms to the principle of utility, it ought to be done. So this act ought to be done if its tendency to augment the community's happiness is greater than its tendency to diminish it.

59. This act would conform to the principle of utility just in case its tendency to augment the happiness of the community would be greater than its

tendency to diminish it. If this action conforms to the principle of utility, it is permissible. So this act is permissible if its tendency to augment the community's happiness is greater than its tendency to diminish it.

60. If you were to go to Yale, you'd become a very successful lawyer. If you promised Yale you would go, you should. You have no obligation to become a very successful lawyer. Therefore, you didn't promise Yale you'd go there unless you're a fool.

Notes

[1] The first discussions of deontic logic emerged in the fourteenth century, when writers first noted the similarities between the deontic concepts of obligation, permission, and forbiddance and the modal notions of necessity, possibility, and impossibility. Ernst Mally published the first modern work on deontic logic in 1926. But most contemporary research on practical and moral reasoning stems from the Scandinavian logician George Henrik von Wright's "Deontic Logic," *Mind* 60 (1951): 1–15.

[2] This principle was formulated in the fourteenth century by logicians such as Rosetus.

[3] W. D. Ross, *The Right and The Good* (New York: Oxford University Press, 1930): 19–20.

12

QUANTIFIERS
AND
MODALITY

In Chapters 9, 10, and 11 of this text, we added to classical sentential logic a number of nontruth-functional connectives: □, expressing necessity; ◇, expressing possibility; ⊰, for the strict conditional; ⊱, for the strict biconditional; □→ and ◇→, for *would* and *might* counterfactual conditionals; O, expressing obligation; and P, expressing permission. In this chapter, we add these connectives to classical quantificational logic.

Combining quantifiers and nontruth-functional connectives produces a logical system with tremendous power and sophistication.[1] The combination also gives rise to many deep philosophical and linguistic issues. Since the early 1960s, problems arising from quantified modal logic have played a large role in philosophy and linguistics. The question of how to combine modalities and quantifiers relates intimately to many traditional metaphysical questions concerning essences and objects. It is also a crucial part of the larger question of how to represent meanings of natural-language sentences.

12.1 SYSTEM QS5

In a sense, combining modalities and quantifiers requires nothing new. In previous chapters, we developed theories of nontruth-functional connectives and of quantifiers. We already have rules governing all these logical opera-

346

tors. If we lump the rules together, we obtain a relatively simple system of quantified modal logic, which we can call QS5 (for "quantified S5"). We'll call the language of QS5 QML. The system QS5 has more power than its name suggests. It incorporates not only S5's analysis of modal connectives but also theories of counterfactuals and deontic connectives.

We do need to specify how to extend one concept to quantificational logic. Atomic formulas contain no connectives or quantifiers; a formula is modally closed just in case each of its atomic subformulas lies within the scope of a modal connective: $\Box$, $\Diamond$, $\rightarrow\!\!3$, or $8\!\!\leftarrow$. Formulas such as $\Box Fa$, $\Diamond Gb \rightarrow \Diamond Ga$, and $\Diamond \exists x Fx$ thus count as modally closed. What remains unclear is whether the same is true of such formulas as $\forall x \Box Fx$, $\exists y \Diamond Gy$, and $\forall x \exists y \Box Fxy$. Fx, Gy, and Fxy are *atomic pieces* of these formulas, but they are not themselves formulas. Consequently, they aren't atomic subformulas. For the purposes of defining QS5, we'll extend the notion of modal closure as follows:

DEFINITION. A formula is *modally closed in QS5* iff each of its atomic pieces appears within the scope of a modal connective.

A formula is modally closed in QS5, that is, just in case every atomic subportion of it lies within the scope of $\Box$, $\Diamond$, $\rightarrow\!\!3$, or $8\!\!\leftarrow$. This definition has the effect of making $\forall x \Box Fx$, $\exists y \Diamond Gy$, and $\forall x \exists y \Box Fxy$ modally closed.

Earlier we established a semantics for modal logic by speaking of possible worlds. In Chapter 6, we counted $\Box \mathscr{A}$ true on an interpretation just in case $\mathscr{A}$ was true, on that interpretation, in every possible world. In sentential logic, an interpretation of a formula was simply an assignment of truth values to the sentence letters in it. In modal sentential logic, we needed to consider the truth values of formulas in different possible worlds. An interpretation of a formula, therefore, became an assignment of truth value to its sentence letters in each possible world.

In Chapter 10, we introduced a semantics for quantificational logic. An interpretation of a formula of quantificational logic consisted of a domain together with an interpretation function assigning elements of the domain to constants and assigning sets of n-tuples of members of the domain to n-ary predicates.

To interpret the formulas of QS5, we must combine these approaches. Interpretations of formulas of quantified modal logic consist of a set of domain–interpretation function pairs. Each pair is a quantificational interpretation of the usual form and corresponds, intuitively, to a possible world. If our language contains counterfactuals, then we need to add a relation of world proximity so that we can talk meaningfully about the worlds closest to a given world. If our language contains deontic connectives, then we also need a function assigning a set of ideal worlds to each possible world.

There would be no reason to worry further about the combination of quantifiers and modalities, except that QS5 commits us to some highly controversial principles. QS5 counts valid a number of inferences that seem questionable in natural language. In the next section, therefore, we'll consider ways of revising rules to reflect our intuitive conception of validity more accurately.

Before discussing the controversial features of QS5, however, we need to consider a point of symbolization. Many English sentences that contain both quantifier and modal expressions are ambiguous. Consider, for instance, the following sentences, each of which we might symbolize in more than one way.

(1) a. All cyclists are necessarily two-legged.
   b. $\forall x(Cx \rightarrow \Box Tx)$
   c. $\Box \forall x(Cx \rightarrow Tx)$

(2) a. Some people may disapprove.
   b. $\exists x(Px \,\&\, \Diamond Dx)$
   c. $\Diamond \exists x(Px \,\&\, Dx)$

(3) a. Every unit might join the revolt.
   b. $\forall x(Ux \rightarrow \Diamond Jx)$
   c. $\Diamond \forall x(Ux \rightarrow Jx)$

(4) a. Some analyses are necessarily wrong.
   b. $\exists x(Ax \,\&\, \Box Wx)$
   c. $\Box \exists x(Ax \,\&\, Wx)$

These symbolizations are not, in general, equivalent.

In each example, formula c has a modal connective with the entire formula as its scope. Aquinas called statements of this form *de dicto,* because they seem to attribute necessity to a statement rather than to any specific object's possession of a specific property. (1)c, for instance, symbolizes 'It's a necessary truth that cyclists are two-legged'; (2)c symbolizes 'It's possible that some people will disapprove'; (3)c, 'It's possible that every unit will join the revolt'; and (4)c, 'It's a necessary truth that some analyses are wrong.' In English, it's easiest to single out this sense of modal statements by prefixing sentences with *it's necessary that, it's possible that, it's necessarily true that,* and so on.

In formula b of each example, in contrast, the quantifier has the entire formula as its scope. Aquinas called statements of this form *de re,* because they attribute necessity to a thing's having a property. (1)b, for instance, symbolizes 'Every cyclist has the property of being necessarily two-legged'; (2)b, 'Some people are such that they may disapprove'; (3)b, 'Every unit is such that it might join the revolt'; and (4)b, 'Some analyses have the property of being necessarily wrong.' *De re* statements represent the modal properties of particular objects; (4)b makes a representation about some analyses, and specifies that they are necessarily wrong. We can specify *de re* readings in natural language by using the phrases *such that* or *have the property that* (or *of*).

The readings in b and c of each example are not equivalent: The formulas in b represent actual objects and their modal properties, while the formulas in C represent some or all possible circumstances. (1)b, $\forall x(Cx \rightarrow \Box Tx)$, represents the assertion that every cyclist has the property of being necessarily two-legged. This assertion is probably false; in an unfortunate accident, cyclists can lose their legs as readily as anyone else. A cyclist who loses a leg may no longer be a cyclist, but that result is irrelevant to (1)b. That formula requires that everyone who is a cyclist in the *actual* world be two-legged in *every* world. (1)c, in contrast, represents the assertion that it's necessarily true that cyclists have two legs. This, too, is probably false, but for reasons very different from those applying to (1)b. Circus bears, for example, ride cycles and have four legs.

Similarly, (4)b and c are quite different. $\exists x(Ax \,\&\, \Box Wx)$ symbolizes the assertion that some analyses have the property of being necessarily wrong; perhaps they contradict themselves or contain mathematical falsehoods. $\Box \exists x(Ax \,\&\, Wx)$, however, represents the assertion that it's necessary that some analyses be wrong. Perhaps the analyses contradict each other. If so, then some must be wrong. But it could still be a contingent matter whether any given analysis is correct or incorrect.

To make the distinction between *de re* and *de dicto* modalities more precise, let's say that a modal formula containing quantifiers is *de re* just in case all its quantifiers appear outside the scope of any of its modal connectives, and *de dicto* just in case all its modal connectives appear outside the scope of any of its quantifiers. Modal quantified formulas that are neither *de re* nor *de dicto*, such as $\exists x \Box \forall y Fxy$ and $\Diamond \forall x \Box Fx$, are *mixed*.

The Barcan Formula

Several controversial features of QS5 involve the relationship between *de re* and *de dicto* modalities. The first questionable formula that is valid in QS5 is the *Barcan formula:*

(5) $\forall x \Box Fx \rightarrow \Box \forall x Fx$

This formula symbolizes 'If everything is necessarily F, then it's necessary that everything is F.'[2] In other words, in certain simple cases, *de re* modal statements imply corresponding *de dicto* statements. We can use semantic tableaux or natural deduction to see that the formula is valid in QS5.

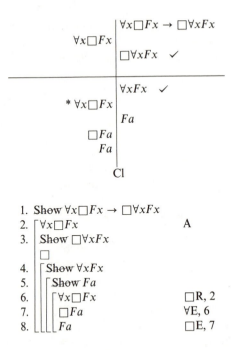

1. Show $\forall x \Box Fx \rightarrow \Box \forall x Fx$
2. $\ulcorner \forall x \Box Fx$ A
3. $\;\;$ Show $\Box \forall x Fx$
   $\;\;\Box$
4. $\;\;\;\ulcorner$ Show $\forall x Fx$
5. $\;\;\;\;\ulcorner$ Show Fa
6. $\;\;\;\;\;\ulcorner \forall x \Box Fx$ $\Box$R, 2
7. $\;\;\;\;\;\;\Box Fa$ $\forall$E, 6
8. $\;\;\;\;\;\;Fa$ $\Box$E, 7

The crucial step, in both methods, is transferring the information in $\forall x \Box Fx$ from one world to another. The fact that $\forall x \Box Fx$ counts as modally closed in QS5 amounts to the assumption that the formula has the same truth value in each possible world. We can reiterate the formula into a modal proof or allow it to survive a world shift, because its truth in one world suffices to establish its truth in any other world.

Is it right that the Barcan formula counts as valid in QS5? $\forall x \Box Fx$ represents 'Everything is necessarily F.' If $\forall x \Box Fx$ is true in the actual world on a particular interpretation, everything in the actual world must be necessarily F. That is, everything in our world must be F in every world, or, at least, in every world in which those things exist. Suppose that F means 'pays homage to Xerxes.' Then $\forall x \Box Fx$ means 'Everyone must pay homage to Xerxes.'

$\Box \forall x Fx$, in contrast, represents 'It's necessarily true that everything is F.' If this formula is true, then in every world everything must be F. If $\Box \forall x Fx$ is false, then there must be a world where something is not F. If we read F as 'pays homage to Xerxes,' $\Box \forall x Fx$ means 'It's a necessary truth that everyone pay homage to Xerxes.'

Could $\forall x \Box Fx$ be true while $\Box \forall x Fx$ is false? For instance, might it be the case that everyone must pay homage to Xerxes, but that it's not a necessary truth that everyone pay homage to Xerxes? If it's not a necessary truth, then somebody, in some world, doesn't pay homage to Xerxes. In other words, it's possible for somebody to be stronger than Xerxes. But if the former formula is true, then everybody in the actual world does pay homage to Xerxes in every world; nobody in this world has the capacity to be stronger than Xerxes.

The person stronger than Xerxes doesn't exist in the actual world. This is no contradiction, unless no worlds contain entities that don't exist in our world. The Barcan formula, then, amounts to a principle that prohibits additions to the domains of possible worlds.

Viewed in this light, the Barcan formula should not be valid, at least in many realms of application. We ordinarily think that there might have been things that don't, as it happens, actually exist. Consider, for example, each of these sentences:

(6) a. If the leaders of Western European nations hadn't imposed national boundaries on the countries of Eastern Europe but had instead allowed their inhabitants to negotiate boundaries, there would be many more distinct nations in Eastern Europe than there are today.
b. If the government didn't pay farmers to limit production, U.S. farms would produce much more than they do under the current price support system.
c. If the birth control pill hadn't been invented, there would be many more children in elementary schools than there actually are.

Each seems plausible. But in QS5, each is supposed absurd. On the face of it, at least, this supposition seems unreasonable. Surely the world could have contained more things than it actually does. If so, then the Barcan formula shouldn't be valid in general. But QS5 may nevertheless be adequate when we're willing to restrict our discussion to what exists in the actual world.

We can see the same point by considering a formula involving possibility and the existential quantifier. The following is equivalent to the Barcan formula:

(7) $\Diamond \exists x F x \rightarrow \exists x \Diamond F x$

In QS5, (7) counts as valid. But suppose that it's possible to breed a pig that can fly. This fact doesn't imply that there is anything in our world now that could be a flying pig.

The Converse Barcan Formula

The second puzzling consequence of QS5 is the *converse Barcan formula:*

(8) $\Box \forall x F x \rightarrow \forall x \Box F x$

This formula represents 'If it's necessary that everything is F, then everything is necessarily F.' In other words, in certain simple cases, *de dicto* statements imply corresponding *de re* statements. Intuitively, this formula seems more plausible than the Barcan formula itself. If, in every possible world, everything is F, then it's natural to think that everything in the actual world is F in every world. But the principle seems suddenly very implausible if Fx is an existential claim. Everything in the domain of a possible world exists in that world. So, the antecedent of the conditional, with $\exists y\ x = y$ substituted for Fx, is valid. In QS5,

(9) $\Box \forall x \exists y\ x = y$

is a valid formula. ($\forall x \exists y\ x = y$ is valid in quantificational logic, and so it must be true in every possible circumstance, that is, in every possible world. But then $\Box \forall x \exists y\ x = y$ is true.) If we accept the converse Barcan formula, consequently, we have to count the consequent as valid as well. The consequent, with F expressing existence, is 'Everything that exists in our world exists in every possible world':

(10) $\forall x \Box \exists y\ x = y$.

It's easy to see that (10) is valid in QS5.

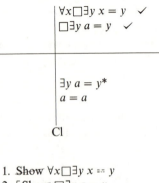

1. Show $\forall x \Box \exists y\ x = y$
2. ⌈ Show $\Box \exists y\ a = y$
       □
3. ⌈ Show $\exists y\ a = y$
4. ⌈ $a = a$ =I
5. ⌊ $\exists y\ a = y$ ∃I, 4

But does everything in the actual world exist necessarily? Perhaps God necessarily exists; perhaps numbers exist necessarily. But, in many contexts, we wouldn't want to assume that *everything* necessarily exists. Sentences such as these, for example, seem contingent:

(11) a. If my parents had never met, I might not have existed.
      b. None of those things would have happened if you had followed my advice.
      c. Without their kind assistance, the book would never have been written.
      d. If the union had struck, the company would have produced many fewer sausages.

In QS5, however, we suppose that sentences such as these are false. The converse Barcan formula has the effect of leading us to suppose that no worlds have fewer objects than our own. So, QS5 is adequate only when we're willing to grant this assumption.

We can see this as well by examining a formula equivalent to the converse Barcan formula:

(12) $\exists x \Diamond Fx \rightarrow \Diamond \exists x Fx$

This formula symbolizes 'If something is possibly F, then it's possible for something to be F.' Like the converse Barcan formula, (12) sounds reasonable. But this appearance evaporates if we replace F with something symbolizing nonexistence. The formula then symbolizes 'If there's anything that might not have existed, then it's possible for there to be something that doesn't exist.' It seems obviously true that our world contains things that could have failed to exist, but it seems strange and even contradictory to say that, in some possible world, there exist things that don't exist. Indeed, in QS5 $\Diamond \exists x \neg \exists y \, x = y$ is contradictory, since $\exists x \neg \exists y \, x = y$ is contradictory in quantification theory. So (12), too, implies that everything has necessary existence; if anything might not have existed, it asserts, then a contradiction follows.

The Barcan formula and its converse are valid only if all possible worlds have the same domain. If domains can expand, then the Barcan formula fails. But if domains can contract, the converse Barcan formula fails. To formulate a semantics for QS5 we therefore need to require that all possible worlds contain exactly the same objects. In some contexts—for instance, within a mathematical theory—this assumption may be reasonable. As we'll see below, it also seems reasonable in much moral and practical reasoning.

Similar Formulas

A third questionable consequence of QS5 is the validity of a pair of equivalent formulas:

(13) a. $\Diamond \forall x Fx \rightarrow \forall x \Diamond Fx$
  b. $\exists x \Box Fx \rightarrow \Box \exists x Fx$

These, like the converse Barcan formula, have the effect of requiring necessary existence. (13)a symbolizes 'If it's possible for everything to be F, then everything has the potential to be F'; (13)b, 'If something has a necessary property, then it's necessarily true that something has that property.' Both principles seem plausible. But (13)a loses its luster of reasonableness when we construe F as representing necessary existence. (13)a then becomes

(14) $\Diamond \forall x \Box \exists y \, x = y \rightarrow \forall x \Diamond \Box \exists y \, x = y$

And, since in S5 $\Diamond \Box \mathscr{A}$ is equivalent to $\Box \mathscr{A}$, (14) amounts to

(15) $\Diamond \forall x \Box \exists y \, x = y \rightarrow \forall x \Box \exists y \, x = y$

(15) symbolizes 'If there is any possible world containing only things that exist necessarily, then everything exists necessarily.' So, imagine a possible world where God, who exists necessarily, decides not to create anything else. In that world, what exists exists necessarily. If such a world is possible, then the antecedent of (15) is true. But it doesn't follow that Lupe, Lupe's logic text, my car, and all other things exist necessarily.

Another equivalent pair of formulas look very much like the formulas we've been discussing, but they are not valid in QS5. Indeed, they aren't valid in any reasonable system of logic combining quantifiers and modalities.

(16) a. $\forall x \Diamond Fx \rightarrow \Diamond \forall x Fx$
   b. $\Box \exists x Fx \rightarrow \exists x \Box Fx$

(16)a symbolizes 'If everything has the potential of being F, then it's possible for everything to be F at once.' This principle is certainly false. Before the beginning of the baseball season, it might be true that any team could win the World Series; each team might have the potential of winning the championship. But that in no way means that it's possible that every team will win. Similarly, the head coach of an NFL team once declared that any team in the NFL can beat any other team on any given Sunday. His proclamation doesn't imply that there could be a season in which every team beat every other team on every Sunday.

(16)b symbolizes 'If it's necessary for there to be an F, then something in particular is necessarily F.' Thus, if it's necessary for someone to win and for someone to lose, then somebody has to win—in the strong sense that this person cannot lose—and someone else has to lose—again, in the strong sense that he or she cannot win. But this is absurd. It's perfectly consistent to hold that someone has to lose without holding that any particular person's defeat is inevitable.

Despite their similarity to the Barcan formula, its converse, and (13)a and b, then, (16)a and b are not valid, even in QS5. They fail whether the domains or possible worlds expand, contract, or remain constant.

Problems

Symbolize the following in QML. If a sentence is ambiguous, list each plausible symbolization.

1. Some new species will come along. They may be wiser (Admiral Rickover).

2. All things are conquerable (David Ambrose).

3. I couldn't possibly tell you. I have never been a man (Claire Booth Luce, on being asked whether she found being a woman a disadvantage).

4. Were I to await perfection, my book would never be finished (Tai T'ung).

▶ **5.** Paper cannot wrap up a fire (Chinese proverb).

   6. We must all take time to do enough thinking to formulate our own conclusions (Thomas J. Watson).

   7. Every experience can be a learning experience (Roberto Goizueta).

   8. But the fact is that some fat is natural, necessary and probably hereditary (*Connoisseur*).

   9. Prudence, like experience, must be paid for (Richard Sheridan).

▶ **10.** Important principles may and must be flexible (Abraham Lincoln).

   11. No business can long prosper in an unhealthy environment, in a polarized society that locks millions out of the mainstream (George Weissman).

   12. The Fed may still do us in. It may help some, though, to know that it didn't mean to (Lindley H. Clark, Jr.).

   13. We are in great haste to construct a magnetic telegraph from Maine to Texas; but Maine and Texas, it may be, have nothing important to communicate (Henry David Thoreau).

   14. He cannot be good that knows not why he is good (Thomas Fuller).

▶ **15.** You can never plan the future by the past (Edmund Burke).

   16. One need not hope in order to act; or succeed in order to persevere (William of Orange).

   17. You cannot dream yourself into a character; you must hammer and forge one yourself (James A. Froude).

   18. Who rides a tiger cannot dismount (Chinese proverb).

   19. What a dull world it would be if every imaginative maker of legends were stigmatized as a liar (Heywood Broun).

   20. Every country can produce good men (Gotthold Lessing).

   21. The person who can laugh with life has developed deep roots with confidence and faith (Democritus).

   22. No one can take away from you what you have stored inside your head (Roberto Goizueta).

   23. It is sometimes necessary to play the fool to avoid being deceived by clever men (La Rochefoucauld).

   24. You can never ride on the wave that went out yesterday (John Wanamaker).

▶ 25. In the infancy of society every author is necessarily a poet (Percy Bysshe Shelley).

26. Human beings do not do all the evil of which they are capable (Henri de Montherlant).

27. Hot-air ballooning is the only sport where you can be going full throttle and still get passed by a butterfly (William Lewis).

28. It may be those who do most, dream most (Stephen Leacock).

29. A man who cannot tolerate small ills can never accomplish great things (Chinese proverb).

30. A man can't have his head full of odds and ends and succeed (Philip D. Armour).

31. A tale which may content the minds of learned men and grave philosophers (Gascoyne, used by Robert Southey as the motto to *The Story of the Three Bears*).

32. We could wake up one morning and find another 1939 Hitler-Stalin pact with a new group in power. The whole balance of power could shift overnight (Henry Jackson).

33. What one does not say does not have to be explained (Sam Rayburn).

34. You can't hit what you can't see (Joe Tinker, speaking of Rube Marquard).

▶ 35. By working faithfully eight hours a day, you may eventually get to be boss and work twelve hours a day (Robert Frost).

36. Those who bring sunshine to the lives of others cannot keep it from themselves (James Barrie).

37. A man in earnest finds means, or if he cannot find, creates them (William Ellery Channing).

38. If the politicians want to have central planning and command, they cannot have dynamism and life (George Gilder).

39. Without kindness there can be no true joy (Thomas Carlyle).

▶ 40. He who cannot change the very fabric of his thought will never be able to change reality (Anwar Sadat).

41. No one can misunderstand a boy like his own mother (Norman Douglas).

42. Every year, if not every day, we have to wager our salvation upon some prophecy based upon imperfect knowledge (Oliver Wendell Holmes, Jr.).

43. Only an optimist can win in playing the game of business (J. P. Morgan).

44. People who acted meanly or cynically or dishonorably in political life found that they could not look Mendès-France in the eye any longer (Ved Mehta).

▶ 45. You can be sincere and still be stupid (Charles F. Kettering).

46. He that cannot be angry is no man (Thomas Dekker).

47. One can always be kind to people about whom one cares nothing (Oscar Wilde).

48. There is no power on earth that can neutralize the influence of a high, pure, simple and useful life (Booker T. Washington).

49. So long as man is capable of self-renewal he is a living being (Henri Frederic Amiel).

50. Thinking is one thing no one has been able to tax (Charles F. Kettering).

51. We can do anything we want to do if we stick to it long enough (Helen Keller).

52. A good tree cannot bear bad fruit, or a poor tree good fruit (Matthew 7:18–19).

53. Not both love and majesty can be together enthroned (Ovid).

54. ... the ox you gore may be your own (Michael K. Evans, on tax reform).

▶ 55. If I can put a Maxim's in Peking, I can put a Maxim's on the moon! (Pierre Cardin).

56. It is impossible to live without brains, either one's own or borrowed (Baltasar Gracian).

57. Every human being on the face of the earth could be housed in the state of Texas in one-story single family homes, each with a front- and backyard (Thomas Sowell).

58. Men that cannot entertain themselves want somebody, though they care for nobody (George Savile).

59. The only man who can't change his mind is a man who hasn't got one (Edward Noyes Wescott).

▶ 60. To make yourself understood you have to speak plain and write plain (William Feather).

61. One cannot both be in love and be wise (P. Syrus).

62. All the discontented people I know are trying to be something they are not, to do something they cannot do (David Grayson).

63. Unless there be correct thought, there cannot be any action, and when there is correct thought, right action will follow (Henry George).

▶ 64. Business must sell itself to the public to preserve itself (Edward L. Bernays).

65. Anyone who works would be amazed at what he could accomplish and, more importantly, leave out, if he had only two days in which to do what usually took him a week (Jonathan Gathorne-Hardy).

66. In Victorian England, if a young lady wanted to be considered properly spiritual, dainty and refined, she could not be caught visibly enjoying food or wine (George Lang).

67. Man must work.... But he may work grudgingly or he may work gratefully; he may work as a man or he may work as a machine (Henry Giles).

68. Trade could not be managed by those who manage it if it had much difficulty (Samuel Johnson).

69. A man who is not a fool can rid himself of every folly but vanity (Jean-Jacques Rousseau).

▶ 70. People who cannot find time for recreation are obliged sooner or later to find time for illness (John Wanamaker).

71. The right man can make a good job out of any job (William Feather).

72. It [the political centerpoint] may, for example, go somewhat further right, even as everyone is diving for the center (Ben J. Wattenberg).

73. No man can be adaptable who is not accustomed to self-reliance and independence of action (James Truslow Adams).

74. If there were no bad people, there would be no good lawyers (Charles Dickens).

▶ 75. Fortune can take away one's goods, but it cannot take away one's heart (Seneca).

76. We can't always please, but we can avoid offending (Arnold Glasow).

77. The economic defects of Soviet-style Communism that have long since seemed chronic and incurable may prove terminal (Strobe Talbott).

78. No man can climb out beyond the limitations of his own character (Robespierre).

79. You can buy mediocrity for gold, but excellence cannot be purchased. It must be achieved (Harry G. Mendelson).

80. Neither homes nor lands nor heaps on heaps of gold and silver can banish fevers from the body nor cares from the mind (Horace).

81. Politicians are people who resolve through linguistic processes conflicts that would otherwise have to be solved by force (S. I. Hayakawa).

82. All singers have this fault—among friends, if asked to sing, they cannot; not asked, cannot desist (Horace).

83. One can live in the shadow of an idea without grasping it (Elizabeth Bowen).

84. Real difficulties can be overcome; it is only the imaginary ones that are unconquerable (Theodore N. Vail).

▶ 85. If Mr. Newman doesn't send me an MX missile, I'm going to report him to the postal service people for fraud (William F. Buckley, Jr., on sending for a "nuclear war prevention kit").

86. A man may be sharper than another, but not than all others (La Rochefoucauld).

87. It would hardly be necessary to talk sense if one spoke well enough (Quentin Crisp).

88. Only people who possess firmness can possess genuine gentleness (La Rochefoucauld).

89. You can't lie to life. You may deceive your teacher about what you know, but you can't deceive life. What you haven't learned leaves a hole that nothing but that learning can fill and no amount of covering over can disguise (Edward R. Sims).

▶ 90. He who cannot forgive others destroys the bridge over which he himself must pass (George Herbert).

91. Money is not required to buy one necessity of the soul (Henry David Thoreau).

92. To attain happiness in another world we need only to believe something, while to secure it in this world, we must do something (Charlotte Perkins Gilman).

93. No man can be really big who does not read widely outside his own field (Theodore N. Vail).

94. But if we had a classless society, no one could *afford* to run for president (*Punch* cartoon).

▶ 95. You can't do anything about the length of your life, but you can do something about its width and depth (H. L. Mencken).

96. In a country with inflation like Argentina's, you can lose money if you have a flat tire on the way to the bank (Sam Ayoub).

97. All the problems of the world could be settled easily if men were only willing to think (Nicholas Murray Butler).

98. The ideal must be high; the purpose strong, worthy and true; or the life will be a failure (Orison Swett Marden).

99. One great, strong, unselfish soul in every community could actually redeem the world (Elbert Hubbard).

100. The most precious thing a parent can give a child is a lifetime of happy memories (Frank Tyger).

101. If all creatures were immortal and preserved the same functions and abilities throughout their endless lives, they would preempt the possibility of improvement or adaptation (Jonathan Miller).

102. To me the meanest flower that blows can give thoughts that do often lie too deep for tears (William Wordsworth).

103. Nothing will ever be attempted if all possible objections must first be overcome (Samuel Johnson).

104. Any man can learn to do anything that any other man has done if he will apply himself (Charles M. Schwab).

105. Under a world sugar glut, prices have sagged, and the Administration must now either impose quotas on sugar imports, pushing up the price to U.S. consumers, or buy close to $300 million worth of surplus sugar that it will not be able to sell any time soon (*Business Week*).

106. Ye can lade a man up to th' university, but ye can't make him think (Finley Peter Dunne).

107. Every man who knows how to read has it in his power to magnify himself, to multiply the ways in which he exists, to make his life full, significant and interesting (Aldous Huxley).

108. The character inherent in the American people has done all that has been accomplished; and it would have done somewhat more, if the government had not sometimes got in its way (Henry David Thoreau).

Use QS5 to evaluate the following arguments and answer the following questions.

109. Thomas Hobbes said, "A man's conscience and his judgment is the same thing, and as the judgment, so also the conscience, may be erroneous." Hobbes appears to be arguing:

> A man's conscience and his judgment is the same thing.
> A man's judgment may be erroneous.
> ∴ A man's conscience may be erroneous.

Is this argument valid?

"Change does not necessarily assure progress, but progress implacably requires change. Education is essential to change for education creates both wants and the ability to satisfy them" (Henry Steele Commager). Which of the following follow from Commager's statement? Which contradict it?

▶ **110.** It's possible for there to be change without progress.

111. It's possible for there to be progress without change.

112. It's possible for there to be education without change.

113. It's possible for there to be change without education.

114. It's possible for there to be education without progress.

115. It's possible for there to be progress without education.

Virgil's *Aeneid* contains the line, "They can because they think they can." We might try to symbolize this in several different ways (where *a* stands for the group Virgil discusses, *Tx* represents '*x* thinks *x* can,' and *Fx* represents '*x* does').

▶ **116.** $\Diamond Fa \dashv 3\ Ta$*   **117.** $Ta \dashv 3\ \Diamond Fa$*

118. $Ta\ \varepsilon 3\ \Diamond Fa$*   **119.** $Ta\ \Box \to\ \Diamond Fa$*

120. $\Diamond Fa\ \Box \to\ Ta$*   **121.** $\neg Ta\ \Box \to\ \neg \Diamond Fa$*

122. $\neg \Diamond Fa\ \Box \to\ \neg Ta$*

123. $(Ta\ \Box \to\ \Diamond Fa)\ \&\ (\neg Ta\ \Box \to\ \neg \Diamond Fa)$*

124. $(\Diamond Fa\ \Box \to\ Ta)\ \&\ (\neg \Diamond Fa\ \Box \to\ \neg Ta)$*

125. $(Ta\ \Box \to\ \Diamond Fa)\ \&\ (\Diamond Fa\ \Box \to\ Ta)$*

126. $(\neg \Diamond Fa\ \Box \to\ \neg Ta)\ \&\ (\neg Ta\ \Box \to\ \neg \Diamond Fa)$*

127. $(Ta\ \&\ \Diamond Fa)\ \&\ (\neg Ta\ \Box \to\ \Diamond Fa)$*

128. $(Ta\ \&\ \Diamond Fa)\ \&\ (\neg \Diamond Fa\ \Box \to\ \neg Ta)$*

Which of these imply (a) that they couldn't, if they didn't think they could ($\neg Ta\ \Box \to\ \neg \Diamond Fa$)? (b) that they can ($\Diamond Fa$)? (c) that they think they can because they can? (d) that they can because it's possible for them to think they can? (e) that it's necessarily true that they can because they think they can? **Assuming that (a) and (b) should follow, and (c), (d), and (e) should not, which seems like the best symbolization?

Quine writes: "Mathematicians may conceivably be said to be necessarily rational and not necessarily two-legged; and cyclists necessarily two-legged and not necessarily rational. But what of an individual who counts among his eccentricities both mathematics and cycling?"[3]

129. Can there be a mathematician who is also a cyclist, if we interpret Quine's assumptions as *de dicto*?

130. Can there be a mathematician who is also a cyclist, if we interpret Quine's assumptions as *de re*?

Say that the predicate F is *essential* iff $\Box\forall x(Fx \to \Box Fx)$, *weakly contingent* iff $\Diamond\exists xFx$ & $\Diamond\exists x\neg Fx$, and *strongly contingent* iff $\Box\forall x(\Diamond Fx$ & $\Diamond\neg Fx)$. Prove or refute the following claims.

131. Self-identity is an essential predicate.

132. If F is essential, not-F is essential.

133. If F is weakly contingent, not-F is weakly contingent.

134. If F is strongly contingent, not-F is, too.

135. If F is strongly contingent, then F is weakly contingent.

136. If F is essential, then F is not strongly contingent.

137. If F is essential, then F is not weakly contingent.

138. If F is not essential, then F is strongly contingent.

139. If F is not essential, then F is weakly contingent.

▶ **140.** If F is essential, possibly-F is essential.

141. If F is essential, necessarily-F is essential.

142. Possibly-F is always essential.

143. Necessarily-F is always essential.

144. If F and G are essential, F-or-G is essential.

145. If F and G are essential, F-and-G is essential.

146. If F and G are essential, F-only-if-G is essential.

147. If F and G are essential, F-iff-G is essential.

148. If F and G are essential, F-if-it-were-the-case-that-G is essential.

149. If F and G are weakly contingent, F-or-G is weakly contingent.

▶ **150.** If F and G are weakly contingent, F-and-G is weakly contingent.

151. If F and G are weakly contingent, F-if-it-were-the-case-that-G is weakly contingent.

152. If F and G are strongly contingent, F-or-G is strongly contingent.

153. If F and G are strongly contingent, F-and-G is strongly contingent.

154. If F and G are strongly contingent, F-if-it-were-the-case-that-G is strongly contingent.

155. If F-or-G is essential, both F and G are essential.

156. If F-and-G is essential, both F and G are essential.

157. If *F*-or-*G* is weakly contingent, both *F* and *G* are weakly contingent.

158. If *F*-or-*G* is weakly contingent, either *F* or *G* is weakly contingent.

159. If *F*-and-*G* is weakly contingent, both *F* and *G* are weakly contingent.

▶ **160.** If *F*-and-*G* is weakly contingent, either *F* or *G* is weakly contingent.

161. If *F*-or-*G* is strongly contingent, both *F* and *G* are strongly contingent.

162. If *F*-and-*G* is strongly contingent, both *F* and *G* are strongly contingent.

163. If *F*-or-*G* is strongly contingent, either *F* or *G* are strongly contingent.

164. If *F*-and-*G* is strongly contingent, either *F* or *G* are strongly contingent.

12.2 Free Logic

In this section we'll develop a new logic, which we'll call FS5 (for "free S5"), to solve the problems of the last section. To devise a logical system which counts the Barcan formula and its relatives invalid, we need to alter QS5 in several ways.

Modal Closure

A crucial step in showing the Barcan formula valid involves reiterating a quantified formula into a modal proof (or allowing such a formula to cross a world shift line). If we reject the assumption that all possible worlds contain the same objects, however, we have no reason to allow such steps. Formulas such as $\exists x \Box Fx$, $\forall y \Diamond Fy$, and $\exists z \Diamond Gz$ need not maintain the same truth value in every possible world. Consequently, we won't any longer count these formulas as modally closed. We'll adapt our original definition of modal closure.

> **Definition.** A formula is *modally closed in FS5* iff each of its quantifiers and each of its atomic subformulas occur within the scope of a modal connective.

We'll continue, then, to consider $\Box \forall x Fx$, $\exists x Fx \dashv Ga$, and $\Diamond Fa \vee (Gb \mathbin{\text{\rotatebox[origin=c]{180}{ε}}} Hc)$ modally closed. But $\forall x \Box Fx$, $\exists x \Diamond Fx$, etc., are not modally closed in FS5; they contain quantifiers that fall outside the scope of any modal connective.

Quantifier Rules

To eradicate fully the assumption that all worlds share the same objects, we must do more than alter our conception of modal closure. To see why, observe that, even with our revision to QS5, we can show that everything exists necessarily:

$$\forall x \square \exists y \; x = y \quad \checkmark$$
$$\square \exists y \; a = y \quad \checkmark$$

$$\exists y \; a = y*$$
$$a = a$$

Cl

1. ~~Show~~ $\forall x \square \exists y \; x = y$
2. ~~Show~~ $\square \exists y \; a = y$
   $\square$
3. $a = a$ =I
4. $\exists y \; a = y$ ∃I, 3

Where do our rules go wrong? In the deduction argument, step 4 seems to be the root of the problem. Nothing in the deduction indicates that a exists in the world we're discussing at that point. Similarly, in the tableau above, we substitute a for y, moving from the information that nothing is a in a certain world—that is, that a doesn't exist in that world—to the conclusion that a isn't a. The questionable tableau and deduction steps both involve rules for the existential quantifier. Apparently, then, to revise QS5 to allow possible worlds to contain different objects, we must change our quantifier rules.

Specifically, we need to change our quantifier rules to account for terms in our language that refer to objects that don't exist in some possible worlds. Standard quantification theory commits us to the existence of an object corresponding to every constant or function term in the language: $\exists x \; x = a$ and $\exists x \; x = f(a)$ are valid. QS5 correspondingly counts $\square \exists x \; x = a$ and $\square \exists x \; x = f(a)$ valid.

Free logic liberates us from this commitment. Free logic is useful for eliminating classical logic's commitment to the existence of a denotation for every closed term in the language. In modal logic, as we've seen, that commitment extends to a more dubious assumption that every term in the language denotes a necessarily existing object.[4]

To formulate free quantification rules, we'll first define a useful abbreviation.

DEFINITION. Where ℓ is any constant or function term, E!ℓ iff $\exists x \; x = \ell$.

We can read E!a as 'a exists.' The symbol E!—called "E shriek"—acts as a singulary predicate expressing existence.

What is the difference between the functions of ∃ and E!? E! acts as a predicate, while ∃ is a quantifier. ∃, in other words, functions much as the English determiners such as *some, a,* and *an* do. E! functions like the verb *exist*. We can define E! in terms of ∃, but not vice versa. In classical quantification theory and in QS5, there is no point to introducing E! at all, since everything under discussion exists. Formulas of the form ∃x x = t play little role in analyzing arguments within these theories. In FS5, however, we want to be able to discuss things that don't exist in certain possible worlds. The issue of what exists thus becomes important in FS5.

Given the predicate E!, we can easily see how to revise the rule of existential introduction. We can move from an instance with a term t to its corresponding existential generic if we know that t exists. So we can formulate the rule as follows:

Free Existential Introduction (F∃I)

n. $\mathscr{A}[t/v]$
m. $E!t$
p. $\exists v \mathscr{A}$ F∃I, n, m

This is the same as the classical rule, except that it requires an additional premise of the form $E!t$.

Notice that this rule eliminates the assumption that all terms denote something. In classical quantificational logic, ∃x x = c is valid. But the proof fails when we replace ∃I with F∃I: We cannot move from $c = c$ to ∃x x = c unless we've already established E!c.

In fact, F∃I also eliminates our logic's assumption that something exists. Russell fretted that the system of *Principia Mathematica* contained this assumption; although ∃x x = x is classically valid, the existence assumption it carries, Russell thought, should not be a part of logic. The validity of ∃x x = x corresponds to a requirement that the domain—the universe of discourse—be nonempty. Classical logic has been called *exclusive* because it excludes the possibility of a null domain.

Free logic, in contrast, is *inclusive:* It allows for the possibility that nothing exists. We cannot derive ∃x x = x from $c = c$ unless we already have E!c; but E!c simply states that c exists. Interpretations of free quantificational logic are similar to those of classical logic in that they consist of a set of objects (a domain) and an interpretation function linking the constants, predicates, and function symbols of the language to the objects. The only differences are that, in free logic, the domain may be empty, and some constants may not denote.

Very similar in character is the free version of universal exploitation. We can instantiate a universally quantified formula with any term, t, provided that we have established $E!t$.

Free Universal Exploitation (F∀E)

n $\forall v \mathscr{A}$
m. $\underline{E!\ell}$
p. $\mathscr{A}[\ell/v]$ F∀E, n, m

The rule of free existential exploitation, like the classical rule, allows us to derive from an existentially quantified formula an instance with a new constant. So, from $\exists x Fx$ we can derive Fc, provided that c is new to the proof. But, if there is an F, we can conclude not only that some c is F but that some existing c is F. So we can also derive E!c.

Free Existential Exploitation (F∃E)

n. $\underline{\exists v \mathscr{A}}$
m. $\mathscr{A}[c/v] \ \& \ E!c$ F∃E, n

Here c must be a constant new to the proof.

To see how these rules work, consider:

$\forall x(Fx \rightarrow Gx)$
$\exists x(Hx \ \& \ \neg Gx)$
$\therefore \ \exists x(Hx \ \& \ \neg Fx)$

This is valid in both classical and free logic, as this proof shows.

1.	$\forall x(Fx \rightarrow Gx)$	A
2.	$\exists x(Hx \ \& \ \neg Gx)$	A
3.	Show $\exists x(Hx \ \& \ \neg Fx)$	
4.	$(Ha \ \& \ \neg Ga) \ \& \ E!a$	F∃E, 2
5.	$E!a$	&E, 4
6.	$Ha \ \& \ \neg Ga$	&E, 4
7.	$Fa \rightarrow Ga$	F∀E, 1, 5
8.	$\neg Ga$	&E, 6
9.	$\neg Fa$	→E*, 7, 8
10.	Ha	&E, 6
11.	$Ha \ \& \ \neg Fa$	&I, 10, 9
12.	$\exists x(Hx \ \& \ \neg Fx)$	F∃I, 11, 5

Exploiting the universal quantification in $\forall x(Fx \rightarrow Gx)$ and introducing the existential quantification in $\exists x(Hx \ \& \ \neg Fx)$ both required the presence of E!a. Fortunately, we obtained E!a from applying free existential exploitation to $\exists x(Hx \ \& \ \neg Gx)$.

Finally, we need to revise our method of universal proof. To show that everything is *F*, we need to show that any arbitrarily chosen thing is *F*. But we needn't prove an arbitrary instance of our universally quantified formula, because some terms denote nothing. Instead, we must prove an arbitrary instance with a denoting constant: a constant that refers to an existing object. In other words, to show that everything is *F*, we need to show that everything that exists is *F*.

Free Universal Proof

$$n. \quad \text{Show } \forall v \mathcal{A}$$
$$n + m. \quad \lceil \text{Show E}!c \to \mathcal{A}[c/v]$$

Here *c* must be a constant new to the proof.

A proof of the validity of the following argument form illustrates the use of free universal proof.

1.	$\forall x(Fx \to Gx)$	A
2.	$\forall x(Hx \to \neg Gx)$	A
3.	Show $\forall x(Fx \to \neg Hx)$	
4.	$\lceil$ Show E!$a \to (Fa \to \neg Ha)$	
5.	$\lceil$ E!a	ACP
6.	Show $Fa \to \neg Ha$	
7.	$\lceil Fa$	ACP
8.	$Fa \to Ga$	F∀E, 1, 5
9.	Ga	→E, 7, 8
10.	$Ha \to \neg Ga$	F∀E, 2, 5
11.	$\neg Ha$	→E*, 9, 10

These three rules, the new method of universal proof, and the definition of E!, together with classical rules for sentential connectives, constitute a system of free logic. Free logic forms a subsystem of classical logic: Everything valid in free logic is valid in classical logic. But some formulas and argument forms that are valid in classical logic are invalid in free logic.

When we combine free quantificational logic with modal connectives, we obtain a system of free quantified modal logic, FS5. Everything valid in FS5 is also valid in QS5, but not vice versa.

We can formulate FS5 by means of tableau rules. Free universal left and existential left correspond closely to free universal and existential exploitation, respectively. Free universal right is very similar to free universal proof; free existential right looks very much like free existential introduction.

Given a universally quantified formula on the left, we can write any instance of that formula on the left, provided that we instantiate with a term that denotes something. So, we can move from a universal formula to an instance involving a term, t, only if we already have E!t on the left.

Free Universal Left (F∀L)

$$
\begin{array}{c|}
\text{E!}t \\
*\ \forall v \mathscr{A} \\
\mathscr{A}[t/v]
\end{array}
$$

If an existentially quantified formula appears on the left, we assume it's true. So, we can write an instance of the formula on the left, using a new constant, c. Further, we can write E!c on the left, to say that the new constant refers to an existing object.

Free Existential Left (F∃L)

$$
\begin{array}{c|}
\checkmark\quad \exists v \mathscr{A} \\
\mathscr{A}[c/v] \\
\text{E!}c
\end{array}
$$

Here c must be a constant new to the branch.

If a universally quantified formula appears on the right, we assume it to be false. That is, some instance of it must be false, and so we can write an instance with a new constant on the right. And, just as with free existential left, we can write a formula on the left indicating that the new constant denotes an existing entity.

Free Universal Right (F∀R)

$$
\begin{array}{|c}
\forall v \mathscr{A} \quad \checkmark \\
\mathscr{A}[c/v]
\end{array}
$$
$$
\text{E!}c\ \big|
$$

Here c must be a constant new to the branch.

Finally, if we find an existentially quantified formula on the right, we can record any instance of it on the right, provided that we instantiate with a term that denotes something.

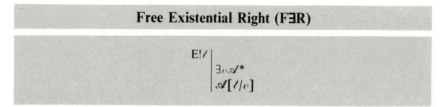

Free Existential Right (F∃R)

$$\begin{array}{c|c} E!\prime \\ \hline & \exists_v\mathcal{A}* \\ & \mathcal{A}[\prime/v] \end{array}$$

To illustrate these rules, let's consider some simple argument forms and their corresponding tableaux.

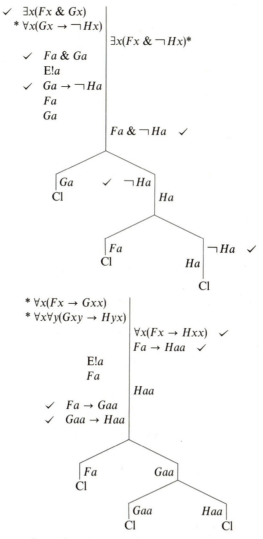

These tableaux close; the argument forms with which they begin are valid in free logic. To see some tableaux that don't close, let's return to the Barcan formula and its relatives. We can begin a tableau to test the validity of the Barcan formula in free logic as follows:

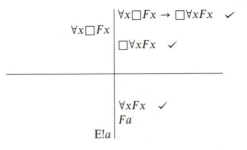

$$\forall x \Box Fx \quad \begin{array}{|l} \forall x \Box Fx \rightarrow \Box \forall x Fx \quad \checkmark \\[1em] \Box \forall x Fx \end{array}$$

Now, we seem to face a choice between applying free universal left and applying necessity right. In fact, however, only the latter is feasible. Free universal left allows us to write an instance of a universally quantified formula, provided that the appropriate existence formula appears earlier on the left. Without a formula of the form E!*t*, we can't apply any rule to $\forall x \Box Fx$; we must turn to the formula on the right, using $\Box$R. We thus generate a world shift line.

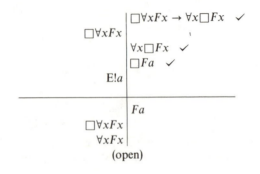

Having applied free universal right, we have *Fa* on the right and E!*a* on the left. Now it's tempting to return to $\forall x \Box Fx$; we have the existence formula we needed. But we've obtained it too late. $\forall x \Box Fx$ is not modally closed in FS5, and so it doesn't survive the world shift. Because no further rules can apply, the tableau terminates, remaining open.

The tableau for the Barcan formula tells us not only that the formula isn't valid but also why it isn't valid. There may be a possible world in which something exists that isn't *F*, even though everything in our world is necessarily *F*. The appearance of E!*a* only below the world shift line is a clue that the possibility of adding objects to the domain invalidates the Barcan formula.

The tableau for the converse Barcan formula is similar:

$$\Box \forall x Fx \quad \begin{array}{|l} \Box \forall x Fx \rightarrow \forall x \Box Fx \quad \checkmark \\[0.5em] \forall x \Box Fx \quad \checkmark \\ \Box Fa \quad \checkmark \end{array}$$

E!*a*

$$\begin{array}{|l} Fa \end{array}$$

$\Box \forall x Fx$
$\forall x Fx$

(open)

We begin the tableau by applying →R and free ∀R. We then apply □R, drawing a world shift line. □∀xFx is modally closed, and so it survives the shift. We then apply □L and seem on the verge of closing the tableau. But we can apply free universal left only if we have available the appropriate existence formula. At first glance, we appear to have that formula; E!a appears above on the left. But the formula is no longer live. It appears above the world shift line. Since our rules all apply only to live formulas, the tableau ends. Because the tableau remains open, the converse Barcan formula is invalid in FS5.

Again, the tableau shows why FS5 reckons the formula invalid. Something—say a—that exists in our world may fail to exist in some other worlds. Even if everything that exists is F, in every possible world, it doesn't follow that a is F in every world. There may be a world in which a isn't F. Such a world, however, must be a world in which F doesn't exist. Thus, the possibility that we can subtract objects from the domain is responsible for the invalidity of the converse Barcan formula in FS5.

A third formula that is valid in QS5 but invalid in FS5 is ◇∀xFx → ∀x◇Fx.

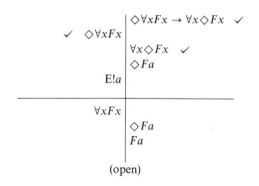

(open)

We generate the world shift line by applying ◇L. ◇Fa survives the shift because it's modally closed. Applying ◇R, we obtain Fa on the right. In QS5, we could apply ∀L to ∀xFx to obtain Fa on the left, closing the tableau. But in free logic, the rule for universal formulas on the right applies only if E!a appears live on the left. Just as in the tableau for the converse Barcan formula, E!a appears above, but on the other side of a world shift line. Once again, the possibility of deleting an object from the domain invalidates a formula.

Since FS5 is a subsystem of QS5, it seems natural to ask what principles we would have to add to FS5 to yield QS5. There's a simple answer: We'd have to add one rule, to the effect that, given any term *t*, at any point in a proof or on a tableau, we can assert E!*t*. In other words, we'd have to assume that every term in the language denoted an object, no matter what possible world we happened to be discussing. This assumption captures the highly objectionable thesis that every term in the language denotes something that exists necessarily.

Problems

Symbolize the following sentences, using E! where appropriate.

1. God is, and all is well (John Greenleaf Whittier).

2. To fail to love is not to exist at all (Mark van Doren).

3. There may now exist great men for things that do not exist (Jakob Burckhardt).

4. There is no God, and Mary is His mother (Robert Lowell).

▶ **5.** I think, therefore I am (René Descartes).

Use tableaux to determine whether these argument forms are valid in FS5.

6. $\exists x(Fx \,\&\, Ga)$; ∴ $\exists xFx \,\&\, Ga$ **7.** $\exists xFx \,\&\, Ga$; ∴ $\exists x(Fx \,\&\, Ga)$

8. $\exists x(Fx \rightarrow Gb)$; ∴ $\forall xFx \rightarrow Gb$ **9.** $\forall xFx \rightarrow Gb$; ∴ $\exists x(Fx \rightarrow Gb)$

10. $\forall x(Fx \rightarrow Ha)$; ∴ $\exists xFx \rightarrow Ha$ **11.** $\exists xFx \rightarrow Ha$; ∴ $\forall x(Fx \rightarrow Ha)$

12. $\forall xFx$; ∴ $\exists xFx$ **13.** $\exists x\forall yFxy$; ∴ $\forall y\exists xFxy$

14. $\forall x\exists y(Fx \lor Gy)$; ∴ $\exists y\forall x(Fx \lor Gy)$

▶ **15.** $\exists x(Fx \,\&\, \Diamond Gx)$; $\forall x(Gx \dashv_3 \Box Hx)$; ∴ $\exists x(Fx \,\&\, \Box Hx)$

16. $\forall x(Fx \dashv_3 Gx)$; $\forall x(Gx \dashv_3 Hx)$; ∴ $\forall x(\neg Hx \dashv_3 \neg Fx)$

17. $\Box\exists x(Fx \,\&\, Gax)$; $\Box\forall x\forall y(\Diamond Gxy \rightarrow \Box Gyx)$; ∴ $\exists x(Fx \,\&\, Gxa)$

18. $\forall x(Fx \dashv_3 Gx)$; $\exists x\Diamond(Fx \,\&\, Hx)$; ∴ $\exists x\Diamond(Gx \,\&\, Hx)$

19. $\forall x(Fx \dashv_3 \Diamond Gx)$; $\Box\forall x(Gx \dashv_3 \Box Hx)$; ∴ $\exists xFx \rightarrow \exists x\Box Hx$

▶ **20.** $\Box\forall x\forall y(Rxy \rightarrow Ryx)$; ∴ $\forall x\forall y(Rxy \dashv_3 Ryx)$

21. $\Box\forall x(\Diamond Fx \rightarrow \Box Hx)$; $\Diamond\exists x\Diamond\neg Hx$; ∴ $\Diamond\exists x\neg\Diamond Fx$

Using natural deduction, show that these argument forms are valid in FS5.

22. $\exists x(Fx \,\&\, Gx)$; $\forall x(Fx \rightarrow Hx)$; ∴ $\exists x(Hx \,\&\, Gx)$

23. $\forall x(Fx \rightarrow Gxx)$; $\forall x(Hax \rightarrow \neg\exists yGxy)$; ∴ $\neg\exists x(Fx \,\&\, Hax)$

24. $\exists x\forall y(Fy \rightarrow Gxy)$; $\forall x\forall y(Gyx \rightarrow \exists zGxz)$; ∴ $\forall x(Fx \rightarrow \exists yGxy)$

▶ **25.** $\exists x(Fx \,\&\, \Box Gx)$; $\forall x(Fx \rightarrow Hx)$; ∴ $\exists x(Hx \,\&\, \Box Gx)$

26. $\Box\forall x(Fx \rightarrow Gx)$; $\Diamond\forall x(Gx \rightarrow Hx)$; ∴ $\Diamond\forall x(Fx \rightarrow Hx)$

27. $\Diamond\exists x(Fx \,\&\, Gx)$; $\Box\forall x(Fx \rightarrow Hxx)$; ∴ $\Diamond\exists x(Gx \,\&\, \exists yHxy)$

28. $Ga \dashv3 \exists y Fya$; $\Box\forall x\forall y(Fxy$ &3 $Fyx)$; $\therefore \Diamond(Ga\&E!a) \dashv3 \Diamond\exists z Faz$

29. $\Box\forall x\forall y(Fxy \rightarrow (Gyx \& Hxx))$; $\Box\exists x\forall y Fxy$; $\therefore \Box\exists x(Hxx \& Gxx)$

▶ **30.** $\Diamond\exists x(\Box Fx \& Gx)$; $\Box\forall x(Gx \dashv3 \Diamond\exists y Hxy)$; $\therefore \Diamond\exists x\Diamond\exists y(Hxy \& \Box Fx)$

12.3 IDENTITY AND DESCRIPTIONS

Many philosophical puzzles concerning quantified modal logic pertain to identity. In QS5 and FS5, all true identity statements that don't contain descriptions are necessarily true. The formula

(17) $\forall x\forall y(x = y \rightarrow \Box x = y)$

is valid, as the following tableau and deduction (in FS5) show:

$$
\begin{array}{ll}
& \forall x\forall y(x = y \rightarrow \Box x = y) \quad \checkmark \\
& a = b \rightarrow \Box a = b \quad \checkmark \\
E!a & \\
E!b & \\
*\ a = b & \\
& \Box a = b \\
& \Box a = a \\
\hline
& a = a \\
& \text{Cl}
\end{array}
$$

1.	Show $\forall x\forall y(x = y \rightarrow \Box x = y)$	
2.	Show $E!a \rightarrow (E!b \rightarrow (a = b \rightarrow \Box a = b))$	
3.	$E!a$	ACP
4.	Show $E!b \rightarrow (a = b \rightarrow \Box a = b)$	
5.	$E!b$	ACP
6.	Show $a = b \rightarrow \Box a = b$	
7.	$a = b$	ACP
8.	Show $\Box a = a$	
	$\Box$	
9.	$a = a$	$= I$
10.	$\Box a = b$	$= E, 7, 8$

These demonstrations are circuitous because $a = b$ isn't modally closed. It can't cross a world shift line or reiterate into a modal proof.

In fact, we can strengthen this result: In FS5 and QS5, all identity statements not involving descriptions are necessarily true or necessarily false. The formula

(18) $\forall x\forall y(\Box x = y \vee \Box x \neq y)$

is valid in FS5, as these show:

$$\forall x \forall y (\Box x = y \vee \Box x \neq y) \quad \checkmark$$
$$\Box a = b \vee \Box a \neq b \quad \checkmark$$

E!a
E!b

$$\Box a = b$$
$$\Box a \neq b \quad \checkmark$$

* a = b

$$a \neq b \quad \checkmark$$

$$\Box a = b$$
$$\Box a = a \quad \checkmark$$

$$a = a$$

CI

1. Show $\forall x \forall y (\Box x = y \vee \Box x \neq y)$
2. Show E!a $\rightarrow$ (E!b $\rightarrow$ ($\Box a = b \vee \Box a \neq b$))
3. E!a   ACP
4. Show E!b $\rightarrow$ ($\Box a = b \vee \Box a \neq b$)
5. E!b   ACP
6. Show $\Box a = b \vee \Box a \neq b$
7. $\neg(\Box a = b \vee \Box a \neq b)$ AIP
8. $\neg \Box a = b \ \& \ \neg \Box a \neq b$ $\neg \vee$, 7
9. $\neg \Box a = b$ &E, 8
10. $\neg \Box a \neq b$ &E, 8
11. Show $\Box a \neq b$
    $\Box$
12. Show $a \neq b$
13. $a = b$ AIP
14. $\neg \Box a = b$ $\Box$R, 9
15. $\neg \Box a = a$ =E, 13, 14
16. Show $\Box a = a$
    $\Box$
17. $a = a$ =I

In FS5 and QS5, then, no identity statements without descriptions are contingent; all are either necessarily true or necessarily false.

FS5 and QS5 reflect the view, therefore, that, if a name refers, it refers to the same thing in every possible world. Kripke, who first argued for this position, calls names *rigid designators*.[5]

Definition. A term, ℓ, is a *rigid designator* iff it refers to the same object in every possible world—iff, that is, $\Box \forall x (x = \ell \rightarrow \Box x = \ell)$.

A name such as *Daniel Bonevac* appears to pick out the same man, no matter what circumstance we might happen to be discussing.[6]

Descriptions such as 'the president of the United States' or 'the first man to circumnavigate the globe alone,' in contrast, typically refer to different objects in different possible circumstances. 'The president of the United States' referred to Ronald Reagan in 1985, Gerald Ford in 1975, Lyndon Johnson in 1965, and Dwight Eisenhower in 1955. But it could have referred to Walter Mondale, Richard Nixon, Barry Goldwater, and Adlai Stevenson in those years, respectively, had some elections and investigations turned out differently. Similarly, 'the first man to circumnavigate the globe alone' refers to Joshua Slocum in the actual world, but it might have referred to Vasco da Gama, Ferdinand Magellan, Commodore Perry, Lord Nelson, or Gerald Massey if things had been somewhat different. If Magellan had sailed around the world without companions, we might say, he'd have been the first man to circumnavigate the globe alone. But when we say, "If Magellan had . . . ," we are talking about possible circumstances in which Magellan himself did something. Names, then, designate rigidly, but descriptions generally do not.

Because descriptions generally aren't rigid designators, statements involving them engender a further complication. Consider this argument:[7]

(19) The number of planets is nine. Nine is necessarily greater than seven. Therefore, the number of planets is necessarily greater than seven.

The conclusion of this argument is ambiguous. On one reading of its conclusion, the argument seems invalid. Nine is necessarily greater than seven, and there are nine planets, but surely there could have been fewer than seven planets.

We can symbolize this in one of three ways, depending on whether we take the description 'the number of the planets' as a singular term or a quantifier:

(20) a. $\imath x P x = 9$
$\Box 9 > 7$
$\therefore \Box \imath x P x > 7$
b. $(\tau x\colon Px)x = 9$
$\Box 9 > 7$
$\therefore \Box(\tau x\colon Px)x > 7$
c. $(\tau x\colon Px)x = 9$
$\Box 9 > 7$
$\therefore (\tau x\colon Px)\Box x > 7$

The last of these argument forms, (20)c, is clearly valid. If there are nine planets, and nine is necessarily greater than seven, then that number (namely, nine) must be necessarily greater than seven. But this doesn't imply that it's a necessary truth that the number of planets is greater than seven; there might have been only six planets in our solar system. So, (20)b ought to be invalid.

(20)a, so long as we maintain our policy concerning the necessity of atomic identity statements, is valid. It represents the same reading of the

argument as (20)c. To contrast the alternative readings of the conclusion of (19), therefore, let's restrict our attention to (20)b and c, taking the descriptions as quantifiers.

To analyze arguments involving descriptions in FS5, we need to develop a free logic for descriptions. Luckily, this isn't hard. If we know, for example, that the F is G, we know that an existing thing is G, and that it's the unique F. Thus, the deduction rule for exploiting descriptions is:

Free Description Exploitation (FτE)

$$
\begin{array}{ll}
\text{n.} & (\tau v \colon \mathscr{A})\mathscr{B} \\
\text{n + p.} & \overline{\text{E!}c \,\&\, \mathscr{B}[c/v] \,\&\, \forall v(v = c \leftrightarrow \mathscr{A})} \qquad \text{FτE, n}
\end{array}
$$

Here c must be a constant new to the proof.

Closely related is a tableau rule for descriptions on the left:

Free Description Left (FτL)

$$
\begin{array}{r|}
\checkmark \;\; (\tau v \colon \mathscr{A})\mathscr{B} \; \\
\text{E!}c \; \\
\mathscr{B}[c/v] \; \\
\forall v(v = c \leftrightarrow \mathscr{A}) \;
\end{array}
$$

Here c must be a constant new to the branch.

To introduce a description into a proof, we need to know that there is a unique $\mathscr{A}$, and that it is $\mathscr{B}$:

Free Description Introduction (FτI)

$$
\begin{array}{ll}
\text{n.} & \text{E!}t \\
\text{m.} & \mathscr{B}[t/v] \\
\text{q.} & \underline{\forall v(v = t \leftrightarrow \mathscr{A})} \\
\text{p.} & (\tau v \colon \mathscr{A})\mathscr{B} \qquad \text{FτI, n}
\end{array}
$$

Here t may be any term.

Closely related, again, is a tableau rule for descriptions on the right, which says that, if it's false that the $\mathscr{A}$ is $\mathscr{B}$, then either nothing is $\mathscr{A}$, or the $\mathscr{A}$ isn't $\mathscr{B}$, or something other than the $\mathscr{A}$ is $\mathscr{B}$.

Free Description Right (FτR)

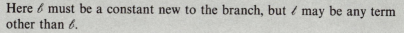

Here ℓ must be a constant new to the branch, but ℓ may be any term other than ℓ.

Verifying that (20)c is valid isn't difficult:

1. $(\tau x: Px)x = 9$
2. $\Box 9 > 7$
3. Show $(\tau x: Px)\Box x > 7$
4. $\lceil$ E!a & $a = 9$ & $\forall x(x = a \leftrightarrow Px)$ FτE, 1
5. $|$ $a = 9$ &E, 4
6. $|$ $\Box a > 7$ =E, 5, 2
7. $|$ $\forall x(x = a \leftrightarrow Px)$ &E, 4
8. $|$ E!a &E, 4
9. $\lfloor$ $(\tau x: Px)\Box x > 7$ FτI, 6, 7, 8

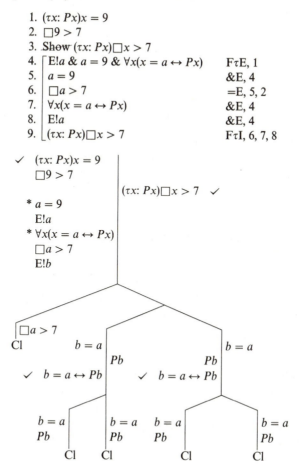

(20)b, however, is invalid—even if we further assume that 9 exists necessarily—as this tableau illustrates:

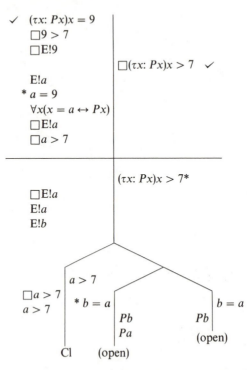

Nothing but $\Box a > 7$ and $\Box 9 > 7$ survives the world shift, and neither of these can help to close the open branches. The open branches reflect the possibility that, in another possible world, something other than 9 is the number of the planets.

Problems

Evaluate these arguments in either QS5 or FS5.

1. The man who might buy our house is fat. Everyone who might buy our house is wealthy. Thus, some man is fat and wealthy.

2. It's impossible that the youth who robbed the store is Manny, for Manny is tall, but the robber must have been short.

3. Rocky is the fighter who is impossible to beat. Hence, Rocky cannot be beaten.

4. Jane is the debater who can outargue everyone. Therefore, someone can outargue Jane.

▶ 5. Cicero is Tully. Cicero could speak more effectively than anyone else in Rome. Cataline was a Roman. Therefore, it was possible for Tully to speak more effectively than Cataline.

6. Hamlet doesn't exist, but Olivier does. So, Olivier couldn't possibly be Hamlet.

7. It's possible for anyone to fear Dracula; necessarily, Dracula is afraid only of me. So, necessarily, I am Dracula.

8. Cicero is Tully. So, it's necessarily true that the person who is Cicero is the person who is Tully.

9. If the king of France is bald, he needs a wig. If he's not bald, he needs a barber. So, the king of France needs a barber or a wig.

▶ 10. Pegasus is a winged horse. But it's impossible for a horse to have wings. So, Pegasus couldn't exist.

Which of the following are implied by the assertion that ℓ and ℓ' are rigid designators?

11. $\forall x(\Diamond x = \ell \rightarrow x = \ell)$

12. $\forall x(\Box x = \ell \vee \neg \Diamond x = \ell)$

13. $\ell = \ell' \rightarrow \Box \ell = \ell'$

14. $\Diamond \ell = \ell' \rightarrow \ell = \ell'$

▶ 15. $\ell \neq \ell' \rightarrow \Box \ell \neq \ell'$

16. $\Diamond \ell \neq \ell' \rightarrow \ell \neq \ell'$

17. $\Box F\ell \dashv \Box F\ell'$

18. $\imath x\, x = \ell$ is a rigid designator.

19. $\Box \ell = \imath x\, x = \ell$

▶ 20. $(\tau x: x = \ell)\Box x = \ell$

21. $(\tau x: x = \ell)\Box Fx \leftrightarrow \Box F\ell$

22. $\Box(\tau x: x = \ell)Fx \leftrightarrow \Box F\ell$

These are some attempted proofs of God's existence. Which are successful in FS5?

23. It's necessarily true that God is the necessarily existing entity. Therefore, God necessarily exists.

24. God is that, the greater than which cannot be conceived. If x exists, and y doesn't, then x is greater than y. Of course, if x is greater than y, then y is not greater than x. So, God exists.

▶ 25. The existence of God is at least possible. If God exists, He exists necessarily. Therefore, God exists necessarily.

26. For every existing thing x, there is another thing y such that, if y didn't exist, then x wouldn't exist either. Therefore, there is something such that, if it didn't exist, nothing would exist.

12.4 QUANTIFICATION AND OBLIGATION

The contrast between *de re* and *de dicto* modal statements reappears in deontic statements. We analyzed *de dicto* statements as having no modal operator within the scope of a quantifier, as in $\Box \forall x Fx$ and $\Diamond \exists x Fx$; *de re* statements, in contrast, had no quantifier within the scope of a modal operator, as in $\forall x \Box F x$ and $\exists x \Diamond F x$. Of course, there are in addition simple modal statements, such as $\Box F a$ and $\Diamond G b$, and complex mixed statements, such as $\Box \forall x \Diamond G x$ and $\exists x \Box \forall y F x y$, that combine the features of *de re* and *de dicto* statements.

When we consider obligation and permission rather than necessity and possibility, we can contrast *de re* and *de dicto* deontic statements in much the same way. *De dicto* deontic statements have no deontic connective within the scope of a quantifier, while *de re* deontic statements have no quantifier within the scope of a deontic connective. So $O\forall x F x$, $P\forall y G y$, $O\exists z F z$ and $P\exists x \exists y H x y$ are all *de dicto;* $\forall x O F x$, $\forall y P G y$, $\exists z O F z$, and $\exists x \exists y P H x y$ are all *de re*. If we interpret F as 'fair,' G as 'generous,' and H as 'hates' and assume that our universe of discourse consists entirely of persons, we can read these formulas as follows:

(21)	*De Dicto*	
a.	$O\forall x F x$	It is obligatory that everyone be fair.
b.	$P\forall y G y$	It is permissible that everyone be generous.
c.	$O\exists z F z$	It is obligatory that someone be fair.
d.	$P\exists x H x x$	It is acceptable that some hate themselves.

(22)	*De Re*	
a.	$\forall x O F x$	Everyone has an obligation to be fair.
b.	$\forall y P G y$	Everyone has permission to be generous.
c.	$\exists z O F z$	Someone has an obligation to be fair.
d.	$\exists x P H x x$	Some have permission to hate themselves.

Other ways of conveying the content of these sentences are ambiguous.

(23) *Ambiguous*
a. It is obligatory for everyone to be fair.
b. Everyone ought to be fair.
c. It's permissible for everyone to be generous.
d. Everyone may be generous.
e. It's obligatory for someone to be fair.
f. Someone should be fair.
g. It's acceptable for some to hate themselves.
h. Some may hate themselves.

All these sentences are ambiguous. Each may be taken to correspond to either a *de re* and a *de dicto* meaning, that is, to a sentence in (21) or to a sentence in (22). (23)d and h are ambiguous in another way as well; we may read *may* as deontic, but we may also read it as expressing epistemic possibility. (That is, 'Some may hate themselves,' on one reading, is equivalent to 'For all I know, it's possible that some hate themselves'; see Chapter 9.)

A traditional distinction in moral theory corresponds closely to the contrast between *de re* and *de dicto* statements of obligation. *Perfect* obligations, the English philosopher John Stuart Mill said, are "duties in virtue of which a correlative *right* resides in some person or persons"; they are those incumbent on a particular person, to perform or refrain from performing a particular action.[8] Suppose that Jack loaned Jim $10; then Jim has an obligation to repay Jack. This is a perfect obligation, for it's Jim that has the obligation; *he* must repay Jack. Jack, correlatively, has a right to the money. We can think of perfect obligations as specific duties of particular people.

Imperfect obligations, however, do not fall on specific people or do not involve specific actions; they ought to be done to make the world a better place, but nobody in particular has a duty to do them. According to Mill, imperfect obligations are "those in which, though the act is obligatory, the particular occasions of performing it are left to our choice; as in the case of charity or beneficence, which we are indeed bound to practise, but not towards any definite person, nor at any prescribed time."[9] Giving to charity is an imperfect obligation because we are obliged to give to charity, but to no charity in particular. The obligation is indefinite; we are obliged to give to some charities, but there are no specific charities we must support. There are many charities to choose from.

Imperfect obligations, then, are indefinite in the sense that they mean that something ought to be the case, but they do not oblige any specific people to perform any specific actions. It may be imperfectly obligatory that something ought to be done without anyone in particular having an obligation to do it, or without there being any particular action that should be performed.

In quantified deontic logic, we can construe statements of imperfect obligation as *de dicto* statements and statements of perfect obligation as *de re* statements. Consider an ambiguous sentence, such as 'Someone ought to give Jack $10.' This sentence might express a perfect obligation; perhaps someone borrowed $10 from Jack, and the time has come to pay up. In that case, we can symbolize the sentence as $\exists xOGxjm$, where G symbolizes 'give,' j symbolizes 'Jack,' and m symbolizes '$10.' The formula specifies that someone in particular has an obligation to give Jack $10. Perhaps, however, the sentence expresses an imperfect obligation. Jack might really need $10 to buy some shoes, but he doesn't have it; so, someone should give him the money. Nevertheless, no particular person has a duty to give Jack $10. In this case, we can symbolize 'Someone ought to give Jack $10' as $O\exists xGxjm$. This formula symbolizes 'It ought to be the case that someone give $10 to Jack.'

To return to the case of charity, we may symbolize Diane's indefinite obligation to give some money to charity as $O\exists x\exists y(Cx \ \& \ My \ \& \ Gdyx)$, where G again symbolizes 'give,' C symbolizes 'is a charity,' M symbolizes 'is a sum of money,' and d symbolizes 'Diane.' This formula is *de dicto;* it stands for 'It ought to be the case that Diane give a sum of money to some charity.' The amount and the recipient are at her discretion.

To represent that Diane has an obligation to give to a particular charity—because she promised to, or because it is especially worthy—we can write $\exists x(Cx \,\&\, O\exists y(My \,\&\, Gdxy))$. This mixed formula symbolizes that there is a particular charity to which Diane should give some money. It leaves Diane the freedom to choose how much she will donate.

To represent that Diane has an obligation to donate a particular sum of money to charity—because, perhaps, that was a condition of her receiving the money in the first place—we can write $\exists y(My \,\&\, O\exists x(Cx \,\&\, Gdxy))$. This formula symbolizes that there is a sum of money that Diane should give to a charity but leaves to her the decision of which charity or charities to support.

Finally, to represent that Diane has a duty to give a specific sum of money to a specific charity, we can write $\exists x \exists y(Cx \,\&\, My \,\&\, OGdxy)$, which symbolizes that there is a charity to which Diane should give a particular sum of money. This formula is fully *de re*.

Mill believed that perfect and imperfect obligations play very different roles in moral theory. Perfect obligations amount to duties and correspond to rights enjoyed by others; imperfect obligations do not. If we fail to fulfill a perfect obligation, we violate somebody else's rights, and so do them an injustice. If we do not fulfill an imperfect obligation, in contrast, we violate nobody's rights, and do nobody an injustice. Nevertheless, we fail to do something we ought to do and so, do something wrong. Mill's theory holds that the sphere of perfect obligation corresponds precisely to the sphere of justice and injustice, while the sphere of imperfect obligation corresponds to the rest of morality.

Since perfect and imperfect obligations are quite different, we should expect statements expressing them not to be equivalent. It's easy to see that statements of imperfect obligation do not imply corresponding assertions of perfect obligation: $O\exists xFx$ does not imply $\exists xOFx$.

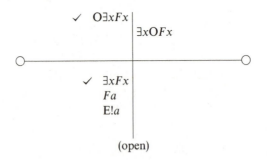

(open)

Notice that we can't begin the tableau by applying F∃R to the formula on the right; we need an existence formula to apply the rule. So, we have no choice but to begin with the obligation formula on the left. Applying OL, we draw an ideal world shift line and obtain $\exists xFx$, which yields in turn Fa and $E!a$ by F∃L. We can go no further; $\exists xOFx$ doesn't survive the world shift.

Even in QS5, imperfect obligations do not imply perfect obligations:

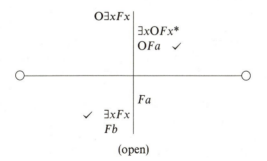

(open)

This tableau, too, is open.

It seems reasonable to think that perfect obligation statements imply assertions of imperfect obligation. This is true in QS5 but incorrect in FS5:

QS5

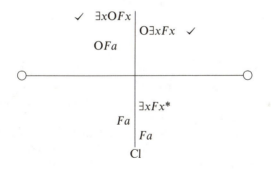

FS5

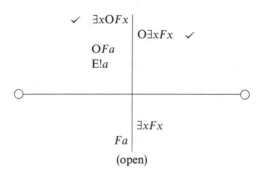

It appears that we could apply a rule to $\exists xFx$ on the right to obtain Fa on the right, closing the branch. But F∃R requires that we already have an appropriate live existence formula. We do have E!a on the left, but it doesn't survive the ideal world shift. The tableau therefore terminates.

In QS5, however, the inference from $\exists xOFx$ to $O\exists xFx$ is valid; standard quantifier rules make no requirements about existence formulas. The FS5 tableau indicates why FS5 counts the inference invalid. The world where a fulfills the perfect obligation may be a world in which a doesn't exist. This sounds bizarre. Indeed, in many contexts, QS5's assumption that the objects in the actual world also exist in any ideal world conforms best with our intuitions. But suppose that the statement of perfect obligation is 'There are events that should never have happened.' This statement is surely true. The corresponding statement of imperfect obligation is 'It ought to be the case that there be events that have never happened,' however, which sounds strange. The statement of perfect obligation doesn't imply that the world would have been a better place had some events both happened and not happened. In FS5, then, perfect obligations do not entail imperfect obligations, although counterexamples are fairly rare.

The logical system this chapter has developed is extremely powerful, as the exercises that follow indicate. But a vast array of natural-language constructions and arguments remain outside its boundaries.

Problems

Symbolize the following in the language of quantified modal logic, using deontic connectives where appropriate. Symbolize imperatives as mixed or *de re* statements of obligation, directed at people. For example, symbolize 'Don't find a fault, find a remedy' (Henry Ford) as $\forall x(Px \rightarrow O\neg \exists y(Ly \,\&\, Fxy)) \,\&\, \forall x(Px \rightarrow O\exists z(Rz \,\&\, Fxz))$. (The formula $\forall x(Px \rightarrow (O\neg \exists y(Ly \,\&\, Fxy) \,\&\, O\exists z(Rz \,\&\, Fxz)))$ is equivalent.) If a sentence refers only to persons, then it's acceptable to limit the domain to persons and omit $(Px \rightarrow \ldots)$.

1. Carthage must be destroyed! (Cato).

2. Fame is something which must be won; honor is something which must not be lost (Arthur Schopenhauer).

3. If you want to succeed you should strike out on new paths rather than travel the worn paths of accepted success (John D. Rockefeller).

4. In everything one must consider the end (Jean de la Fontaine).

▶ 5. One should absorb the color of life, but should never remember its details (Oscar Wilde).

6. One should be a slave to nothing but one's toothbrush (Romaine Brooks).

7. Always treat others as you would like them to treat you (Matthew 7:12).

8. Assume a virtue, if you have it not (William Shakespeare).

9. Don't do anything that will make your conscience uneasy (William Feather).

▶ 10. If any would not work, neither should he eat (II Thessalonians 3:10).

11. Who has the fame to be an early riser may sleep till noon (James Howell).

12. Never mind your happiness; do your duty (**Will Durant**).

13. Be content to act, and leave the talking to others (Baltasar Gracián).

14. Let nothing pass that will advantage you (Cato).

▶ 15. Follow your honest convictions and be strong (William Makepeace Thackeray).

16. Do not wait for extraordinary circumstances to do good actions; try to use ordinary circumstances (Jean Paul Richter).

17. It is necessary to try to surpass one's self always; this occupation ought to last as long as life (Queen Christina).

18. Such strength as a man has he should use (Cicero).

19. Make no little plans; they have no magic to stir men's blood (Daniel Hudson Burnham).

▶ 20. We must cut our coat according to our cloth, and adapt ourselves to changing circumstances (William R. Inge).

21. Defeat should never be a source of discouragement, but rather a fresh stimulus (Robert South).

22. It's not good enough that we do our best; sometimes we have to do what's required (Winston Churchill).

23. Love all, trust a few, do wrong to no one (William Shakespeare). (We cannot adequately symbolize the meaning of *a few;* symbolize *a few* as *some.*)

24. Generosity should never exceed ability (Cicero).

▶ 25. Go confidently in the direction of your dreams. Live the life you have imagined (Henry David Thoreau).

26. Beware of false prophets, which come to you in sheep's clothing, but inwardly they are ravening wolves (Matthew 7:15).

27. Be kind, for everyone you meet is fighting a hard battle (Philo).

28. Always take a job that is too big for you (Harry Emerson Fosdick).

29. Keep neither a blunt knife nor an ill-disciplined looseness of tongue (Epictetus).

▶ 30. If you have done your best and failed, try doing your worst; you might see your error and succeed (Benjamin Lichtenberg).

31. Keep away from people who try to belittle your ambition (Mark Twain).

32. A wise man ought to have money in his head, but not in his heart (Jonathan Swift).

33. If your conscience won't stop you, pray for cold feet (Elmer G. Leterman).

34. We must have courage to bet on our ideas, to take the calculated risk, and to act (Maxwell Maltz).

▶ 35. And all who heard should see them there, and all should cry, Beware! Beware! (Samuel Taylor Coleridge).

36. Talks, speeches, articles and resolutions should all be concise and to the point. Meetings also should not go on too long (Mao Tse-Tung).

37. Never invest in anything that eats or needs repairing (Billy Rose).

38. Diplomatic life entails the obligation to ruin one's liver on behalf of one's country (Sir Oliver Wright).

39. Trust everybody, but yourself most of all (Danish proverb).

▶ 40. Failure . . . should challenge us to new heights of accomplishment, not pull us to new depths of despair (William Arthur Ward).

41. If it be possible, as much as lieth in you, live peaceably with all men (Romans 12:18).

42. A man should never be ashamed to own he has been in the wrong, which is but saying in other words, that he is wiser today than he was yesterday (Alexander Pope).

43. One should . . . be able to see things as hopeless and yet be determined to make them otherwise (F. Scott Fitzgerald).

44. Any book which has as major objective helping you help yourself, fighting back, getting what you deserve, should include a discussion of how to protect yourself against lawyers, especially your own (Roy Cohn).

▶ 45. Never speak evil of anyone if you do not know it for a certainty. If you know it for a certainty, ask yourself, "Why should I tell it?" (Johann Kaspar Lavater).

46. You can't expect to win unless you know why you lose (Benjamin Lipson).

47. A man should work eight hours and sleep eight hours but not the same eight hours (Elmer G. Leterman).

48. A man who cannot command his temper should not think of becoming a man of business (Lord Chesterfield).

49. Let men decide firmly what they will not do, and they will be free to do vigorously what they ought to do (Mencius).

▶ **50.** If you can't win, make the winner look great (Whitt N. Schultz).

51. Don't part with your illusions. When they are gone, you may still exist, but you have ceased to live (Mark Twain).

52. There is no feeling in this world to be compared with self-reliance. Don't sacrifice that to anything else (John D. Rockefeller).

53. It does not do to leave a live dragon out of your calculations, if you live near him (J. R. R. Tolkien).

54. Hold yourself responsible for a higher standard than anybody else expects of you (Henry Ward Beecher).

▶ **55.** The perversion of the mind is only possible when those who should be heard in its defense are silent (Archibald MacLeish).

56. A man may and ought to pride himself more on his will than on his talent (Honoré de Balzac).

57. If you want your judgment to be accepted, express it coolly and without passion (Arthur Schopenhauer).

58. Give your decisions, never your reasons; your decisions may be right, your reasons are sure to be wrong (William Mansfield).

59. Do what you love. Know your own bone; gnaw at it, bury it, unearth it and gnaw it still (Henry David Thoreau).

60. A President ought to be allowed to "hang" two men every year without giving any reason or explanation.... I could get word to twenty or thirty people that they were being considered for the honor (Herbert Hoover).

61. Money, make money; by honest means if you can; if not, by any means make money (Horace).

62. Do not anticipate trouble, or worry about what may never happen. Keep in the sunlight (Benjamin Franklin).

63. The only failure a man ought to fear is failure in cleaving to the purpose he sees to be best (Mary Ann Evans).

64. Luck affects everything; let your hook always be cast; in the stream where you least expect it, there will be fish (Ovid).

65. A conservative is a man who does not believe that anything should be done for the first time (Frank A. Vanderlip).

66. Our duty is to believe that for which we have sufficient evidence, and to suspend our judgment when we have not (John Lubbock).

67. We should make every sacrifice necessary to put usable conventional weapons in the hands of those who would do the fighting (Thomas J. Watson).

David Ben-Gurion, echoing Plato, said, "Courage is a special kind of knowledge: the knowledge of how to fear what ought to be feared and how not to fear what ought not to be feared." We can capture part of what Ben-Gurion had in mind in the formula $\Box\forall x(Cx \leftrightarrow \forall y((OFxy \rightarrow Fxy) \mathbin{\&} (O\neg Fxy \rightarrow \neg Fxy)))$, where Cx symbolizes 'x has courage' and Fxy symbolizes 'x fears y.' Use tableaux to determine, in QS5 or FS5, whether the following are implied by this formula.

68. $\forall x(Cx \leftrightarrow \forall y(Fxy \leftrightarrow OFxy))$ ('Courage is fearing all and only what ought to be feared.')

69. $\Box\forall x\forall y((Cx \mathbin{\&} Fxy) \rightarrow OFxy)$ ('Necessarily, the courageous fear only what they ought to fear.')

▶ **70.** $\Box\forall x\forall y((Cx \mathbin{\&} OFxy) \rightarrow Fxy)$ ('Necessarily, the courageous fear everything they ought to fear.')

71. $\Box\exists xCx \rightarrow \Box\exists x\forall y(OFxy \rightarrow Fxy)$ ('If it's necessary for someone to have courage, it's necessary for someone to fear all they ought to fear.')

72. $\Box\forall x\forall y((Cx \mathbin{\&} Fxy) \rightarrow PFxy)$ ('Necessarily, the courageous fear only what it's permissible for them to fear.')

73. "That which is wrong is never necessary" (B. C. Forbes). We can perhaps formulate Forbes's dictum as $\Box\forall x\forall y(O\neg Dxy \rightarrow \neg\Box Dxy)$, where Dxy symbolizes 'x does y.' Use natural deduction to show that this formula is valid.

"No longer good from taking thought, but led by habit, so that I am not merely able to do right, but cannot help doing right" (Horace). We can contrast being able to do right from being unable to do wrong in the formulas (a) $\forall y(ODay \rightarrow \Diamond Day)$ and (b) $\forall y(ODay \rightarrow \Box Day)$. Use semantic tableaux to answer these questions, in either QS5 or FS5.

74. Does a imply b?

75. Does b imply a?

76. Is either a or b valid?

77. Does either imply $\forall y(ODay \rightarrow Day)$ ('I do what I should')?

78. Does either imply $\forall y(Day \rightarrow PDay)$ ('I do only what I may')?

79. Does either imply $\forall y(\diamond Day \rightarrow ODay)$ ('I am able to do only what I ought')?

▶ **80.** Is b equivalent to $\forall y(O \neg Day \leftrightarrow \square \neg Day)$? ('I am unable to do all and only what I should not do.')

81. Is b equivalent to $\forall y(\diamond Day \leftrightarrow PDay)$? ('I can do all and only what I may do.')

"In Germany, under the law everything is prohibited except that which is permitted. In France, under the law everything is permitted except that which is prohibited. In the Soviet Union, everything is prohibited, including that which is permitted. And in Italy, under the law everything is permitted, especially that which is prohibited" (Newton Minow). These formulas express a simple way of construing Minow's point. Germany: $\forall x \forall y(O \neg Dxy \vee PDxy)$; France: $\forall x \forall y(\neg O \neg Dxy \rightarrow PDxy)$; Soviet Union: $\forall x \forall y O \neg Dxy$; Italy: $\forall x \forall y PDxy$. Using tableaux in FS5 or QS5, answer these questions.

82. Are any of these valid?

83. Does Germany's principle contradict that of any other country?

84. Does Italy's principle contradict that of any other country?

85. Does the truism $\exists x \exists y \square Dxy$ contradict any country's principle?

86. Does the truism $\exists x \exists y \neg \diamond Dxy$ contradict any country's principle?

A more sophisticated approach to Minow's point would be to let Lx symbolize 'x is a law,' $O_z Dxy$ symbolize 'x ought to do y according to z,' and $P_z Dxy$ symbolize 'x may do y according to z.' Assume that the usual logical rules hold for O_z and P_z. Then the countries' principles become: Germany: $\forall x \forall y(O \neg Dxy \vee \exists z(Lz \ \& \ P_z Dxy))$; France: $\forall x \forall y(PDxy \vee \exists z(Lz \ \& \ O_z \neg Dxy))$; Soviet Union: $\forall x \forall y O \neg Dxy$; Italy: $\forall x \forall y PDxy$. Use natural deduction in QS5 or FS5 to show the following.

87. The Soviet Union's principle entails Germany's.

88. Italy's principle entails France's.

89. Germany's principle implies that, if there are no laws, the Soviet Union's principle holds.

▶ **90.** France's principle implies that, if there are no laws, Italy's principle holds.

91. If we adopt the schema $O\mathscr{A} \leftrightarrow \exists z(Lz \ \& \ O_z \mathscr{A})$—'You ought to do something just in case there's a law that directs it'—then France's principle follows.

92. If we adopt the schema $O\mathscr{A} \leftrightarrow \forall z(Lz \rightarrow O_z \mathscr{A})$—'You ought to do something just in case every law directs it'—then Germany's principle follows.

93. Say that a law is *negative* iff it prohibits someone from doing something. If we adopt the schema $O\mathscr{A} \leftrightarrow \exists z(Lz \ \& \ O_z\mathscr{A})$, then Italy's principle implies that no laws are negative.

94. Say that a law is *positive* iff it permits someone to do something. If we adopt the schema $O\mathscr{A} \leftrightarrow \forall z(Lz \rightarrow O_z\mathscr{A})$, then the Soviet Union's principle implies that no laws are positive.

Robert Benchley joked that "Anyone can do any amount of work provided it isn't the work he is supposed to do at that moment." We can symbolize Benchley's statement as $\forall x\forall y((Px \ \& \ Wy) \rightarrow (\neg ODxy \rightarrow \Diamond Dxy))$. Use tableaux to determine which of these follow.

95. $\forall x\forall y((Px \ \& \ Wy) \rightarrow (\neg \Diamond Dxy \rightarrow ODxy))$ ('Everyone is supposed to do the work that he or she is incapable of doing.')

96. $\forall x\forall y((Px \ \& \ Wy) \rightarrow (ODxy \rightarrow \neg \Diamond Dxy))$ ('Nobody can do the work that he or she is supposed to do.')

97. $\forall x\forall y((Px \ \& \ Wy) \rightarrow \neg \Diamond(Dxy \ \& \ ODxy))$ ('It's impossible for anyone to do the work he or she is supposed to do.')

98. $\forall x\forall y((Px \ \& \ Wy) \rightarrow \Diamond Dxy)$ ('Anyone can do any amount of work.')

"So act that your principle of action might safely be made a law for the whole world." Immanuel Kant formulated this principle, called the *categorical imperative*. Where Fxy symbolizes 'x follows principle of action y,' we might symbolize Kant's imperative as (a) $\Box\forall x\forall y(OFxy \rightarrow \forall zOFzy)$; (b) $\Box\forall x\forall y(\forall zOFzy \rightarrow OFxy)$; (c) $\Box\forall x\forall y(OFxy \leftrightarrow \forall zOFzy)$; (d) $\Box\forall x\forall y(OFxy \rightarrow O\forall zFzy)$; (e) $\Box\forall x\forall y(O\forall zFzy \rightarrow OFxy)$; or (f) $\Box\forall x\forall y(OFxy \leftrightarrow O\forall zFzy)$.
Use tableaux to answer these questions in QS5 or FS5.

99. Are any of a–f valid?

100. Do any of the others imply a? If so, which?

101. Do any of the others imply c? If so, which?

102. Do any of the others imply d? If so, which?

103. Do any of the others imply f? If so, which?

104. Do any of a–f imply $\Box\forall x\forall y(OFxy \rightarrow \Diamond\forall zFzy)$? ('So act that it would be possible for everyone to follow your principle of action.') If so, which?

105. Do any of a–f imply $\Box\forall x\forall y(OFxy \rightarrow P\forall zFzy)$? ('So act that it would be permissible for everyone to follow your principle of action.')

106. Do any of a–f imply $\forall y(\exists xOFxy \rightarrow \forall xOFxy$? ('If anyone has an obligation to follow a principle of action, then everyone has an obligation to follow that principle.')

107. Do any of a–f imply $\forall y(O \exists x Fxy \rightarrow O \forall x Fxy)$? ('Any principle of action that ought to be followed by anyone ought to be followed by everyone.)

This is Ambrose Bierce's definition of *compromise:* "Compromise: Such an adjustment of conflicting interests as gives each adversary the satisfaction of thinking he has got what he ought not to have and is deprived of nothing except what was justly his due." It seems that, if nobody is deceived, then Bierce's definition takes one of these forms:

    a. $\Box \forall x \forall y (Cxy \leftrightarrow \forall z((O \neg Hxz \leftrightarrow Hxz) \ \& \ (O \neg Hyz \leftrightarrow Hyz)))$
    b. $\Box \forall x \forall y (Cxy \leftrightarrow \forall z((\neg OHxz \leftrightarrow Hxz) \ \& \ (\neg OHyz \leftrightarrow Hyz)))$
    c. $\Box \forall x \forall y (Cxy \leftrightarrow \forall z((O \neg Hxz \rightarrow Hxz) \ \& \ (OHxz \rightarrow \neg Hxz) \ \&$
       $(O \neg Hyz \rightarrow Hyz) \ \& \ (OHyz \rightarrow \neg Hyz)))$
    d. $\Box \forall x \forall y (Cxy \leftrightarrow \forall z((O \neg Hxz \rightarrow Hxz) \ \& \ (OHxz \leftrightarrow \neg Hxz) \ \&$
       $(O \neg Hyz \rightarrow Hyz) \ \& \ (OHyz \leftrightarrow \neg Hyz)))$

108. Do any of a–d imply $\forall x \forall y (Cxy \rightarrow \forall z(Hxz \lor Hyz))$?*

109. Do any of a–d imply $\forall x \forall y (Cxy \rightarrow \forall z(OHxz \lor OHyz))$?*

110. Do any of a–d imply $\forall x (\exists y Cxy \rightarrow \forall z(OHxz \lor O \neg Hxz))$?*

111. Do any of a–d imply $\forall x \forall y (Cxy \rightarrow \neg \exists z(Hxz \ \& \ Hyz))$?*

112. Do any of a–d imply $\forall x \forall y (Cxy \rightarrow \neg \exists z(OHxz \ \& \ OHyz))$?*

113. Do any of a–d imply $\forall x \forall y (Cxy \rightarrow \neg \exists z(O \neg Hxz \ \& \ O \neg Hyz))$?*

114. Do any of a–d imply $\forall x \forall y (Cxy \rightarrow \forall z((OHxz \ \& \ OHyz) \rightarrow \neg (Hxz \lor Hyz)))$?*

115. If we assume $\Box \forall x (\exists y Cxy \rightarrow \forall z(OHxz \lor O \neg Hxz))$, is a equivalent to b, c, or d?*

116. If we assume $\Box \forall x (\exists y Cxy \rightarrow \forall z(OHxz \lor O \neg Hxz))$, is b equivalent to a, c, or d?*

117. If we assume $\forall x \forall y (Cxy \rightarrow \forall z(Hxz \lor Hyz))$, do any of a–d imply $\forall x \forall y (Cxy \rightarrow \neg \exists z(OHxz \ \& \ OHyz))$?*

118. Do any of a–d contradict $\exists x \exists y (Cxy \ \& \ \forall z((Hxz \leftrightarrow OHxz) \ \& \ (Hyz \leftrightarrow OHyz)))$?*

The Golden Rule commands, "Always treat others as you would like them to treat you" (Matthew 7:12). This is the New English Bible translation. In the Revised Standard Version, it is, "So whatever you wish that men would do to you, do so to them"; in the King James Bible, "Therefore all things whatsoever ye would that men should do to you, do ye even so unto them." Assuming the rule applies to all possible circumstances, it seems most natural to symbolize Jesus' command in the Sermon on the Mount as

    a. $\Box \forall x \forall y \forall z (Wxyzx \rightarrow ODxzy)$

where $Wxyzw$ symbolizes 'x wants y to do z to w' and where $Dxyz$ symbolizes 'x does y to z.' But we can develop many variants of this principle:

b. $\Box\forall x\forall y\forall z(Wxyzx \rightarrow \mathrm{P}Dxzy)$ ('You may treat others as you would like them to treat you.')

c. $\Box\forall x\forall y\forall z(\mathrm{O}Dxzy \rightarrow Wxyzx)$ ('You have an obligation to treat others in a certain way only if you would like them to treat you that way.')

d. $\Box\forall x\forall y\forall z(\mathrm{P}Dxzy \rightarrow Wxyzx)$ ('You are permitted to treat others in a certain way only if you would like them to treat you that way.')

In either QS5 or FS5:

119. Are any of a–d valid?*

120. Do any of the above imply $\Box\forall x\forall y\forall z(\Box Dxzy \rightarrow Wxyzx)$? ('If it's necessary for you to treat others in a certain way, then you would like them to treat you that way.')*

121. Do any of the above imply a?*

122. Do any of the above imply $\Box\forall x\forall y\forall z(Wxyzx \rightarrow \Diamond Dxzy)$? ('It's possible for you to treat others as you would like them to treat you.')*

123. Do any of the above imply $\Box\forall x\forall y\forall z(\neg Wxyzx \rightarrow \neg\mathrm{O}Dxzy)$? ('You're forbidden to treat others as you would not want them to treat you.')*

124. Do any of the above imply $\Box\forall x\forall y\forall z(\mathrm{O}\neg Dxzy \rightarrow \neg Wxyzx)$? ('You would not like others to treat you as you are forbidden to treat them.')*

125. Do any of the above imply $\Box\forall x\forall y\forall z(\mathrm{P}\neg Dxzy \rightarrow \neg Wxyzx)$? ('You would not like others to treat you as you are permitted not to treat them.')*

According to a–d, what can we conclude from each of the following? What do these results indicate about a–d as formalizations of the Golden Rule?

126. Juan would like Maria to kiss him.**

127. Maria wouldn't like Juan to kiss her.**

128. Juan would like to lose weight. (Symbolize as "Juan would like (himself) to cause himself to weigh less.)**

129. Juan doesn't want (himself) to be self-destructive.**

130. It's necessarily true that everyone would like to be happy.**

131. Sue should help Fred.**

132. Jill may hire Jack.**

133. Al shouldn't see Lisa.**

134. Harriet has no obligation to lend money to Ken.**

135. Gary can stop Fran.**

136. Donna can't forgive Sam.**

137. It's necessary for Ed to repay Wendy.**

138. It's OK for Rikki to indulge herself.**

139. Ted ought to take care of himself.**

140. It's necessary for all of us to look out for our own survival.**

Notes

[1] Aristotle was the first logician to study combinations of modality and quantification; he developed a theory of modal syllogisms. Medieval logicians made significant progress in understanding the complexities of such combinations. In the 1940s and 1950s, American logicians such as Ruth Barcan Marcus, Rudolf Carnap, and W. V. Quine used the tools of symbolic logic to advance this work. The American logician Saul Kripke formulated a rigorous semantics for modal logic in 1959.

[2] Ruth Barcan Marcus proposed this principle in "A Functional Calculus of the First Order Based on Strict Implication," *Journal of Symbolic Logic* 11 (1946): 1–16.

[3] W. V. Quine, *Word and Object* (Cambridge: MIT Press, 1960): 199.

[4] Saul Kripke first recognized that, for this reason, any adequate quantified modal logic should be a free logic. See S. Kripke, "Semantical Considerations on Modal Logic," *Acta Philosophica Fennica* 16 (1963): 83–94. The American logicians Henry S. Leonard and Karel Lambert developed free logic in the late 1950s and 1960s; see H. Leonard, "The Logic of Existence," *Philosophical Studies* 7 (1956): 49–64.

[5] See S. Kripke, *Naming and Necessity* (Cambridge: Harvard University Press, 1972, 1980).

[6] Many names, of course, help us pick out individuals only with the help of contextual information. There are many John Smiths and Mary Joneses.

[7] W. V. Quine made this example famous in "Reference and Modality," in *From a Logical Point of View* (Cambridge: Harvard University Press, 1961).

[8] John Stuart Mill, *Utilitarianism* (New York: New American Library, 1962): 305.

[9] John Stuart Mill, *Utilitarianism* (New York: New American Library, 1962): 305.

BIBLIOGRAPHY

Alban, M. J. "Independence of the Primitive Symbols of Lewis' Calculi of Propositions," *Journal of Symbolic Logic* 8 (1943): 25–26.

Anderson, Alan Ross, and Belnap, Nuel D., Jr. *Entailment: The Logic of Relevance and Necessity*, Volume I (Princeton: Princeton University Press, 1975).

Beth, E. W. "Semantic Entailment and Formal Derivability," in K. J. J. Hintikka (ed.), *Philosophy of Mathematics* (Oxford: Oxford University Press, 1969): 9–41.

Church, Alonzo. "A Note on the Entscheidungsproblem," *Journal of Symbolic Logic* 1 (1936): 40–41. (Correction on 101–2.)

Feys, R. "Les logiques nouvelles de modalités," *Revue Néoscolastique de Philosophie* 40 (1937): 517–533.

Gentzen, Gerhard. "An Investigation into Logical Deduction," in M. Szabo (ed.), *The Collected Papers of Gerhard Gentzen* (Amsterdam: North-Holland, 1969).

Grice, H. P. "Logic and Conversation," in P. Cole and J. L. Morgan (eds.) *Syntax and Semantics 3: Speech Acts* (New York: Academic Press, 1975): 45–58.

Hallden, S. "On the Decision Problem of Lewis's Calculus S5," *Norsk Mathematisk Tidsskrift* 31 (1949): 89–94.

Harper, William L. "A Sketch of Some Recent Developments in the Theory of Conditionals," in W. L. Harper, R. Stalnaker, and G. Pearce (eds.), *Ifs* (Dordrecht: D. Reidel, 1981): 3–40.

Hintikka, K. J. J. "Form and Content in Quantification Theory," *Acta Philosophica Fennica* 8 (1955): 7–55.

Kalish, Donald; Montague, Richard; and Mar, Gary. *Logic: Techniques of Formal Reasoning* (New York: Harcourt-Brace-Jovanovich, 1964, 1980).

Kneale, William, and Kneale, Martha. *The Development of Logic* (Oxford: Clarendon Press, 1962, 1984).

Kripke, Saul. *Naming and Necessity* (Cambridge: Harvard University Press, 1972, 1980).

Kripke, Saul. "Semantical Considerations on Modal Logic," *Acta Philosophica Fennica* 16 (1963): 83–94.

Lakoff, George. "Linguistics and Natural Logic," in Donald Davidson and Gilbert Harman (eds.), *Semantics of Natural Language* (Dordrecht: D. Reidel, 1972): 545–665.

Leonard, Henry S. "The Logic of Existence," *Philosophical Studies* 7 (1956): 49–64.

Lewis, C. I., and Langford, C. H. *Symbolic Logic* (New York: Dover, 1932).

Lewis, David. *Counterfactuals* (Cambridge: Harvard University Press, 1973).

Lewis, David. "Counterfactuals and Comparative Possibility," in W. L. Harper, R. Stalnaker, and G. Pearce (eds.), *Ifs* (Dordrecht: D. Reidel, 1981): 57–86.

Marcus, Ruth Barcan. "A Functional Calculus of the First Order Based on Strict Implication," *Journal of Symbolic Logic* 11 (1946): 1–16.

McCawley, James D. "A Program for Logic," in Donald Davidson and Gilbert Harman (eds.), *Semantics of Natural Language* (Dordrecht: D. Reidel, 1972): 498–544.

McCawley, James D. *What Linguists Have Always Wanted to Know About Logic** (Chicago: University of Chicago Press, 1980).

Mill, John Stuart. *Utilitarianism* (New York: New American Library, 1962).

Parry, W. T. "Modalities in the Survey System of Strict Implication," *Journal of Symbolic Logic* 4 (1939): 131–54.

Prior, Arthur N. *Formal Logic* (Oxford: Oxford University Press, 1962).

Quine, Willard van Orman. *Methods of Logic* (Cambridge: Harvard University Press, 1950, 1982).

Quine, Willard van Orman. "Reference and Modality," in *From a Logical Point of View* (Cambridge: Harvard University Press, 1961).

Quine, Willard van Orman. *Word and Object* (Cambridge: MIT Press, 1960).

Ross, W. D. *The Right and the Good* (New York: Oxford University Press, 1930).

Russell, Bertrand. "On Denoting," in R. C. Marsh (ed.), *Logic and Knowledge* (New York: G. P. Putnam's Sons, 1956): 39–56.

Sobocinski, B. "Note on a Modal System of Feys-von Wright," *The Journal of Computing Systems* 1 (1953): 171–78.

Sobocinski, B. "Remarks about the Axiomatizations of Certain Modal Systems," *Notre Dame Journal of Formal Logic* 5 (1964): 71–80.

Stalnaker, Robert. "A Defense of Conditional Excluded Middle," in W. L. Harper, R. Stalnaker, and G. Pearce (eds.), *Ifs* (Dordrecht: D. Reidel, 1981): 87–106.

Stalnaker, Robert. *Inquiry* (Cambridge: Bradford Books, 1985).

Stalnaker, Robert. "A Theory of Conditionals," in W. L. Harper, R. Stalnaker, and G. Pearce (eds.), *Ifs* (Dordrecht: D. Reidel, 1981): 41–56.

Thomason, Richmond. "A Fitch-Style Formulation of Conditional Logic," *Logique et Analyse* 52 (1970): 397–412.

von Wright, G. H. "Deontic Logic," *Mind* 60 (1951): 1–15.

von Wright, G. H. *An Essay on Modal Logic* (Amsterdam: North-Holland, 1951).

ANSWERS TO SELECTED PROBLEMS

Chapter 1
BASIC CONCEPTS OF LOGIC

1.1 Arguments

5. Most criminals believe that their chances of being caught and punished are small.
   ∴ The perceived costs of a life of crime are low.

10. Swedish is an Indo-European language, but Finnish isn't.
    ∴ Finnish is more difficult for English-speakers to learn than Swedish.

15. The earth receives radiant heat from the sun and loses heat to outer space by its own radiative emissions.
    The energy received undergoes many transformations.
    In the long run, no appreciable fraction of this energy is stored on the earth, and there is no persistent trend toward higher or lower temperatures.
    ∴ The energy the earth receives from the sun is roughly equal to the energy it loses to outer space by its own radiative emissions.

1.2 Validity

5. Invalid. Suppose the Mets are contenders, but Strawberry doesn't hit 30 home runs.

10. Valid.   15. Valid.

20. Invalid. The patient may die even if we operate.

25. Invalid. Jack might devote most of the next few weeks to securing the sale without understanding how important it is.

30. Valid.

35. This argument is ambiguous. *Few* has both a relative and an absolute meaning. If we take *few* as an absolute, the argument is, roughly:
    Few people are mathematics students who take courses in logic.
    All accounting majors take courses in logic.
    ∴ Few people are accounting majors who are students of mathematics.
    This is valid. If we assign *few* a relative meaning, the argument is invalid. If there are many more mathematics students than accounting majors, then the group of accounting majors who are also mathematics students may be small relative to the group of mathematics students but large relative to the group of accounting majors.

1.3 Implication and Equivalence

5. Not equivalent; (a) implies (b).   10. Not equivalent. Neither implies the other.

15. Not equivalent. Neither implies the other.

20. (b) implies (a). Whether (a) also implies (b) is controversial.

25. Nothing.   30. Neither follows from nor implies.

35. Does not follow; whether it implies the conditional is controversial.

40. Neither follows from nor implies.

45. Not equivalent.   **50.** Factive.

55. Not factive, in general, but factive under some conditions.

60. Factive.   **65.** Not factive.   **70.** Factive.

1.4 Logical Properties of Sentences

5. Valid.   **10.** Contingent.   **15.** Contingent.   **20.** Contingent.   **25.** Valid.

30. Contingent.   **35.** Valid.   **40.** Contradictory.

55. These problems pose no problem for such definitions. The argument is valid. Pseudo-Scot was wrong to think that the conclusion is contradictory. There are surely possible circumstances in which no sentence is negative. In such circumstances, we couldn't articulate the argument. But this fact has no bearing on its validity.

1.5 Sets of Sentences

5. Satisfiable.   **10.** Satisfiable.   **15.** Satisfiable. (There may be no actresses.)

20. Controversial. I find this set satisfiable on one reading, but contradictory if *ought* means "really, in the final analysis, ought."

25. True. Contradictory sets cannot be true. So it can never happen that all the members of such a set are true while anything else is false.

30. True. The premises can never all be true; so they can't all be true while the conclusion is false.

35. False. Consider the set {John is fat, John is not fat}.

40. True. If a sentence is consistent with the premises of a valid argument, there is a circumstance in which all are true. Because the argument is valid, its conclusion must be true in this circumstance as well. So there must be a circumstance in which the sentence and the conclusions are true.

45. True. Valid sentences are true in all circumstances, because there cannot be a circumstance in which they are false but something else is true.

Chapter 2
SENTENCES

2.1 Sentence Connectives

5. Not truth-functional.   **10.** Not truth-functional.   **15.** Not truth-functional.

A	May A
T	T
F	?

A	Should A
T	?
F	?

A	Maybe A
T	T
F	?

20. Controversial. Only the second row of this table is uncomplicated.

A	B	If A, then B
T	T	
T	F	F
F	T	
F	F	

2.2 A Sentential Language

5. (c)   **10.** (c)   **15.** (c)

2.3 Truth Functions

5. The falsity of the antecedent sentence, 'the moon smiles,' does not make the entire conditional true automatically.

10. The predicate 'are utter opposites' applies to the conjoined subject 'Fame and rest': The sentence can't be analyzed as 'Fame is an utter opposite and rest is an utter opposite.' Thus, *and* does not act as a sentence connective here.

2.4 Symbolization

5. p: You get simple beauty; q: You get nothing other than beauty; r: You get about the best thing God invents. $(p \mathbin{\&} q) \to r$.

10. p: You get what you want; q: You seriously wanted it; r: You tried to bargain over the price. $\neg p \to (\neg q \vee r)$.

15. p: The light shineth in the darkness; q: The darkness comprehendeth it. $p \mathbin{\&} \neg q$.

20. p: Happiness is the end of life; q: Character is the end of life. $\neg p \mathbin{\&} q$.

25. p: The mind itself, like other things, must sometimes be unbent; q: The mind is weakened; r: The mind is broken. $p \vee (q \vee r)$.

30. p: Action always brings happiness; q: There is no happiness without action. $\neg p \mathbin{\&} q$.

35. p: The spirit of business adventure is dulled; q: This country will cease to hold the foremost position in the world. $p \to q$.

40. p: Arguments are to be avoided; q: Arguments are always vulgar; r: Arguments are often convincing. $p \mathbin{\&} (q \mathbin{\&} r)$.

2.5 Validity

5. Correct. Satisfiable sets do not imply contradictions, because, on an interpretation making all the members of such a set true, any contradiction will be false.

10. Incorrect. Every formula implies any valid formula.

15. Incorrect. No member of $\{p, \neg p\}$ implies a contradiction, but the set is not satisfiable.

20. Correct. If $\neg \mathscr{A}$ implies $\mathscr{A}$, there is no interpretation on which $\neg \mathscr{A}$ is true. But $\neg \mathscr{A}$ is true just when $\mathscr{A}$ is false. So, there is no interpretation on which $\mathscr{A}$ is false; thus, $\mathscr{A}$ is valid.

25.

p	q	r	$(p \leftrightarrow \neg r) \to (r \to q)$
T	T	T	T F FT T T T T

30.

p	q	r	$\neg((q \vee \neg p) \to \neg r)$
F	T	T	T T T TF F FT

35.

p	q	r	$(p \to \neg r) \to \neg(p \mathbin{\&} q)$
T	F	F	T T TF T T T F F

40. Correct. The argument form $p \therefore p$ is valid, but it becomes invalid if its premise is left out. So, $p \therefore p$ is exact. And its conclusion, p, is contingent.

45. Correct. If $\mathscr{A}$ exactly implied $\neg \mathscr{A}$, then the argument form $\mathscr{A} \therefore \neg \mathscr{A}$ would be exact. So, $\mathscr{A} \therefore \neg \mathscr{A}$ would be valid, but $\neg \mathscr{A}$ itself would not be valid. Thus, $\mathscr{A}$ would not be contradictory. But $\mathscr{A} \therefore \neg \mathscr{A}$ is valid only when $\mathscr{A}$ is contradictory. So, there is no such $\mathscr{A}$.

2.6 Truth Tables for Formulas

5.

p	$p \to p$
T	T T T
F	F T F

Valid.

10.

p	q	$p \to (q \mathbin{\&} p)$
T	T	T T T T T
T	F	T F F F T
F	T	F T T F F
F	F	F T F F F

Contingent.

15.

p	q	$((p \mathbin{\&} \neg q) \to q) \to \neg p$
T	T	T F FT T T F FT
T	F	T T TF F F T FT
F	T	F F FT T T T TF
F	F	F F TF T F T TF

Contingent.

20.

p q	(p → q) → (p & q)
T T	T T T **T** T T T
T F	T F F **T** T F F
F T	F T T **F** F F T
F F	F T F **F** F F F

Contingent.

25.

p q r	(p ∨ (q & r)) ↔ ((p ∨ q) & (p & r))
T T T	T T T T T **T** T T T T T T T
T T F	T T T F F **F** T T T F T F F
T F T	T T F F T **T** T T F T T T T
T F F	T T F F F **F** T T F F T F F
F T T	F T T T T **F** F T T F F F T
F T F	F F T F F **T** F T T F F F F
F F T	F F F F T **T** F F F F F F T
F F F	F F F F F **T** F F F F F F F

2.7 Other Uses of Truth Tables

5. *p*: I'm mistaken; *q*: I'm a fool.

¬p → q
q → p
∴ p

p q	(¬p → q)	(q → p)	p
T T	FT **T** T	**T** T T	T
T F	FT **T** F	**F** T T	T
F T	TF **T** T	**T** F F	F
F F	TF **F** F	**F** T F	F

Valid.

10. *p*: Socrates died; *q*: Socrates died while he was living; *r*: Socrates died while he was dead.

p → (q ∨ r)
¬q
¬r
∴ ¬p

p q r	(p → (q ∨ r))	¬q	¬r	¬p
T T T	T **T** T T T	FT	FT	FT
T T F	T **T** T T F	FT	TF	FT
T F T	T **T** F T T	TF	FT	FT
T F F	T **F** F F F	TF	TF	FT
F T T	F **T** T T T	FT	FT	TF
F T F	F **T** T T F	FT	TF	TF
F F T	F **T** F T T	TF	FT	TF
F F F	F **T** F F F	TF	TF	TF

Valid.

15. *p*: We ought to philosophize; *q*: We ought not philosophize.

p ∨ q
p → p
q → p
∴ p

p q	(p ∨ q)	(p → p)	(q → p)	p
T T	T **T** T	T **T** T	T **T** T	T
T F	T **T** F	T **T** T	F **T** T	T
F T	F **T** T	F **T** F	T **F** F	F
F F	F **F** F	F **T** F	F **T** F	F

Valid.

20.

p q	p ∨ q	¬q	p
T T	T **T** T	FT	T
T F	T **T** F	TF	T
F T	F **T** T	FT	F
F F	F **F** F	TF	F

Valid.

30.
```
p q     ¬(p ↔ q)     ¬p ↔ ¬q
T T     F T T T      FT T FT
T F     T T F F      FT F TF
F T     T F F T      TF F FT
F F     F F T F      TF T TF
```
Not equivalent; neither implies the other.

35.
```
p q     p & q     (p ∨ q) & (p ↔ q)
T T     T T T     T T T T T T T
T F     T F F     T T F F T F F
F T     F F T     F T T F F F T
F F     F F F     F F F F F T F
```
Equivalent.

45.
```
p q     (p ↔ q)     (p ↔ q) ↔ p
T T     T T T       T T T T T
T F     T F F       T F F F T
F T     F F T       F F T T F
F F     F T F       F T F F F
```
Not equivalent; neither implies the other.

50.
```
p q r     q ↔ (p → (q ∨ ¬r))     ¬(p → (q ∨ r))     r
T T T     T T  T T  T T  FT      F T T  T T T       T
T F F     T T  T T  T T  TF      F T T  T T F       F
T F T     F T  T F  F F  FT      F T T  F T T       T
T F F     F F  T T  F T  TF      T T F  F F F       F
F T T     T T  F T  T T  FT      F F T  T T T       T
F T F     T T  F T  T T  TF      F F T  T T F       F
F F T     F F  F T  F F  FT      F F T  F T T       T
F F F     F F  F T  F T  TF      F F T  F F F       F
```
Unsatisfiable.

55.
```
p q r     ¬(p ↔ ¬q)     p ↔ (r ↔ q)     q → (p ↔ r)
T T T     T T F FT      T T T T T       T T  T T T
T T F     T T F FT      T F F F T       T F  T F F
T F T     F T T TF      T F T F F       F T  T T T
T F F     F T T TF      T T F T F       F T  T F F
F T T     F F T FT      F F T T T       T F  F F T
F T F     F F T FT      F T F F T       T T  F T F
F F T     T F F TF      F T T F F       F T  F F T
F F F     T F F TF      F F F T F       F T  F T F
```
Satisfiable. Interpretations: p q r
                              T T T
                              F F T

60.
```
p q r     ((p → q) → p) → r     ((r → p) → r) → q     ((q → r) → q) → p
T T T     T T T  T T  T T       T T T  T T  T T       T T T  T T  T T
T T F     T T T  T T  F F       F T T  F F  T T       T F F  T T  T T
T F T     T F F  T T  T T       T T T  T T  F F       F T T  F F  T T
T F F     T F F  T T  F F       F T T  F F  T F       F T F  F F  T T
F T T     F T T  F F  T T       T F F  T T  T T       T T T  T T  F F
F T F     F T T  F F  T F       F T F  F F  T T       T F F  T T  F F
F F T     F T F  F F  T T       T F F  T T  F F       F T T  F F  T F
F F F     F T F  F F  T F       F T F  F F  T F       F T F  F F  T F
```
Satisfiable. Interpretations: p q r
                              T T T
                              F F F

70.

p q r	((p ∨ q) & p) → ¬q	((p → (q → r)) & (q → ¬r)) → ¬p
T T T	T T T T T F FT	T T T T T F T F FT **T** FT
T T F	T T T T T F FT	T F T F F F T T TF **T** FT
T F T	T T F T T T TF	T T F T T T F T FT **F** FT
T F F	T T F T T T TF	T T F T F T F T TF **F** FT
F T T	F T T F F T FT	F T T T T F T F FT **T** TF
F T F	F T T F F T FT	F T T F F T T T TF **T** TF
F F T	F F F F F T TF	F T F T T T F T FT **T** TF
F F F	F F F F F T TF	F T F T F T F T TF **T** TF

Does not follow.

Chapter 3
SEMANTIC TABLEAUX

5. (1) True: $p \lor q$, $p \to q$; False: q, p.   Open
   (2) True: $p \lor q$, $p \to q$, q; False: q.   Closed
   The tableau is open.

3.1 Rules for Negation, Conjunction, and Disjunction

5.

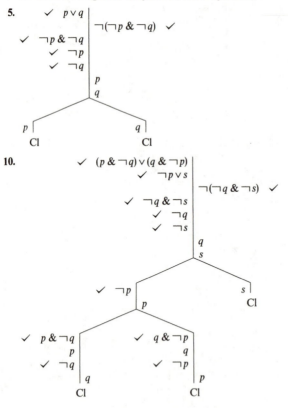

10.

3.2 Rules for the Conditional and Biconditional

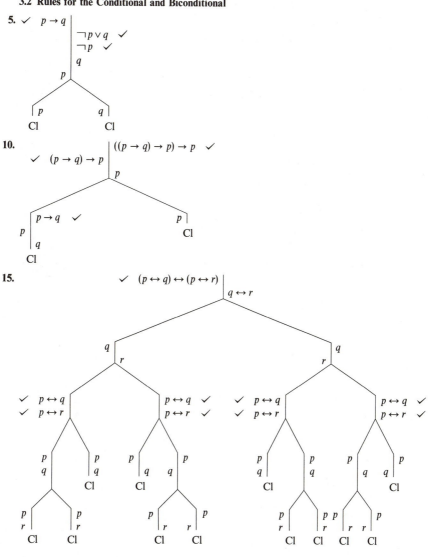

5. ✓ $p \to q$

$\neg p \lor q$ ✓
$\neg p$ ✓
q
p

p q
Cl   Cl

10. $((p \to q) \to p) \to p$ ✓

✓ $(p \to q) \to p$
p

$p \to q$ ✓ p
p Cl
q
Cl

15. ✓ $(p \leftrightarrow q) \leftrightarrow (p \leftrightarrow r)$

$q \leftrightarrow r$

q q
r r

✓ $p \leftrightarrow q$ $p \leftrightarrow q$ ✓   ✓ $p \leftrightarrow q$ $p \leftrightarrow q$ ✓
✓ $p \leftrightarrow r$ $p \leftrightarrow r$ ✓   ✓ $p \leftrightarrow r$ $p \leftrightarrow r$ ✓

p p p p p p p p
q q q q q q q q
      Cl   Cl   Cl   Cl

p p p p p p p p
r r r r r r r r
Cl   Cl   Cl   Cl   Cl   Cl Cl   Cl

3.3 Decision Procedures

10. p: The Soviets march into Poland; q: The Soviets denounce free trade unions in the wrong way; r: The Communist Party's ideal of representing workers is exposed to the Russian people as a sham; s: Poland remains within the Soviet orbit; p_1: Poland drifts from Soviet control. Valid.

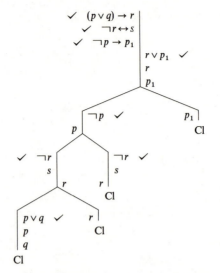

15. *p*: My cat sings opera; *q*: All the lights are out; *r*: I am very insistent; *s*: I howl at the moon.

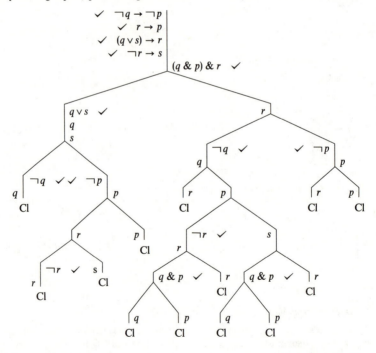

19.

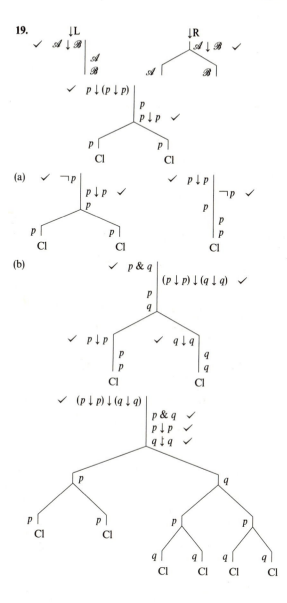

(c)

(d)

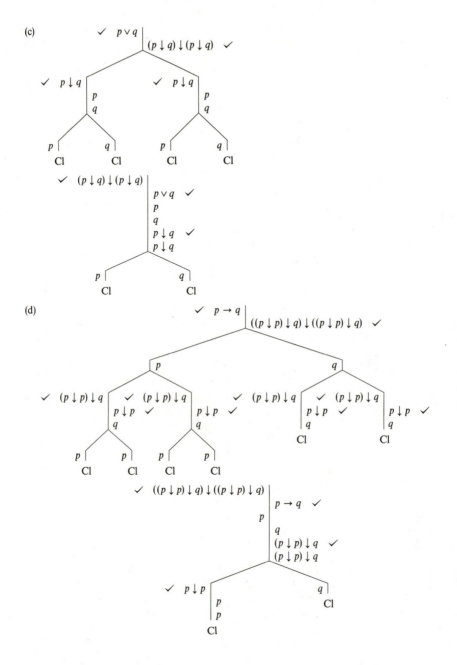

Chapter 4
NATURAL DEDUCTION

4.2 Rules for Negation and Conjunction

5. 1. Show $\neg((p\ \&\ \neg q)\ \&\ (q\ \&\ \neg p))$
2. $(p\ \&\ \neg q)\ \&\ (q\ \&\ \neg p)$ — AIP
3. $p\ \&\ \neg q$ — &E, 2
4. $q\ \&\ \neg p$ — &E, 2
5. p — &E, 3
6. $\neg p$ — &E, 4

15. 1. $p\ \&\ q$ — A
2. $\neg(q\ \&\ \neg r)$ — A
3. $\neg(p\ \&\ \neg s)$ — A
4. Show $r\ \&\ s$
5. Show $\neg\neg r$
6. $\neg r$ — AIP
7. q — &E, 1
8. $q\ \&\ \neg r$ — &I, 7, 6
9. $\neg(q\ \&\ \neg r)$ — R, 2
10. r — $\neg\neg$, 5
11. Show $\neg\neg s$
12. $\neg s$ — AIP
13. p — &E, 1
14. $p\ \&\ \neg s$ — &I, 13, 12
15. $\neg(p\ \&\ \neg s)$ — R, 3
16. s — $\neg\neg$, 11
17. $r\ \&\ s$ — &I, 10, 16

20. 1. $\neg(p\ \&\ r)$ — A
2. $\neg(\neg(p\ \&\ q)\ \&\ \neg p)$ — A
3. r — A
4. Show $\neg s$
5. s — AIP
6. Show $\neg p$
7. p — AIP
8. $p\ \&\ r$ — &I, 7, 3
9. $\neg(p\ \&\ r)$ — R, 1
10. Show $\neg\neg(p\ \&\ q)$
11. $\neg(p\ \&\ q)$ — AIP
12. $\neg(p\ \&\ q)\ \&\ \neg p$ — &I, 11, 6
13. $\neg(\neg(p\ \&\ q)\ \&\ \neg p)$ — R, 2
14. $p\ \&\ q$ — $\neg\neg$, 10
15. p — &E, 4

4.3 Rules for the Conditional and Biconditional

5. 1. $p\rightarrow r$ — A
2. $q\rightarrow\neg r$ — A
3. Show $p\rightarrow\neg q$
4. p — ACP
5. r — $\rightarrow$E, 1, 4
6. Show $\neg q$
7. q — AIP
8. $\neg r$ — $\rightarrow$E, 2, 7
9. r — R, 5

10. 1. $p\rightarrow q$ — A
2. $\neg q$ — A
3. Show $\neg p$
4. p — AIP
5. q — $\rightarrow$E, 1, 4
6. $\neg q$ — R, 2

15. 1. $s\rightarrow(r\ \&\ p)$ — A
2. $q\rightarrow(\neg r\ \&\ \neg p_1)$ — A
3. Show $(q\ \&\ s)\rightarrow p_2$
4. $q\ \&\ s$ — ACP
5. Show $\neg\neg p_2$
6. $\neg p_2$ — AIP
7. q — &E, 4
8. s — &E, 4
9. $r\ \&\ p$ — $\rightarrow$E, 1, 8
10. $\neg r\ \&\ \neg p_1$ — $\rightarrow$E, 2, 7
11. r — &E, 9
12. $\neg r$ — &E, 10
13. p_2 — $\neg\neg$, 5

4.4 Rules for Disjunction

5.

1.	$\neg p \vee \neg r$	A
2.	Show $\neg(p \,\&\, r)$	
3.	$\lceil p \,\&\, r$	AIP
4.	p	&E, 3
5.	r	&E, 3
6.	Show $\neg r \rightarrow \neg r$	
7.	$\lceil \neg r$	ACP
8.	Show $\neg p \rightarrow \neg r$	
9.	$\lceil \neg p$	ACP
10.	Show $\neg r$	
11.	$\lceil r$	AIP
12.	p	R, 4
13.	$\llcorner \neg p$	R, 9
14.	$\neg r$	$\vee$E, 1, 6, 8

15.

1.	$\neg s \vee (s \,\&\, p)$	A
2.	$(s \rightarrow p) \rightarrow r$	A
3.	Show r	
4.	$\lceil$ Show $s \rightarrow p$	
5.	$\lceil s$	ACP
6.	Show $\neg s \rightarrow p$	
7.	$\lceil \neg s$	ACP
8.	Show $\neg\neg p$	
9.	$\lceil \neg p$	AIP
10.	s	R, 5
11.	$\llcorner \neg s$	R, 7
12.	p	$\neg\neg$, 8
13.	Show $(s \,\&\, p) \rightarrow p$	
14.	$\lceil s \,\&\, p$	ACP
15.	$\llcorner p$	&E, 14
16.	p	$\vee$E, 1, 6, 13
17.	$\llcorner r$	$\rightarrow$E, 2, 4

20.

1.	$p \rightarrow q$	A
2.	$r \rightarrow p$	A
3.	Show $\neg r \vee q$	
4.	$\lceil$ Show $\neg\neg(\neg r \vee q)$	
5.	$\lceil \neg(\neg r \vee q)$	AIP
6.	Show $\neg q$	
7.	$\lceil q$	AIP
8.	$\neg r \vee q$	$\vee$I, 7
9.	$\llcorner \neg(\neg r \vee q)$	R, 5
10.	Show $\neg r$	
11.	$\lceil r$	AIP
12.	p	$\rightarrow$E, 2, 11
13.	q	$\rightarrow$E, 1, 12
14.	$\llcorner \neg q$	R, 6
15.	$\llcorner \neg r \vee q$	$\vee$I, 10
16.	$\llcorner \neg r \vee q$	$\neg\neg$, 4

30. 1. $s \rightarrow p$ A
     2. $(s \,\&\, p) \rightarrow q$ A
     3. $r \rightarrow s_1$ A
     4. $r \vee s$ A
     5. Show $q \vee s_1$
     6. ⌈Show $r \rightarrow (q \vee s_1)$
     7. ⌈ r ACP
     8. │ s_1 →E, 3, 7
     9. ⌊ $q \vee s_1$ ∨I, 8
    10. Show $s \rightarrow (q \vee s_1)$
    11. ⌈ s ACP
    12. │ p →E, 1, 11
    13. │ $s \,\&\, p$ &I, 11, 12
    14. │ q →E, 2, 13
    15. ⌊ $q \vee s_1$ ∨I, 14
    16. ⌊$q \vee s_1$ ∨E, 4, 6, 10

35. 1. $\neg p \vee (q \,\&\, r)$ A
     2. $(r \vee \neg q) \rightarrow (s \,\&\, q_1)$ A
     3. $(q_1 \,\&\, q_2) \vee \neg(q_1 \vee q_2)$ A
     4. Show $\neg p \vee q_2$
     5. ⌈Show $\neg p \rightarrow (\neg p \vee q_2)$
     6. ⌈ $\neg p$ ACP
     7. ⌊ $\neg p \vee q_2$ ∨I, 6
     8. Show $(q \,\&\, r) \rightarrow (\neg p \vee q_2)$
     9. ⌈ $q \,\&\, r$ ACP
    10. │ r &E, 9
    11. │ $r \vee \neg q$ ∨I, 10
    12. │ $s \,\&\, q_1$ →E, 2, 11
    13. │ q_1 &E, 12
    14. │ Show $(q_1 \,\&\, q_2) \rightarrow (q_1 \,\&\, q_2)$
    15. │ ⌊$q_1 \,\&\, q_2$ ACP
    16. │ Show $\neg(q_1 \vee q_2) \rightarrow (q_1 \,\&\, q_2)$
    17. │ ⌈ $\neg(q_1 \vee q_2)$ ACP
    18. │ │ Show $\neg\neg(q_1 \,\&\, q_2)$
    19. │ │ ⌈ $\neg(q_1 \,\&\, q_2)$ AIP
    20. │ │ │ $q_1 \vee q_2$ ∨I, 13
    21. │ │ ⌊ $\neg(q_1 \vee q_2)$ R, 17
    22. │ ⌊ $q_1 \,\&\, q_2$ ¬¬, 18
    23. │ $q_1 \,\&\, q_2$ ∨E, 3, 14, 16
    24. │ q_2 &E, 23
    25. ⌊ $\neg p \vee q_2$ ∨I, 24
    26. ⌊$\neg p \vee q_2$ ∨E, 1, 5, 8

4.5 Derivable Rules

5. p: The President pursues arms limitation talks; q: The European left will acquiesce to the placement of additional nuclear weapons in Europe; r: The President gets the foreign policy mechanism working more harmoniously.

1. $q \to p$ A
2. $\neg q \to (s \lor r)$ A
3. Show $\neg r \to (\neg p \to s)$
4. $\neg r$ ACP
5.   Show $\neg p \to s$
6. $\neg p$ ACP
7. $\neg q$ $\to$E*, 1, 6
8. $s \lor r$ $\to$E, 2, 7
9. s $\lor$E*, 8, 4

28.
1. $\neg(p \leftrightarrow q)$ A
2. Show $p \leftrightarrow \neg q$
3. Show $p \to \neg q$
4. p ACP
5.   Show $\neg q$
6. q AIP
7.   Show $p \to q$
8. p ACP
9. q R, 6
10.   Show $q \to p$
11. q ACP
12. p R, 4
13. $p \leftrightarrow q$ $\leftrightarrow$I, 7, 10
14. $\neg(p \leftrightarrow q)$ R, 1
15. Show $\neg q \to p$
16. $\neg q$ ACP
17.   Show $\neg\neg p$
18. $\neg p$ AIP
19.   Show $p \to q$
20. p ACP
21.   Show $\neg\neg q$
22. $\neg q$ AIP
23. p R, 20
24. $\neg p$ R, 18
25. q $\neg\neg$, 21
26.   Show $q \to p$
27. q ACP
28.   Show $\neg\neg p$
29. $\neg p$ AIP
30. q R, 27
31. $\neg q$ R, 16
32. p 77, 28
33. $p \leftrightarrow q$ $\leftrightarrow$I, 19, 26
34. $\neg(p \leftrightarrow q)$ R, 1
35. p 77, 17
36. $p \leftrightarrow \neg q$ $\leftrightarrow$I, 3, 15

38.
1. $(p \mathbin{\&} \neg r) \leftrightarrow (s \lor \neg q)$ A
2. $s_1 \mathbin{\&} ((\neg s \mathbin{\&} \neg r) \to p)$ A
3. $(s_1 \to q) \lor (s_1 \to r)$ A
4. $((p \mathbin{\&} q) \mathbin{\&} s) \to r$ A
5. Show $q \mathbin{\&} r$
6. s_1 &E, 2
7. $(\neg s \mathbin{\&} \neg r) \to p$ &E, 2
8.   Show q
9. $\neg q$ AIP
10. $s_1 \mathbin{\&} \neg q$ &I, 6, 9
11. $\neg(s_1 \to q)$ $\neg\to$, 10
12. $s_1 \to r$ $\lor$E*, 3, 11
13. r $\to$E, 12, 6
14. $s \lor \neg q$ $\lor$I, 9
15. $p \mathbin{\&} \neg r$ $\leftrightarrow$E, 1, 14
16. $\neg r$ &E, 15
17.   Show r
18. $\neg r$ AIP
19. $s_1 \mathbin{\&} \neg r$ &I, 6, 18
20. $\neg(s_1 \to r)$ $\neg\to$, 19
21. $s_1 \to q$ $\lor$E*, 3, 20
22. q $\to$E, 21, 6
23.   Show p
24. $\neg p$ AIP
25. $\neg(\neg s \mathbin{\&} \neg r)$ $\to$E*, 7, 24
26. $\neg\neg s \lor \neg\neg r$ $\neg$&, 25
27. $s \lor r$ $\neg\neg$, 26
28. s $\lor$E*, 27, 18
29. $s \lor \neg q$ $\lor$I, 28
30. $p \mathbin{\&} \neg r$ $\leftrightarrow$E, 1, 29
31. p &E, 30
32. $p \mathbin{\&} \neg r$ &I, 23, 18
33. $s \lor \neg q$ $\leftrightarrow$E, 1, 32
34. $\neg\neg q$ $\neg\neg$, 22
35. s $\lor$E*, 33, 34
36. $p \mathbin{\&} q$ &I, 23, 22
37. $(p \mathbin{\&} q) \mathbin{\&} s$ &I, 36, 35
38. r $\to$E, 4, 37
39. $q \mathbin{\&} r$ &I, 8, 17

Chapter 5
QUANTIFIERS

(When the universe is unspecified, it consists of objects.)

5.2 Categorical Sentence Forms

5. Bx: x is big; Mx: x is a man; Dx: x is a dreamer. $\forall x((Bx \,\&\, Mx) \to Dx)$.

10. Dx: x is delusion. $\forall x Dx$.

5.3 Polyadic Predicates

5. Fx: x is finite; Rxy: x reveals y; a: Infinitude. $\forall x(Fx \to Rxa)$. Alternatively, we could take "infinitude" as "something infinite" and so symbolize Roethke's statement as $\forall x(Fx \to \exists y(\neg Fy \,\&\, Rxy))$.

10. Mx: x is a man; Nx: To do nothing is in x's power. $\forall x(Mx \to Nx)$. It is tempting to use

$$Pxy: \text{To do } y \text{ is in } x\text{'s power}$$

and to symbolize this sentence as $\forall x(Mx \to \neg\exists yPxy)$ or $\neg\exists y\forall x(Mx \to Pxy)$. But the first says that no man has the power to do anything, while the second says that there is nothing that every man has the power to do.

5.4 The Language QL

5. (c)   **10.** (a)   **15.** (b)

20. (c)   **25.** (c)   **30.** (c)

35. (b)   **40.** (c)

45. (a) $\forall xFx \leftrightarrow Gc$, a formula; (b) $\forall xFx \leftrightarrow Gd$, a formula; (c) $\forall xFx \leftrightarrow Fx$, not a formula; (d) $\forall xFx \leftrightarrow Gc$, a formula; (e) $\forall yFy \leftrightarrow Gc$, a formula.

50. (a) $Fxc \leftrightarrow \exists yFcy$, not a formula; (b) $Fxd \leftrightarrow \exists yFdy$, not a formula; (c) $Fxx \leftrightarrow \exists yFdy$, not a formula; (d) $Fxc \leftrightarrow \exists dFdd$, not a formula; (e) $Fyc \leftrightarrow \exists yFdy$, not a formula.

5.5 Symbolization

5. Px: x is a person (i.e., one of us); Dx: x consists (in part) of dust; Rx: x consists (in part) of dreams. $\forall x(Px \to (Dx \,\&\, Rx))$. Alternatively, we could symbolize as follows: Dx: x is dust; Rx: x is a dream; Cxy: x consists of y. $\forall x(Px \to \exists y\exists z(Dy \,\&\, Rz \,\&\, Cxy \,\&\, Cxz))$. Note that neither conveys, however, that we are *only* dust and dreams.

10. Px: x is a person; Vx: x is a great virtue; Cx: x is a great vice; Dx: x is disagreeable; Lx: x is delightful; Hxy: x has y. $\exists x((Px \,\&\, \exists y(Vy \,\&\, Hxy)) \,\&\, Dx) \,\&\, \exists x((Px \,\&\, \exists y(Cy \,\&\, Hxy)) \,\&\, Lx)$.

15. a: Loafing; Nxy: x needs y; Ex: x is an explanation; Cxy: x is an excuse for y. $\neg\exists y(Ey \,\&\, Nay) \,\&\, Caa$.

20. Fx: x falls; Bx: x is built again. $\forall x(Fx \,\&\, Bx)$. Alternatively: Bxy: x builds y again. $\forall x(Fx \,\&\, \exists yByx)$.

25. Tx: x is a time; $Wxyz$: x is a 30-hour week for y at time z; Mx: x is a man; $Dxyz$: x has y to do at time z. $\neg\exists x\exists y\exists z(Tx \,\&\, My \,\&\, Wzyx \,\&\, \exists wDywx)$.

30. Mx: x is a man; Fxy: x is a friend of y. $\forall x(\forall y(My \to Fxy) \to \neg\exists y(My \,\&\, Fxy))$.

35. *Bx*: *x* is one of the Bears; *Gx*: *x* is a good Bear; *Hxy*: *x* harms *y*; *Sxyz*: *x* suspects that *y* would harm *z* at time *t*; *Cx*: *x* is a creature; *Tx*: *x* is a time. ∀*x*(*Bx* → (*Gx* & ¬∃*y*(*Cy* & *Hxy*) & ¬∃*z*∃*w*∃*t*(*Tz* & *Cw* & *Bt* & *Sxwtz*))). Note: *Cx* symbolizes -*body*. In the world of the three Bears, symbolizing -*body* as '*x* is a person' would be too restrictive. Also note the symbolization of 'never suspected that anyone would harm them.' We could take this as saying that no Bear ever suspected that anyone would harm him (or her), but it seems more accurate to render it as saying that no Bear ever suspected that anyone would harm any of the Bears.

Chapter 6
QUANTIFIED TABLEAUX

6.2 Strategies

5. Universe: People; *Mx*: *x* is moral; *Rxy*: *x* respects the dignity of *y*; *a*: Julie. Valid.

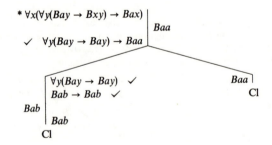

10. Universe: Teams; *a*: Longhorns; *Bxy*: *x* can beat *y*. Valid.

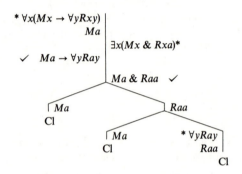

15. *Ax*: *x* is an object of art. Valid.

$$\exists x(Ax \to \forall yAy)*$$
$$Aa \to \forall yAy \checkmark$$
Aa
$$\forall yAy \checkmark$$
$$Ab$$
$$Ab \to \forall yAy \checkmark$$
Ab
$$\forall yAy$$
Cl

25. Follows.

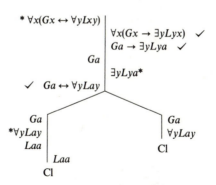

30. Universe: Amok natives; *Pxy: x* is parent of *y*. The fact does not follow from the anthropologist's thesis.

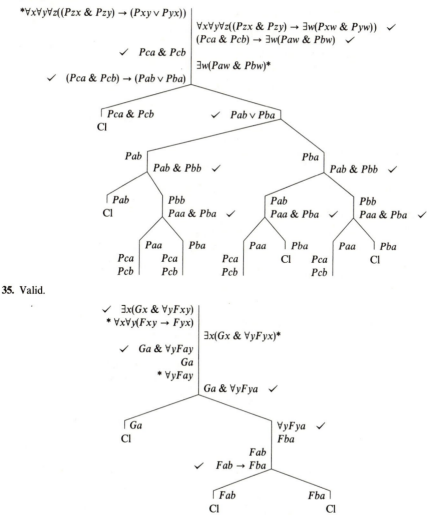

35. Valid.

6.3 Interpretations

5. False.	**10.** True.	**15.** False.
20. True.	**25.** True.	**30.** True.
35. True.	**40.** False.	**45.** True.
50. True.	**55.** False.	**60.** True.
65. False.	**70.** True.	**75.** False.
80. True.	**85.** True.	**90.** False.
95. True.	**100.** False.	**105.** False.
110. True.	**115.** True.	**120.** False.
125. True.	**130.** True.	**135.** False.
140. True.	**145.** False.	**150.** False.
155. True.	**160.** False.	**165.** False.
170. False.	**175.** True.	**180.** False.

6.4 Constructing Interpretations from Tableaux

5. (1) $D = \{a\}$; $\varphi(F) = \emptyset$; $\varphi(R) = \emptyset$.
(2) $D = \{a, b\}$; $\varphi(F) = \emptyset$; $\varphi(R) = \{\langle a, b \rangle\}$.

$$\checkmark \quad Fa \to \exists yRay$$

```
              Fa                    ✓  ∃yRay
                                       Rab
```

10. (1) $D = \{a\}$; $\varphi(F) = \emptyset$.

$$* \; \forall x \neg Fxx$$
$$\checkmark \quad \neg Faa$$
$$Faa$$

20. (1) $D = \{a\}$; $\varphi(R) = \{\langle a, a \rangle\}$.

$$\checkmark \quad \exists x \forall y Rxy$$
$$* \; \forall y Ray$$
$$Raa$$

25. (1) $D = \{a\}$; $\varphi(R) = \emptyset$.
(2) $D = \{a\}$; $\varphi(R) = \{\langle a, a \rangle\}$.

$$* \; \forall x(\neg Rxx \to \forall y(Rxy \to \neg Ryx))$$
$$\checkmark \quad \neg Raa \to \forall y(Ray \to \neg Rya)$$

```
        ┌ ¬Raa  ✓              * ∀y(Ray → ¬Rya)
    Raa │                        ✓  Raa → ¬Raa

                                  ┌ Raa  ✓    ¬Raa
                                      Raa
```

30. (1) $D = \{a\}$; $\varphi(F) = \emptyset$; $\varphi(G) = \{\langle a \rangle\}$.

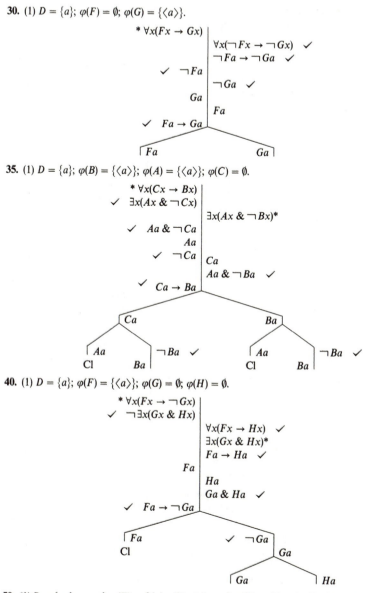

35. (1) $D = \{a\}$; $\varphi(B) = \{\langle a \rangle\}$; $\varphi(A) = \{\langle a \rangle\}$; $\varphi(C) = \emptyset$.

40. (1) $D = \{a\}$; $\varphi(F) = \{\langle a \rangle\}$; $\varphi(G) = \emptyset$; $\varphi(H) = \emptyset$.

50. (1) $D = \{a, b, c, \ldots\}$; $\varphi(F) = \{\langle a \rangle, \langle b \rangle, \langle c \rangle, \ldots\}$; $\varphi(R) = \{\langle a, a \rangle, \langle b, a \rangle, \langle c, b \rangle, \langle d, c \rangle, \ldots\}$.
(2) $D = \{a\}$; $\varphi(F) = \{\langle a \rangle\}$; $\varphi(R) = \{\langle a, a \rangle\}$.

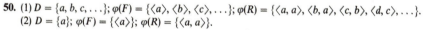

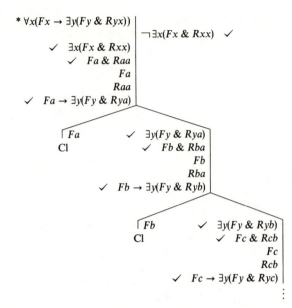

Chapter 7
QUANTIFIED NATURAL DEDUCTION

7.1 Deduction Rules for Quantifiers

5. *Cx*: *x* is coherent; *Bxy*: *x* baffles *y*; *a*: Me; *b*: This course.

1.	¬∃x(Cx & Bxa)	A
2.	Bba	A
3.	S̶h̶o̶w̶ ¬Cb	
4.	⌈ Cb	AIP
5.	│ Cb & Bba	&I, 4, 2
6.	│ ∃x(Cx & Bxa)	∃I, 5
7.	└ ¬∃x(Cx & Bxa)	R, 1

10. *Cx*: *x* is a computer program used for processing natural language; *Px*: *x* is written in PROLOG; *Rx*: *x* relies heavily on the notion of a list.

1.	∃x(Cx & Px)	A
2.	¬∃x(Px & Rx)	A
3.	S̶h̶o̶w̶ ¬∀x(Cx → Rx)	
4.	⌈ ∀x(Cx → Rx)	AIP
5.	│ Ca & Pa	∃E, 1
6.	│ Ca → Ra	∀E, 4
7.	│ Ca	&E, 5
8.	│ Ra	→E, 6, 7
9.	│ Pa	&E, 5
10.	│ Pa & Ra	&I, 9, 8
11.	│ ∃x(Px & Rx)	∃I, 10
12.	└ ¬∃x(Px & Rx)	R, 2

15. Universe: People; *Ax*: *x* is an analyst; *Bx*: *x* insists that we are in the middle of a historic bull market; *Cx*: *x* says the market will soon collapse; *Rx*: *x* is recommending only utility stocks; *Mx*: *x* believes that M1 controls the direction of the economy.

1.	$\exists x(Ax \ \& \ Bx)$	A
2.	$\exists x(Ax \ \& \ Cx)$	A
3.	$\forall x(Cx \rightarrow Rx)$	A
4.	$\neg\exists x(Mx \ \& \ Bx)$	A
5.	Show $\exists x(Ax \ \& \ Rx) \ \& \ \exists x(Ax \ \& \ \neg Mx)$	
6.	$\lceil Aa \ \& \ Ba$	$\exists$E, 1
7.	$Ab \ \& \ Cb$	$\exists$E, 2
8.	$Cb \rightarrow Rb$	$\forall$E, 3
9.	Cb	&E, 7
10.	Rb	$\rightarrow$E, 8, 9
11.	Ab	&E, 7
12.	$Ab \ \& \ Rb$	&I, 11, 10
13.	$\exists x(Ax \ \& \ Rx)$	$\exists$I, 12
14.	Aa	&E, 6
15.	Show $\neg Ma$	
16.	$\lceil Ma$	AIP
17.	Ba	&E, 6
18.	$Ma \ \& \ Ba$	&I, 16, 17
19.	$\exists x(Mx \ \& \ Bx)$	$\exists$I, 18
20.	$\lfloor\neg\exists x(Mx \ \& \ Bx)$	R, 4
21.	$Aa \ \& \ \neg Ma$	&I, 14, 15
22.	$\exists x(Ax \ \& \ \neg Mx)$	$\exists$I, 21
23.	$\lfloor\exists x(Ax \ \& \ Rx) \ \& \ \exists x(Ax \ \& \ \neg Mx)$	&I, 13, 22

7.2 Universal Proof

5. *Sx*: *x* is stupid; *Dx*: *x* is difficult; *Cxy*: *x* can do *y*; *Bxyz*: *x* can do *y* better than *z* can; *a*: You; *b*: Me.

1.	$\forall x(Sx \rightarrow \neg Dx)$	A
2.	$\forall x(Cax \rightarrow Sx)$	A
3.	$\forall x(\neg Dx \rightarrow Bbxa)$	A
4.	Show $\forall x(Cax \rightarrow Bbxa)$	
5.	$\lceil$ Show $Cac \rightarrow Bbca$	
6.	$\lceil Cac$	ACP
7.	$Cac \rightarrow Sc$	$\forall$E, 2
8.	Sc	$\rightarrow$E, 7, 6
9.	$Sc \rightarrow \neg Dc$	$\forall$E, 1
10.	$\neg Dc$	$\rightarrow$E, 9, 8
11.	$\neg Dc \rightarrow Bbca$	$\forall$E, 3
12.	$\lfloor Bbca$	$\rightarrow$E, 11, 10

10. Universe: People; *Fx*: *x* is a Frenchman; *Sx*: *x* is a Socialist; *Cx*: *x* is a Communist; *Axy*: *x* fears *y*.

1.	$\forall x(Fx \rightarrow \forall y(Sy \rightarrow Axy))$	A
2.	$\forall x(Sx \rightarrow \forall y(Axy \rightarrow Cy))$	A
3.	Show $\forall x((Fx \ \& \ Sx) \rightarrow Cx)$	

4.	⌐Show $(Fa \& Sa) \rightarrow Ca$	
5.	⌐$Fa \& Sa$	ACP
6.	Fa	&E, 5
7.	Sa	&E, 5
8.	$Fa \rightarrow \forall y(Sy \rightarrow Aay)$	$\forall$E, 1
9.	$\forall y(Sy \rightarrow Aay)$	$\rightarrow$E, 8, 6
10.	$Sa \rightarrow Aaa$	$\forall$E, 9
11.	Aaa	$\rightarrow$E, 10, 7
12.	$Sa \rightarrow \forall y(Aay \rightarrow Cy)$	$\forall$E, 2
13.	$\forall y(Aay \rightarrow Cy)$	$\rightarrow$E, 12, 17
14.	$Aaa \rightarrow Ca$	$\forall$E, 13
15.	Ca	$\rightarrow$E, 14, 11

20. $Cxyz$: x chooses y over z; Px: x is a person; $Fxyz$: x prefers y to z; Dx: x is a dictator; Vx: x has veto power; g: Government.

1.	$\forall x \forall y(Cgxy \leftrightarrow \neg Cgyx)$	A
2.	$\forall x(Px \rightarrow (Vx \leftrightarrow \forall y \forall z(Cgyz \rightarrow \neg Fxzy)))$	A
3.	$\forall x(Px \rightarrow (Dx \leftrightarrow \forall y \forall z(Fxyz \rightarrow Cgyz)))$	A
4.	Show $\forall x((Px \& Vx) \rightarrow Dx)$	
5.	⌐Show $(Pa \& Va) \rightarrow Da$	
6.	⌐$Pa \& Va$	ACP
7.	Pa	&E
8.	$Pa \rightarrow (Va \leftrightarrow \forall y \forall z((Cgyz \rightarrow \neg Fazy))$	$\forall$E, 2
9.	Va	&E, 6
10.	$Va \leftrightarrow \forall y \forall z(Cygz \rightarrow \neg Fazy)$	$\rightarrow$E, 8, 7
11.	$\forall y \forall z(Cgyz \rightarrow \neg Fazy)$	$\leftrightarrow$E, 10, 9
12.	$Pa \rightarrow (Da \leftrightarrow \forall y \forall z(Fazy \rightarrow Cgyz))$	$\forall$E, 3
13.	$Da \leftrightarrow \forall y \forall z(Fayz \rightarrow Cgyz)$	$\rightarrow$E, 12, 7
14.	Show $\forall y \forall z(Fayz \rightarrow Cgyz)$	
15.	⌐Show $\forall z(Fabz \rightarrow Cgbz)$	
16.	⌐Show $Fabc \rightarrow Cgbc$	
17.	⌐$Fabc$	ACP
18.	$\forall z(Cgcz \rightarrow \neg Fazc)$	$\forall$E, 11
19.	$Cgcb \rightarrow \neg Fabc$	$\forall$E, 18
20.	$\neg Cgcb$	$\rightarrow$E*, 19, 17
21.	$\forall y(Cgby \leftrightarrow \neg Cgyb)$	$\forall$E, 1
22.	$Cgbc \leftrightarrow \neg Cgcb$	$\forall$E, 21
23.	$Cgbc$	$\leftrightarrow$E, 22, 20
24.	Da	$\leftrightarrow$E, 13, 14

35.

1.	$\forall x(Mx \rightarrow Lx)$	A
2.	$\forall x(Sx \rightarrow Mx)$	A
3.	$\exists x Sx$	A
4.	Show $\exists x(Sx \& Lx)$	
5.	⌐Sa	$\exists$E, 3
6.	$Sa \rightarrow Ma$	$\forall$E, 2
7.	Ma	$\rightarrow$E, 6, 5
8.	$Ma \rightarrow La$	$\forall$E, 1
9.	La	$\rightarrow$E, 8, 7
10.	$Sa \& La$	&I, 5, 9
11.	$\exists x(Sx \& Lx)$	$\exists$I, 10

40.

1.	$\forall x(Mx \rightarrow \neg Lx)$	A
2.	$\forall x(Sx \rightarrow Mx)$	A
3.	Show $\neg \exists x(Lx \& Sx)$	
4.	⌐$\exists x(Lx \& Sx)$	AIP
5.	$La \& Sa$	$\exists$E, 4
6.	La	&E, 5
7.	Sa	&E, 5
8.	$Ma \rightarrow \neg La$	$\forall$E, 1
9.	$Sa \rightarrow Ma$	$\forall$E, 2
10.	Ma	$\rightarrow$E, 9, 7
11.	$\neg La$	$\rightarrow$E, 8, 10

45. Px: x is a person; Cx: x is a circle; Fx: x is a figure; Dxy: x draws y.

1. $\forall x(Cx \rightarrow Fx)$ A
2. Show $\forall x(Px \rightarrow (\exists y(Cy \,\&\, Dxy) \rightarrow \exists x(Fz \,\&\, Dxz)))$
3.   Show Pa $\rightarrow (\exists y(Cy \,\&\, Day) \rightarrow \exists z(Fz \,\&\, Daz))$
4. Pa ACP
5.   Show $\exists y(Cy \,\&\, Day) \rightarrow \exists z(Fz \,\&\, Daz)$
6. $\exists y(Cy \,\&\, Day)$
7. $Cb \,\&\, Dab$ $\exists$E, 6
8. $Cb \rightarrow Fb$ $\forall$E, 1
9. Cb &E, 7
10. Fb $\rightarrow$E, 8, 9
11. Dab &E, 7
12. $Fb \,\&\, Dab$ &I, 10, 11
13. $\exists z(Fz \,\&\, Daz)$ $\exists$I, 12

55.
1. Show $\forall x\forall yFxy \leftrightarrow \forall y\forall xFxy$
2. Show $\forall x\forall yFxy \rightarrow \forall y\forall xFxy$
3. $\forall x\forall yFxy$ ACP
4.   Show $\forall y\forall xFxy$
5.   Show $\forall xFxa$
6.   Show Fba
7. $\forall yFby$ $\forall$E, 3
8. Fba $\forall$E, 7
9.   Show $\forall y\forall xFxy \rightarrow \forall x\forall yFxy$
10. $\forall y\forall xFxy$ ACP
11.   Show $\forall x\forall yFxy$
12.   Show $\forall yFcy$
13.   Show Fcd
14. $\forall xFyd$ $\forall$E, 10
15. Fcd $\forall$E, 14
16. $\forall x\forall yFxy \leftrightarrow \forall y\forall xFxy$ $\leftrightarrow$I, 2, 9

60.
1. $\forall x\exists y(Gy \rightarrow Fx)$ A
2. Show $\exists y\forall x(Gy \rightarrow Fx)$
3. $\neg\exists y\forall x(Gy \rightarrow Fx)$ AIP
4.   Show $\forall y\exists x(Gy \,\&\, \neg Fx)$
5.   Show $\exists x(Ga \,\&\, \neg Fx)$
6. $\neg\exists x(Ga \,\&\, \neg Fx)$ AIP
7.   Show $\forall x(Ga \rightarrow Fx)$
8.   Show $Ga \rightarrow Fb$
9. Ga ACP
10.   Show Fb
11. $\neg Fb$ AIP
12. $Ga \,\&\, \neg Fb$ &I, 9, 11
13. $\exists x(Ga \,\&\, \neg Fx)$ $\exists$I, 12
14. $\neg\exists x(Ga \,\&\, \neg Fx)$ R, 6
15. $\exists y\forall x(Gy \rightarrow Fx)$ $\exists$I, 7
16. $\neg\exists y\forall x(Gy \rightarrow Fx)$ R, 3
17. $\exists x(Gb \,\&\, \neg Fx)$ $\forall$E, 4
18. $Gb \,\&\, \neg Fc$ $\exists$E, 17
19. $\exists y(Gy \rightarrow Fc)$ $\forall$E, 1
20. $Gd \rightarrow Fc$ $\exists$E, 19
21. $\exists x(Gd \,\&\, \neg Fx)$ $\forall$E, 4
22. $Gd \,\&\, \neg Fe$ $\exists$E, 21
23. Gd &E, 22
24. Fc $\rightarrow$E, 20, 23
25. $\neg Fc$ &E, 18

65.
 1. $\exists xFx \vee \exists xGx$ A
 2. $\forall x(Fx \rightarrow Gx)$ A
 3. Show $\exists xGx$
 4. ⌈Show $\exists xGx \rightarrow \exists xGx$
 5. ⌈$\exists xGx$ ACP
 6. Show $\exists xFx \rightarrow \exists xGx$
 7. ⌈$\exists xFx$ ACP
 8. | Fa $\exists$E, 7
 9. | $Fa \rightarrow Ga$ $\forall$E, 2
 10. | Ga $\rightarrow$E, 9, 8
 11. ⌊$\exists xGx$ $\exists$I, 10
 12. ⌊$\exists xGx$ $\vee$E, 1, 6, 4

7.3 Derivable Rules for Quantifiers

5.
 1. $\forall x(\exists yFxy \rightarrow \exists y\neg Gy)$ A
 2. $\exists x\exists yFxy$ A
 3. $\forall x(Gx \leftrightarrow \neg Hx)$ A
 4. Show $\exists xHx$
 5. ⌈Fab $\exists$E^2, 2
 6. | $\exists yFay \rightarrow \exists y\neg Gy$ $\forall$E, 1
 7. | $\exists yFay$ $\exists$I, 5
 8. | $\exists y\neg Gy$ $\rightarrow$E, 6, 7
 9. | $\neg Gc$ $\exists$E, 8
 10. | $Gc \leftrightarrow \neg Hc$ $\forall$E, 3
 11. | Hc $\leftrightarrow$E*, 10, 9
 12. ⌊$\exists xHx$ $\exists$I, 11

15.
 1. $\forall x(Fxx \rightarrow Hx)$ A
 2. $\exists xHx \rightarrow \neg\exists yGy$ A
 3. Show $\forall x(Gx \rightarrow \neg\exists zFzz)$
 4. ⌈Show $Ga \rightarrow \neg\exists zFzz$
 5. | ⌈Ga ACP
 6. | | Show $\neg\exists zFzz$
 7. | | ⌈$\exists zFzz$ AIP
 8. | | | Fbb $\exists$E, 7
 9. | | | $Fbb \rightarrow Hb$ $\forall$E, 1
 10. | | | Hb $\rightarrow$E, 9, 8
 11. | | | $\exists xHx$ $\exists$I, 10
 12. | | | $\neg\exists yGy$ $\rightarrow$E, 2, 11
 13. ⌊⌊⌊$\exists yGy$ $\exists$I, 5

25.
 1. $\exists x\forall y\neg Fxy$ A
 2. Show $\exists x\forall y\forall z(Fxz \rightarrow Fzy)$
 3. ⌈$\forall y\neg Fay$ $\exists$E, 1
 4. | Show $\forall y\forall z(Faz \rightarrow Fzy)$
 5. | ⌈Show $Fac \rightarrow Fcb$
 6. | | ⌈Fac ACP
 7. | | | $\neg Fac$ $\forall$E, 3
 8. | | ⌊Fcb I, 6, 7
 9. ⌊$\exists x\forall y\forall z(Fxz \rightarrow Fzy)$ $\exists$I, 4

30.
 1. $\exists x(Fx \,\&\, \forall y(Ty \rightarrow Gy))$ A
 2. $\forall x(Fx \rightarrow (\exists y(Ay \,\&\, Gy) \rightarrow Bxx))$ A

3. $\exists z(Az \ \& \ Tz)$ A
4. Show $\exists xBxx$
5. ⌈ $Fa \ \& \ \forall y(Ty \rightarrow Gy)$ $\exists$E, 1
6. │ Fa &E, 5
7. │ $\forall y(Ty \rightarrow Gy)$ &E, 5
8. │ $Ab \ \& \ Tb$ $\exists$E, 3
9. │ Tb &E, 8
10. │ $Tb \rightarrow Gb$ $\forall$E, 7
11. │ Gb $\rightarrow$E, 10, 9
12. │ Ab &E, 8
13. │ $Ab \ \& \ Gb$ &I, 12, 11
14. │ $\exists y(Ay \ \& \ Gy)$ $\exists$I, 13
15. │ $Fa \rightarrow (\exists y(Ay \ \& \ Gy) \rightarrow Baa)$ $\exists$E, 2
16. │ $\exists y(Ay \ \& \ Gy) \rightarrow Baa$ $\rightarrow$E, 15, 6
17. │ Baa $\rightarrow$E, 16, 14
18. ⌊ $\exists xBxx$ $\exists$I, 17

Chapter 8
IDENTITY, FUNCTION SYMBOLS, AND DESCRIPTIONS

8.1 Identity

5. p: This poem; r: The reader. $p = r \ \& \ r = p$.

10. Universe: Actions; Bxy: x is better than y; Px: x procures the greatest happiness for the greatest number. $\forall x(Px \rightarrow \forall y(y \neq x \rightarrow Bxy))$.

15. Universe: Cases and counsels; Dx: x is a case of difficulty; Hx: x is a case when hopes are small; Cx: x is counsel; $Bxyz$: x is bolder than y in case z; $Sxyz$: x is safer than y in case z. $\forall x((Dx \ \& \ Hx) \rightarrow \forall y(\forall z(z \neq y \rightarrow Byzx) \rightarrow \forall w(w \neq y \rightarrow Sywx)))$.

20. Universe: People; a: Lyndon; Pxy: x pities y. $\forall x(Pxa \rightarrow x = a)$ or $\forall x(Pxa \leftrightarrow x = a)$.

25. Universe: People; a: Lyndon; Pxy: x pities y. $\forall z(\forall y(Pzy \rightarrow \forall x(Pyx \rightarrow x = a)) \rightarrow z = a)$ or $\forall z(\forall y(Pzy \rightarrow \forall x(Pyx \leftrightarrow x = a)) \leftrightarrow z = a)$.

8.2 Tableau Rules for Identity

5. Universe: People; a: Frank; e: Joe; Px: x played in the tournament; Wx: x won.

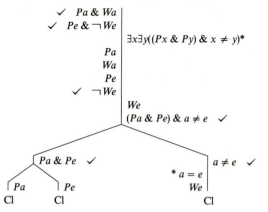

15.

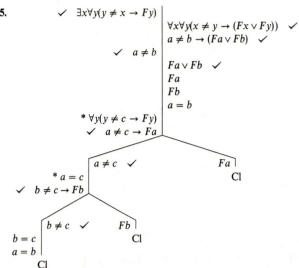

25. Universe: Numbers; *Gxy*: *x* is greater than *y*. Not valid.

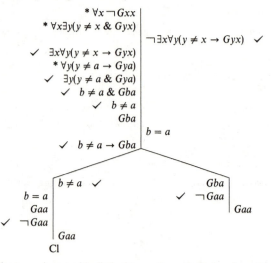

35. Universe: People; *Ixy*: *x* is more idealistic than *y*; *Fxy*: *x* is farther from the problem than *y*. Yes.

* $\forall x\forall y(Ixy \leftrightarrow Fxy)$

$\forall x(\forall y(y \neq x \rightarrow Fxy) \rightarrow \forall z(z \neq x \rightarrow Ixz))$ ✓
$\forall y(y \neq a \rightarrow Fay) \rightarrow \forall z(z \neq a \rightarrow Iaz)$ ✓

* $\forall y(y \neq a \rightarrow Fay)$

$\forall z(z \neq a \rightarrow Iaz)$ ✓
$b \neq a \rightarrow Iab$ ✓

✓ $b \neq a$

Iab
$b = a$

✓ $b \neq a \rightarrow Fab$

	$b \neq a$ ✓		Fab
$b = a$			✓ $Iab \leftrightarrow Fab$
Cl			

	Iab		Iab
	Fab		Fab
	Cl		Cl

8.3 Deduction Rules for Identity

5.

1.	$\exists x(\forall y(y = x \leftrightarrow Fy)$	A
2.	~~Show~~ $\exists x(Fx \& (Gx \rightarrow Ha)) \leftrightarrow (\exists x(Fx \& Gx) \rightarrow Ha)$	
3.	$\ulcorner \forall y(y = d \leftrightarrow Fy)$	$\exists$E, 1
4.	$\quad$ ~~Show~~ $\exists x(Fx \& (Gx \rightarrow Ha)) \rightarrow (\exists x(Fx \& Gx) \rightarrow Ha)$	
5.	$\quad \ulcorner \exists x(Fx \& (Gx \rightarrow Ha))$	ACP
6.	$\quad\quad$ ~~Show~~ $\exists x(Fx \& Gx) \rightarrow Ha$	
7.	$\quad\quad \ulcorner \exists x(Fx \& Gx)$	ACP
8.	$\quad\quad\quad Fb \& Gb$	$\exists$E, 7
9.	$\quad\quad\quad Fb$	&E, 8
10.	$\quad\quad\quad b = d \leftrightarrow Fb$	$\forall$E, 3
11.	$\quad\quad\quad b = d$	$\leftrightarrow$E, 10, 9
12.	$\quad\quad\quad Fd$	=E, 11, 9
13.	$\quad\quad\quad Fc \& (Gc \rightarrow Ha)$	$\exists$E, 5
14.	$\quad\quad\quad Fc$	&E, 13
15.	$\quad\quad\quad c = d \leftrightarrow Fc$	$\forall$E, 3
16.	$\quad\quad\quad c = d$	$\leftrightarrow$E, 15, 14
17.	$\quad\quad\quad Gc \rightarrow Ha$	&E, 13
18.	$\quad\quad\quad Gb$	&E, 8
19.	$\quad\quad\quad Gd$	=E, 11, 18
20.	$\quad\quad\quad Gc$	=E, 16, 19
21.	$\quad\quad\quad Ha$	$\rightarrow$E, 17, 20
22.	$\quad\quad$ ~~Show~~ $(\exists x(Fx \& Gx) \rightarrow Ha) \rightarrow \exists x(Fx \& (Gx \rightarrow Ha))$	
23.	$\quad\quad \ulcorner \exists x(Fx \& Gx) \rightarrow Ha$	ACP
24.	$\quad\quad\quad d = d$	=I
25.	$\quad\quad\quad d = d \leftrightarrow Fd$	$\forall$E, 3
26.	$\quad\quad\quad Fd$	$\leftrightarrow$E, 25, 24
27.	$\quad\quad\quad$ ~~Show~~ $Ga \rightarrow Ha$	
28.	$\quad\quad\quad \ulcorner Gd$	ACP
29.	$\quad\quad\quad\quad Fd \& Gd$	&I, 26, 28
30.	$\quad\quad\quad\quad \exists x(Fx \& Gx)$	$\exists$I, 29
31.	$\quad\quad\quad\quad Ha$	$\rightarrow$E, 23, 30
32.	$\quad\quad\quad Fd \& (Gd \rightarrow Ha)$	&I, 26, 27
33.	$\quad\quad \llcorner \exists x(Fx \& (Gx \rightarrow Ha))$	$\exists$I, 32
34.	$\llcorner \exists x(Fx \& (Gx \rightarrow Ha)) \leftrightarrow (\exists x(Fx \& Gx) \rightarrow Ha)$	$\leftrightarrow$I, 4, 22

8.4 Function Symbols

5. Universe: People a: Me; $m(x)$: x's mother; Lxy: x loves y; Jx: x could be jivin'. $\forall x(\neg Lxa \lor x = m(a))$ & $Jm(a)$.

10. Universe: Objects and abstractions; c: common sense; Hxy: x holds y; $t(x)$: x's tongue. $Hct(c)$.

15. Mx: x is a man; Hxy: x has y; Ix: x is an inner life; $s(x)$: x's surroundings; $l(x)$: x's slave. $\forall x((Mx \,\&\, \neg\exists y(ly \,\&\, Hxy)) \to x = l(s(x)))$.

20. Universe: Objects and names; $n(x)$: x's name.

$$* \ \forall x(n(n(x)) = n(x))$$

$$\checkmark \ \exists ya = n(y)$$

$$* \ a = n(b)$$
$$n(n(b)) = n(b)$$
$$n(a) = n(b)$$
$$* \ n(a) = a$$
$$a = n(a)$$

$$\forall x(\exists yx = n(y) \to x = n(x)) \ \checkmark$$
$$\exists ya = n(y) \to a = n(a) \ \checkmark$$

$$a = n(a)$$

Cl

25. Universe: Objects and abstractions; Dx: x is demonstrable; $c(x)$: x's contrary; Ixy: x implies y; Cx: x is a contradiction; Vx: x is conceivable; $e(x)$: x's existence; $n(x)$: x's nonexistence.
(i) Not valid.

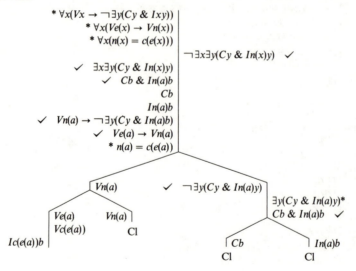

(ii) Valid.

$$\checkmark \quad \neg\exists x\exists y(Cy \ \& \ In(x)y)$$
$$* \ \forall x(\neg Dx \vee \exists y(Cy \ \& \ Ic(x)y))$$
$$* \ \forall x(n(x) = c(e(x)))$$

$\neg\exists xDe(x) \quad \checkmark$

$\checkmark \quad \exists xDe(x)$

$\exists x\exists y(Cy \ \& \ In(x)y)*$

$De(a)$

$\checkmark \quad \neg De(a) \vee \exists y(Cy \ \& \ Ic(e(a))y)$
$* \ n(a) = c(e(a))$

$\checkmark \quad \neg De(a)$ | $\checkmark \quad \exists y(Cy \ \& \ Ic(e(a))y)$
$De(a)$ | $\exists y(Cy \ \& \ In(a)y)$
Cl | $\exists y(Cy \ \& \ Ic(e(a))y$
| Cl

35.
1.	$\exists x\exists y(x \neq y \ \& \ \forall z(z = x \vee z = y))$	A
2.	$\forall x\forall y(f(x) = f(y) \rightarrow x = y)$	A
3.	Show $\forall x\exists y x = f(y)$	
4.	$\quad$ Show $\exists y \ c = f(y)$	
5.	$\quad \quad a \neq b \ \& \ \forall z(z = a \vee z = b)$	$\exists E^2$, 1
6.	$\quad \quad a \neq b$	&E, 5
7.	$\quad \quad \forall z(z = a \vee z = b)$	&E, 5
8.	$\quad \quad f(a) = f(b) \rightarrow a = b$	$\forall E^2$, 2
9.	$\quad \quad f(a) \neq f(b)$	$\rightarrow E^*$, 8, 6
10.	$\quad \quad c = a \vee c = b$	$\forall E$, 7
11.	$\quad \quad f(a) = a \vee f(a) = b$	$\forall E$, 7
12.	$\quad \quad f(b) = a \vee f(b) = b$	$\forall E$, 7
13.	$\quad \quad$ Show $c = a \rightarrow \exists y \ c = f(y)$	
14.	$\quad \quad \quad c = a$	ACP
15.	$\quad \quad \quad$ Show $f(a) = b \rightarrow \exists y \ c = f(y)$	
16.	$\quad \quad \quad \quad f(a) = b$	ACP
17.	$\quad \quad \quad \quad f(b) = a \vee f(b) = f(a)$	$=E$, 16, 12
18.	$\quad \quad \quad \quad f(b) = a$	$\vee E^*$, 17, 9
19.	$\quad \quad \quad \quad f(b) = c$	$=E$, 14, 18
20.	$\quad \quad \quad \quad c = f(b)$	$=E$, 19
21.	$\quad \quad \quad \quad \exists y \ c = f(y)$	$\exists I$, 20
22.	$\quad \quad \quad$ Show $f(a) = a \rightarrow \exists y \ c = f(y)$	
23.	$\quad \quad \quad \quad f(a) = a$	ACP
24.	$\quad \quad \quad \quad c = f(a)$	$=E$, 14, 23
25.	$\quad \quad \quad \quad \exists y \ c = f(y)$	$\exists I$, 24
26.	$\quad \quad \quad \exists y \ c = f(y)$	$\vee E$, 11, 15, 22
27.	$\quad \quad$ Show $c = b \rightarrow \exists y \ c = f(y)$	
28.	$\quad \quad \quad c = b$	ACP
29.	$\quad \quad \quad$ Show $f(b) = a \rightarrow \exists y \ c = f(y)$	
30.	$\quad \quad \quad \quad f(b) = a$	ACP
31.	$\quad \quad \quad \quad f(a) = f(b) \vee f(a) = b$	$=E$, 30, 11
32.	$\quad \quad \quad \quad f(a) = b$	$\vee E^*$, 31, 9
33.	$\quad \quad \quad \quad c = f(a)$	$=E$, 28, 32
34.	$\quad \quad \quad \quad \exists y \ c = f(y)$	$\exists I$, 33
35.	$\quad \quad \quad$ Show $f(b) = b \rightarrow \exists y \ c = f(y)$	
36.	$\quad \quad \quad \quad f(b) = b$	ACP
37.	$\quad \quad \quad \quad c = f(b)$	$=E$, 28, 36
38.	$\quad \quad \quad \quad \exists y \ c = f(y)$	$\exists I$, 37
39.	$\quad \quad \quad \exists y \ c = f(y)$	$\vee E$, 12, 29, 35
40.	$\quad \quad \exists y \ c = f(y)$	$\vee E$, 10, 13, 21

45.

1.	$\forall x \forall y (x < y \leftrightarrow \exists z \, x \oplus z = y)$	A
2.	$\forall x \forall y \, x \oplus y = y \oplus x$	A
3.	$\forall x \, x \oplus x = x$	A
4.	$\forall x \forall y \forall z \, x \oplus (y \oplus z) = (x \oplus y) \oplus z$	A
5.	~~Show~~ $\forall x \forall y ((x < y \,\&\, y < x) \rightarrow x = y)$	
6.	⌈~~Show~~ $(a < b \,\&\, b < a) \rightarrow a = b$	
7.	⌈ $a < b \,\&\, b < a$	ACP
8.	$a < b$	&E, 7
9.	$b < a$	&E, 7
10.	$a < b \leftrightarrow \exists z \, a \oplus z = b$	$\forall$E^2, 1
11.	$b < a \leftrightarrow \exists z \, b \oplus z = a$	$\forall$E^2, 1
12.	$\exists z \, a \oplus z = b$	$\leftrightarrow$E, 10, 8
13.	$\exists z \, b \oplus z = a$	$\leftrightarrow$E, 11, 9
14.	$a \oplus c = b$	$\exists$E, 12
15.	$b \oplus d = a$	$\exists$E, 13
16.	$(a \oplus c) \oplus d = a$	=E, 14, 15
17.	$((b \oplus d) \oplus c) \oplus d = a$	=E, 15, 16
18.	$b \oplus b = b$	$\forall$E, 3
19.	$(((b \oplus b) \oplus d) \oplus c) \oplus d = a$	=E, 17, 18
20.	$(((b \oplus b) \oplus d) \oplus c) \oplus d = ((b \oplus b) \oplus d) \oplus (c \oplus d)$	$\forall$E^3, 4
21.	$((b \oplus b) \oplus d) \oplus (c \oplus d) = a$	=E, 19, 20
22.	$((b \oplus b) \oplus d) \oplus (c \oplus d) = (b \oplus (b \oplus d)) \oplus (c \oplus d)$	$\forall$E^3, 4
23.	$(b \oplus (b \oplus d)) \oplus (c \oplus d) = a$	=E, 21, 22
24.	$b \oplus d = d \oplus b$	$\forall$E^2, 2
25.	$(b \oplus (d \oplus b)) \oplus (c \oplus d) = a$	=E, 23, 24
26.	$c \oplus d = d \oplus c$	$\forall$E^2, 2
27.	$(b \oplus (d \oplus b)) \oplus (d \oplus c) = a$	=E, 25, 26
28.	$b \oplus (d \oplus b) = (b \oplus d) \oplus b$	$\forall$E^3, 4
29.	$((b \oplus d) \oplus b) \oplus (d \oplus c) = a$	=E, 27, 28
30.	$((b \oplus d) \oplus b) \oplus (d \oplus c) = (b \oplus d) \oplus (b \oplus (d \oplus c))$	$\forall$E^3, 4
31.	$(b \oplus d) \oplus (b \oplus (d \oplus c)) = a$	=E, 29, 30
32.	$b \oplus (d \oplus c) = (b \oplus d) \oplus c$	$\forall$E^3, 4
33.	$(b \oplus d) \oplus ((b \oplus d) \oplus c) = a$	=E, 31, 32
34.	$a \oplus (a \oplus c) = a$	=E, 15, 33
35.	$a \oplus (a \oplus c) = (a \oplus a) \oplus c$	$\forall$E^3, 4
36.	$(a \oplus a) \oplus c = a$	=E, 34, 36
37.	$a \oplus a = a$	$\forall$E, 3
38.	$a \oplus c = a$	=E, 36, 37
39.	$b = a$	=E, 14, 38
40.	⌊⌊$a = b$	=E, 39, 39

8.5 Definite Descriptions

5. Universe: Mountains; e: Everest; Txy: x is taller than y.
$$e = \imath x \forall y (y \neq x \rightarrow Txy)$$
$$(\tau x : \forall y (y \neq x \rightarrow Txy)e = x$$

10. Universe: People; Sxy: x saw y; Txy: x talked to y; m: Millie; d: Don.
$$T \imath x Sxmd \qquad (\tau x : Sxm)Txd$$

15. Universe: Things and locations; Px: x is a project; Txy: x threatens y; Bx: x is a building; Cx: x is a city; Oxy: x is older than y; Ixy: x is in y.
$$T \imath x P x \imath y ((By \,\&\, Iy\imath wCw \,\&\, \forall z((Bz \,\&\, Iz\imath wCw \,\&\, z \neq y) \rightarrow Oyz))$$
$$(\tau x : Px)(\tau w : Cw)(\tau y : (By \,\&\, Iyw \,\&\, \forall z((Bz \,\&\, Izw \,\&\, z \neq y) \rightarrow Oyz)))Txy$$

20. Universe: Presidents and states; *b*: James Buchanan; *Px*: *x* has been a president; *a*: Pennsylvania; *Fxy*: *x* is from *y*.

$$b = \imath x((Px \ \& \ Fxa) \ \& \ \forall y((y \neq x \ \& \ Py) \rightarrow \neg Fya))$$
$$(\tau x: ((Px \ \& \ Fxa) \ \& \ \forall y((y \neq x \ \& \ Py) \rightarrow \neg Fya))b = x$$

25. This sentence can't be symbolized, whether we treat descriptions as singular terms or as quantifiers; the two descriptions within it are interdependent. Using this dictionary:

Universe: Things; *Lxy*: *x* loved *y*; *Bx*: *x* is a bird; *Mx*: *x* is a man; *Sxyz*: *x* shot *y* *z*; *b(x)*: *x*'s bow.

We might try

$$(\tau x: Bx \ \& \ (\tau y: (My \ \& \ Syxb(y)))Lxy)Lyx$$
or $(\tau y: My \ \& \ (\tau x: (Bx \ \& \ Lxy))Syxb(y))Lyx$

but these aren't formulas; *Lyx* is not in the scope of the embedded description.

30. *Wxy*: *x* wrote *y*; *a*: Waverley.
(a) $W(\imath x: Wxa)a$ (b) $(\tau x: Wxa)Wxa$

(a) 1. $\forall z(z = b \leftrightarrow Wza)$ A
2. ~~Show~~ *Wba*
3. $\lceil b = b \leftrightarrow Wba$ $\forall$E, 1
4. $\vert b = b$ =I
5. $\llcorner Wba$ $\leftrightarrow$E, 3, 4
Valid.

(b)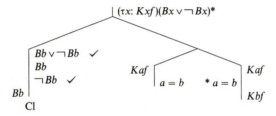

Invalid. (Perhaps nothing, or more than one thing, wrote *Waverley*.)

40. Universe: People and countries; *Bx*: *x* is bald; *Kxy*: *x* is king of *y*; *f*: France.
(a) $B\imath xKxf \lor \neg B\imath xKxf$
(b) Ambiguous: (b1) $(\tau x: Kxf)(Bx \lor \neg Bx)$
     (b2) $(\tau x: Kxf)Bx \lor (\tau x: Kxf)\neg Bx$

(a) 1. $\forall z(z = a \leftrightarrow Kzf)$ A
2. ~~Show~~ $Ba \lor \neg Ba$
3. $\lceil \neg(Ba \lor \neg Ba)$ AIP
4. $\vert \neg Ba \ \& \ Ba$ $\neg\lor$, 3
5. $\vert Ba$ &E, 4
6. $\llcorner \neg Ba$ &E, 4

(b1) Not valid.

(b2) Not valid.

Chapter 9
NECESSITY

9.3 Translation

5. *p*: Rhonda is perceptive; *q*: Rhonda made the error. $\neg \Diamond (p \ \& \ q)$.

10. *p*: The software computes the area; *q*: The software takes an integral; *r*: The software incorporates certain complex approximation techniques. $(p \ \& \ q) \dashv 3 \ (p \dashv 3 \ r)$.

15. *p*: We exert painful effort; *q*: We exert grim energy; *r*: We exhibit resolute courage; *s*: We move on to better things. $s \dashv 3 \ ((p \ \& \ q) \ \& \ r)$.

20. *p*: I love thee so much; *q*: I love honor more than I love thee. $\neg q \dashv 3 \neg \Diamond p$.

9.4 Modal Semantic Tableaux

5.

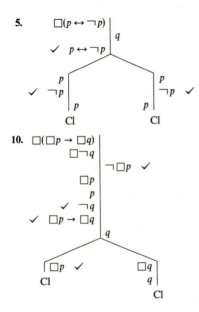

10.

9.5 Other Tableau Rules

10.

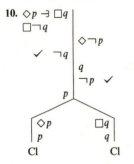

9.6 World Traveling

5. *p*: You die; *q*: You eat tomatoes. $(q \dashv p) \vee (p \dashv q)$. Not valid in S5.

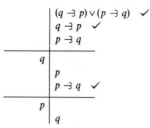

10. *p*: One advocates tax cuts; *q*: One abhors deficits; *r*: One is willing to shrink revenues in the short term. Valid.

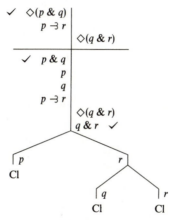

15. *p*: The state university becomes more selective in its admissions policies; *q*: The legislature becomes angry; *r*: There is enough classroom space.

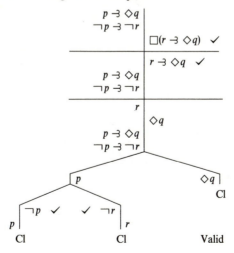

25. Not equivalent.

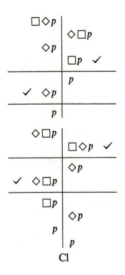

35. Valid in S5.

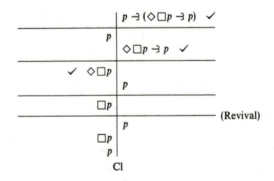

45. Valid in S5.

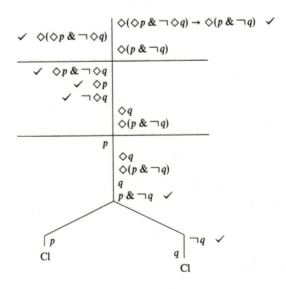

57. Not valid in S5; contradictory when $\mathscr{A}$ is $p \,\&\, \neg p$.

$$\Box \Diamond \Diamond \mathscr{A} \quad \checkmark$$

$$\begin{array}{l}
\Diamond \Diamond \mathscr{A} \\
\Diamond \mathscr{A} \\
\mathscr{A}
\end{array}$$

$$\begin{array}{l}
\Box \Diamond \Diamond \mathscr{A} \\
\checkmark \quad \Diamond \Diamond \mathscr{A}
\end{array}$$

$$\checkmark \quad \Diamond \mathscr{A}$$

$$\mathscr{A}$$

$$\begin{array}{l}
\Box \Diamond \Diamond (p \,\&\, \neg p) \\
\checkmark \quad \Diamond \Diamond (p \,\&\, \neg p)
\end{array}$$

$$\checkmark \quad \Diamond (p \,\&\, \neg p)$$

$$\begin{array}{l}
\checkmark \quad p \,\&\, \neg p \\
p \\
\checkmark \quad \neg p
\end{array}$$

$$p$$

$$\text{Cl}$$

9.7 Modal Deduction

5.
1.	Show $\Diamond p \vee \Diamond \neg p$	
2.	$\neg (\Diamond p \vee \Diamond \neg p)$	AIP
3.	$\neg \Diamond p \,\&\, \neg \Diamond \neg p$	$\neg\vee, 2$
4.	$\neg \Diamond p$	&E, 3
5.	$\neg \Diamond \neg p$	&E, 3
6.	Show $\Box p$	
7.	Show p	
8.	$\neg p$	AIP
9.	$\Diamond \neg p$	$\Diamond$I, 8
10.	$\neg \Diamond \neg p$	$\Box$R, 5
11.	p	$\Box$E, 6
12.	$\Diamond p$	$\Diamond$I, 11

10.
1.	Show $\Diamond \neg p \dashv 3 \neg \Box p$	
2.	$\Diamond \neg p$	A$\Box$P
3.	Show $\neg \Box p$	
4.	$\Box p$	AIP
5.	Show $\neg p \dashv 3 \neg \Box p$	
6.	$\neg p$	A$\Box$P
7.	Show $\neg \Box p$	
8.	$\Box p$	AIP
9.	$\neg p$	R, 6
10.	p	$\Box$E, 8
11.	$\neg \Box p$	$\Diamond$E, 2, 5

20. 1. Show $(p \mathrel{\unicode{8917}} q) \mathrel{\unicode{8888}} \Box(p \leftrightarrow q)$

    ┌□
2. | $p \mathrel{\unicode{8917}} q$ A□P
3. | Show $\Box(p \leftrightarrow q)$

      ┌□
4. | | Show $p \rightarrow q$
5. | | ┌p ACP
6. | | | $p \mathrel{\unicode{8917}} q$ □R, 2
7. | | └q ⫥E, 6, 5
8. | | Show $q \rightarrow p$
9. | | ┌q ACP
10. | | | $p \mathrel{\unicode{8917}} q$ □R, 2
11. | | └p ⫥E, 10, 9
12. └└$p \leftrightarrow q$ ↔I, 4, 8

35. 1. Show $\Diamond(p \mathrel{\unicode{8917}} q) \mathrel{\unicode{8917}} (p \mathrel{\unicode{8917}} q)$
2. ┌Show $\Diamond(p \mathrel{\unicode{8917}} q) \mathrel{\unicode{8888}} (p \mathrel{\unicode{8917}} q)$

      ┌□
3. | | $\Diamond(p \mathrel{\unicode{8917}} q)$ A□P
4. | | Show $(p \mathrel{\unicode{8888}} q) \mathrel{\unicode{8888}} (p \mathrel{\unicode{8888}} q)$

        ┌□
5. | | └$p \mathrel{\unicode{8888}} q$ A□P
6. | └$p \mathrel{\unicode{8917}} q$ $\Diamond$E, 3, 4
7. | Show $(p \mathrel{\unicode{8917}} q) \mathrel{\unicode{8888}} \Diamond(p \mathrel{\unicode{8917}} q)$

      ┌□
8. | | $p \mathrel{\unicode{8917}} q$ A□P
9. | └$\Diamond(p \mathrel{\unicode{8917}} q)$ $\Diamond$I, 8
10. └$\Diamond(p \mathrel{\unicode{8917}} q) \mathrel{\unicode{8917}} (p \mathrel{\unicode{8917}} q)$ ⫥I, 2, 7

45. 1. Show $(p \mathrel{\unicode{8888}} q) \mathrel{\unicode{8888}} ((r \mathrel{\unicode{8888}} p) \mathrel{\unicode{8888}} (r \mathrel{\unicode{8888}} q))$

    ┌□
2. | $p \mathrel{\unicode{8888}} q$ A□P
3. | Show $(r \mathrel{\unicode{8888}} p) \mathrel{\unicode{8888}} (r \mathrel{\unicode{8888}} q)$

      ┌□
4. | | $r \mathrel{\unicode{8888}} p$ A□P
5. | | Show $r \mathrel{\unicode{8888}} q$

        ┌□
6. | | | r A□P
7. | | | $r \mathrel{\unicode{8888}} p$ □R, 4
8. | | | p ⫥E, 7, 6
9. | | | $p \mathrel{\unicode{8888}} q$ □R, 2
10. └└└q ⫥E, 9, 8

50.
1.	Show $(p \mathrel{\&} (q \mathbin{\rightarrow 3} r)) \mathbin{\rightarrow 3} (\Box p \lor \Box \neg p)$	
2.	$p \mathrel{\&} (q \mathbin{\rightarrow 3} r)$	A$\Box$P
3.	Show $\Box p \lor \Box \neg p$	
4.	$\neg(\Box p \lor \Box \neg p)$	AIP
5.	$\neg \Box p \mathrel{\&} \neg \Box \neg p$	$\neg \lor$, 4
6.	$\neg \Box p$	&E, 5
7.	$\neg \Box \neg p$	&E, 5
8.	$\Diamond p$	Df$\Diamond$, 7
9.	Show $p \mathbin{\rightarrow 3} \Box p$	
10.	p	A$\Box$P
11.	$p \mathrel{\&} (q \mathbin{\rightarrow 3} r)$	$\Box$R, 2
12.	$q \mathbin{\rightarrow 3} r$	&E, 11, 10
13.	Show $\Box p$	
14.	$p \mathrel{\&} (q \mathbin{\rightarrow 3} r)$	$\Box$R, 11
15.	$q \mathbin{\rightarrow 3} r$	$\Box$R, 12
16.	p	&E, 14, 15
17.	$\Box p$	$\Diamond$E, 8, 9

55.
1.	$\Diamond p \mathbin{\rightarrow 3} q$	A
2.	Show $p \mathbin{\rightarrow 3} q$	
3.	p	A$\Box$P
4.	$\Diamond p$	$\Diamond$I, 3
5.	$\Diamond p \mathbin{\rightarrow 3} q$	$\Box$R, 1
6.	q	$\mathbin{\rightarrow 3}$E, 5, 4

65.
1.	$\Diamond \Diamond \Box p$	A
2.	Show $\Box p$	
3.	Show $\Diamond \Box p \mathbin{\rightarrow 3} \Box p$	
4.	$\Diamond \Box p$	A$\Box$P
5.	Show $\Box p \mathbin{\rightarrow 3} \Box p$	
6.	$\Box p$	A$\Box$P
7.	$\Box p$	$\Diamond$E, 4, 5
8.	$\Box p$	$\Diamond$E, 1, 3

75.
1.	$\Box \Diamond \Box p$	A
2.	Show $\Diamond \Box \Box p$	
3.	$\Diamond \Box p$	$\Box$E, 1
4.	Show $\Box p \mathbin{\rightarrow 3} \Diamond \Box \Box p$	
5.	$\Box p$	A$\Box$P
6.	Show $\Box \Box p$	
7.	$\Box p$	$\Box$R, 5
8.	$\Diamond \Box \Box p$	$\Diamond$I, 6
9.	$\Diamond \Box \Box p$	$\Diamond$E, 3, 4

85.
1.	$\Box(p \rightarrow q)$	A
2.	Show $\Box p \rightarrow \Box q$	
3.	$\Box p$	ACP
4.	Show $\Box q$	
5.	$\Box p$	$\Box$R, 3
6.	$\Box(p \rightarrow q)$	$\Box$R, 1
7.	p	$\Box$E, 5
8.	$p \rightarrow q$	$\Box$E, 6
9.	q	$\rightarrow$E, 8, 7

90.
1. $\neg(p \rightarrow q)$ ⠀⠀⠀⠀A
2. Show $\Diamond\neg q$
3. ⠀$\neg\Diamond\neg q$ ⠀⠀⠀⠀AIP
4. ⠀$\Box q$ ⠀⠀⠀⠀Df$\Diamond$, 3
5. ⠀Show $p \rightarrow q$
6. ⠀⠀p ⠀⠀⠀⠀A$\Box$P
7. ⠀⠀$\Box q$ ⠀⠀⠀⠀$\Box$R, 4
8. ⠀⠀q ⠀⠀⠀⠀$\Box$E, 7
9. ⠀$\neg(p \rightarrow q)$ ⠀⠀⠀⠀R, 1

100.
1. $\Diamond p \rightarrow q$ ⠀⠀⠀⠀A
2. Show $\Diamond p \rightarrow \Box q$
⠀⠀$\Box$
3. ⠀$\Diamond p$ ⠀⠀⠀⠀A$\Box$P
4. ⠀Show $\Box q$
⠀⠀⠀$\Box$
5. ⠀⠀$\Diamond p$ ⠀⠀⠀⠀$\Box$R, 3
6. ⠀⠀$\Diamond p \rightarrow q$ ⠀⠀⠀⠀$\Box$R, 1
7. ⠀⠀q ⠀⠀⠀⠀$\rightarrow$E, 6, 5

110.
1. $(p \rightarrow s) \vee q$ ⠀⠀⠀⠀A
2. $\Box r \rightarrow \Diamond(p \And \neg s)$ ⠀⠀⠀⠀A
3. Show $\Box r \rightarrow q$
4. ⠀$\Box r$ ⠀⠀⠀⠀ACP
5. ⠀$\Diamond(p \And \neg s)$ ⠀⠀⠀⠀$\rightarrow$E, 2, 4
6. ⠀Show $(p \And \neg s) \rightarrow \neg(p \rightarrow s)$
⠀⠀⠀$\Box$
7. ⠀⠀$p \And \neg s$ ⠀⠀⠀⠀A$\Box$P
8. ⠀⠀Show $\neg(p \rightarrow s)$
9. ⠀⠀⠀$p \rightarrow s$ ⠀⠀⠀⠀AIP
10. ⠀⠀⠀p ⠀⠀⠀⠀&E, 7
11. ⠀⠀⠀$\neg s$ ⠀⠀⠀⠀&E, 7
12. ⠀⠀⠀s ⠀⠀⠀⠀$\rightarrow$E, 9, 10
13. ⠀$\neg(p \rightarrow s)$ ⠀⠀⠀⠀$\Diamond$E, 5, 6
14. ⠀q ⠀⠀⠀⠀$\vee$E*, 1, 13

115.
1. $(p \rightarrow (q \rightarrow r)) \leftrightarrow ((p \rightarrow q) \rightarrow r)$ ⠀⠀⠀⠀A
2. Show $\Box(p \vee r)$
⠀⠀$\Box$
3. ⠀$(p \rightarrow (q \rightarrow r)) \leftrightarrow ((p \rightarrow q) \rightarrow r)$ ⠀⠀$\Box$R, 1
4. ⠀Show $p \vee r$
5. ⠀⠀$\neg(p \vee r)$ ⠀⠀⠀⠀AIP
6. ⠀⠀$\neg p \And \neg r$ ⠀⠀⠀⠀$\neg\vee$, 5
7. ⠀⠀$\neg p$ ⠀⠀⠀⠀&E, 6
8. ⠀⠀$\neg r$ ⠀⠀⠀⠀&E, 6
9. ⠀⠀Show $p \rightarrow (q \rightarrow r)$
10. ⠀⠀⠀p ⠀⠀⠀⠀ACP
11. ⠀⠀⠀$\neg p$ ⠀⠀⠀⠀R, 7
12. ⠀⠀⠀$q \rightarrow r$ ⠀⠀⠀⠀!, 10, 11
13. ⠀⠀$(p \rightarrow q) \rightarrow r$ ⠀⠀⠀⠀$\leftrightarrow$E, 3, 9
14. ⠀⠀$\neg(p \rightarrow q)$ ⠀⠀⠀⠀$\rightarrow$E*, 13, 8
15. ⠀⠀$p \And \neg q$ ⠀⠀⠀⠀$\neg\rightarrow$, 14
16. ⠀⠀p ⠀⠀⠀⠀&E, 15

125. p: Our attempt to prevent a leftist takeover succeeds; r: We reassert our leadership position in the world.

1. $\Box\neg(\neg p \And r)$ ⠀⠀⠀⠀A
2. $\Diamond\neg\Diamond p$ ⠀⠀⠀⠀A
3. Show $\neg\Diamond r$

4.	$\Diamond r$	AIP
5.	Show $r \dashv3 \Box \Diamond p$	
6.	r	A$\Box$P
7.	Show $\Box \Diamond p$	
8.	$\Box \neg (\neg p \ \& \ r)$	$\Box$R, 1
9.	$\neg (\neg p \ \& \ r)$	$\Box$E, 8
10.	$p \lor \neg r$	$\neg$&, 9
11.	p	$\lor$E*, 10, 6
12.	$\Diamond p$	$\Diamond$I, 11
13.	$\Box \Diamond p$	$\Diamond$E, 4, 5
14.	$\neg \Diamond \neg \Diamond p$	Df$\Diamond$, 13

130. s: A recent Supreme Court decision is correct; q: Men and women are to be treated equally under the law; p: We permit the criminal status of an act to depend on the sex of the agent; r: We are willing to accept any sexual discrimination that might benefit society.

1.	$s \dashv3 \Diamond (q \ \& \ p)$	A
2.	$\Diamond p \dashv3 r$	A
3.	Show $s \dashv3 \neg \Box \neg (q \ \& \ r)$	
4.	s	A$\Box$P
5.	$s \dashv3 \Diamond (q \ \& \ p)$	$\Box$R, 1
6.	$\Diamond (q \ \& \ p)$	$\dashv3$E, 5, 4
7.	Show $(q \ \& \ p) \dashv3 \neg \Box \neg (q \ \& \ r)$	
8.	$q \ \& \ p$	A$\Box$P
9.	Show $\neg \Box \neg (q \ \& \ r)$	
10.	$\Box \neg (q \ \& \ r)$	AIP
11.	$\Diamond p \dashv3 r$	$\Box$R, 2
12.	p	&E, 8
13.	$\Diamond p$	$\Diamond$I, 12
14.	r	$\dashv3$E, 11, 13
15.	q	&E, 8
16.	$q \ \& \ r$	&I, 15, 14
17.	$\neg (q \ \& \ r)$	$\Box$E, 10
18.	$\neg \Box \neg (q \ \& \ r)$	$\Diamond$E, 6, 7

135.

1.	Show $\Box p \leftrightarrow (p \ \& \ \neg \nabla p)$	
2.	Show $\Box p \rightarrow (p \ \& \ \neg \nabla p)$	
3.	$\Box p$	ACP
4.	p	$\Box$E, 3
5.	Show $\neg \nabla p$	
6.	∇p	AIP
7.	$\Diamond p \ \& \ \neg \Box p$	Df∇, 6
8.	$\neg \Box p$	&E, 7
9.	$\Box p$	R, 3
10.	$p \ \& \ \neg \nabla p$	&I, 4, 5
11.	Show $(p \ \& \ \neg \nabla p) \rightarrow \Box p$	
12.	$p \ \& \ \neg \nabla p$	ACP
13.	p	&E, 12
14.	$\neg \nabla p$	&E, 12
15.	$\neg (\Diamond p \ \& \ \neg \Box p)$	Df∇, 14
16.	$\neg \Diamond p \lor \Box p$	$\neg$&, 15
17.	$\Diamond p$	$\Diamond$I, 13
18.	$\Box p$	$\lor$E*, 16, 17
19.	$\Box p \leftrightarrow (p \ \& \ \neg \nabla p)$	$\leftrightarrow$I, 2, 11

140.
1. Show $\Box \mathcal{A} \leftrightarrow (\neg \mathcal{A} \dashv (p \,\&\, \neg p))$
2. Show $\Box \mathcal{A} \to (\neg \mathcal{A} \dashv (p \,\&\, \neg p))$
3. $\Box \mathcal{A}$ ACP
4.   Show $\neg \mathcal{A} \dashv (p \,\&\, \neg p)$
5. $\neg \mathcal{A}$ A$\Box$P
6. $\Box \mathcal{A}$ $\Box$R, 3
7. $\mathcal{A}$ $\Box$E, 6
8. $p \,\&\, \neg p$!, 5, 7
9.   Show $(\neg \mathcal{A} \dashv (p \,\&\, \neg p)) \to \Box \mathcal{A}$
10. $\neg \mathcal{A} \dashv (p \,\&\, \neg p)$ ACP
11.   Show $\Box \mathcal{A}$
12.   Show $\mathcal{A}$
13. $\neg \mathcal{A}$ AIP
14. $\neg \mathcal{A} \dashv (p \,\&\, \neg p)$ $\Box$R, 10
15. $p \,\&\, \neg p$ $\dashv$E, 14, 13
16. p &E, 15
17. $\neg p$ &E, 15
18. $\Box \mathcal{A} \leftrightarrow (\neg \mathcal{A} \dashv (p \,\&\, \neg p))$ $\leftrightarrow$I, 2, 9

Chapter 10
COUNTERFACTUALS

10.1 The Meaning of Counterfactuals

5. p: I knew how he would end up; q: I was nicer to the guy. $p \,\Box\!\!\to q$.

10. p: The MX is studied carefully; q: We have the MX in a properly survivable mode; r: We need the MX. $p \,\Box\!\!\to (\neg \Diamond q \,\&\, \neg r)$.

15. p: I object to dying; q: Dying is followed by death. $\neg q \,\Box\!\!\to \neg p$.

20. p: Don't knock the weather; q: Nine-tenths of the people start a conversation; r: The weather changes once in a while. $p \,\&\, (\neg r \,\Box\!\!\to \neg \Diamond q)$.

10.2 Deduction Rules for Counterfactuals

5.
1. Show $\neg \Diamond p \to (p \,\Box\!\!\to q)$
2. $\neg \Diamond p$ ACP
3. Show $p \,\Box\!\!\to q$
4. p A$\Box\!\!\to$P
5. $\Diamond p$ $\Diamond$I, 4
6. $\neg \Diamond p$ $\Box\!\!\to$R1, 2
7. q !, 5, 6

10.
1. Show $(p \,\&\, (p \,\Box\!\!\to q)) \to q$
2. $p \,\&\, (p \,\Box\!\!\to q)$ ACP
3. p &E, 2
4. $p \,\Box\!\!\to q$ &E, 2
5. q $\Box\!\!\to$E, 4, 3

20.
1. Show $((p \:\square\!\!\rightarrow q) \,\&\, (p \:\square\!\!\rightarrow \neg q)) \rightarrow \neg \Diamond p$
2. $(p \:\square\!\!\rightarrow q) \,\&\, (p \:\square\!\!\rightarrow \neg q)$ ACP
3. Show $\neg \Diamond p$
4. $\Diamond p$ AIP
5. $p \:\square\!\!\rightarrow q$ &E, 2
6. $p \:\square\!\!\rightarrow \neg q$ &E, 2
7. Show $p \:\square\!\!\rightarrow \square(q \,\&\, \neg q)$
     $\square\!\!\rightarrow$
8. p A$\square\!\!\rightarrow$P
9. q $\square\!\!\rightarrow$R2, 5(8)
10. $\neg q$ $\square\!\!\rightarrow$R2, 6(8)
11. $\square(q \,\&\, \neg q)$!, 9, 10
12. $\square(q \,\&\, \neg q)$ $\Diamond$E*, 4, 7
13. $q \,\&\, \neg q$ $\square$E, 12
14. q &E, 13
15. $\neg q$ &E, 13

25.
1. $p \mathbin{\boxminus} q$ A
2. Show $(r \:\square\!\!\rightarrow p) \leftrightarrow (r \:\square\!\!\rightarrow q)$
3. Show $(r \:\square\!\!\rightarrow p) \rightarrow (r \:\square\!\!\rightarrow q)$
4. $r \:\square\!\!\rightarrow p$ ACP
5. Show $r \:\square\!\!\rightarrow q$
     $\square\!\!\rightarrow$
6. r A$\square\!\!\rightarrow$P
7. p $\square\!\!\rightarrow$R2, 4(6)
8. $p \mathbin{\boxminus} q$ $\square\!\!\rightarrow$R1, 1
9. q $\boxminus$E, 8, 7
10. Show $(r \:\square\!\!\rightarrow q) \rightarrow (r \:\square\!\!\rightarrow p)$
11. $r \:\square\!\!\rightarrow q$ ACP
12. Show $r \:\square\!\!\rightarrow p$
     $\square\!\!\rightarrow$
13. r A$\square\!\!\rightarrow$P
14. q $\square\!\!\rightarrow$R2, 11(13)
15. $p \mathbin{\boxminus} q$ $\square\!\!\rightarrow$R1, 1
16. p $\boxminus$E, 15, 14
17. $(r \:\square\!\!\rightarrow p) \leftrightarrow (r \:\square\!\!\rightarrow q)$ $\leftrightarrow$I, 3, 10

35.
1. $q \:\square\!\!\rightarrow p$ A
2. $p \mathbin{\prec\!\!\!3} (q \,\&\, r)$ A
3. Show $\Diamond q \rightarrow \Diamond r$
4. $\Diamond q$ ACP
5. Show $q \:\square\!\!\rightarrow \Diamond r$
     $\square\!\!\rightarrow$
6. q A$\square\!\!\rightarrow$P
7. p $\square\!\!\rightarrow$R2, 1(6)
8. $p \mathbin{\prec\!\!\!3} (q \,\&\, r)$ $\square\!\!\rightarrow$R1, 2
9. $q \,\&\, r$ $\prec\!\!\!3$E, 8, 7
10. r &E, 9
11. $\Diamond r$ $\Diamond$I, 10
12. $\Diamond r$ $\Diamond$E*, 4, 5

40.
1. $(p \,\&\, r) \:\square\!\!\rightarrow s$ A
2. $p \:\square\!\!\rightarrow (p \,\&\, q)$ A
3. $(p \,\&\, q) \:\square\!\!\rightarrow r$ A

4. Show $p \; \square \rightarrow s$
5. ⌐Show $p \; \square \rightarrow r$

	⌐$\square \rightarrow$	
6.	p	A$\square \rightarrow$P
7.	$p \; \& \; q$	$\square \rightarrow$R2, 2(6)
8.	Show $(p \; \& \; q) \; \text{⇔} \; (p \; \& \; (p \; \& \; q))$	
9.	⌐Show $(p \; \& \; q) \; \dashv \; (p \; \& \; (p \; \& \; q))$	
	⌐$\square$	
10.	$p \; \& \; q$	A$\square$P
11.	p	&E, 10
12.	$p \; \& \; (p \; \& \; q)$	&I, 11, 10
13.	Show $(p \; \& \; (p \; \& \; q)) \; \dashv \; (p \; \& \; q)$	
	⌐$\square$	
14.	$p \; \& \; (p \; \& \; q)$	A$\square$P
15.	$p \; \& \; q$	&E, 14
16.	$(p \; \& \; q) \; \text{⇔} \; (p \; \& \; (p \; \& \; q))$	⇔I, 9, 13
17.	r	$\square \rightarrow$R2, 3(6, 7, 8)
18.	Show $p \; \square \rightarrow s$	
	⌐$\square \rightarrow$	
19.	p	A$\square \rightarrow$P
20.	r	$\square \rightarrow$R2, 5(19)
21.	s	$\square \rightarrow$R2, 1(19, 20)

45.
1.	$p \; \square \rightarrow q$	A
2.	$q \; \square \rightarrow s$	A
3.	$r \; \square \rightarrow q$	A
4.	$p \; \square \rightarrow r$	A
5.	$q \; \square \rightarrow p$	A
6.	Show $r \; \square \rightarrow s$	
7.	⌐Show $(q \; \& \; p) \; \square \rightarrow p$	
	⌐$\square \rightarrow$	
8.	$q \; \& \; p$	A$\square \rightarrow$P
9.	p	&E, 8
10.	Show $p \; \square \rightarrow (q \; \& \; p)$	
	⌐$\square \rightarrow$	
11.	p	A$\square \rightarrow$P
12.	q	$\square \rightarrow$R1, 1(11)
13.	$q \; \& \; p$	&I, 12, 11
14.	Show $q \; \square \rightarrow r$	
	⌐$\square \rightarrow$	
15.	q	A$\square \rightarrow$P
16.	p	$\square \rightarrow$R2, 5(15)
17.	r	$\square \rightarrow$R2, 4(15, 16, 7, 10)
18.	Show $r \; \square \rightarrow s$	
	⌐$\square \rightarrow$	
19.	r	A$\square \rightarrow$P
20.	s	$\square \rightarrow$R2, 2(3, 14)

55. p: 4 is odd; q: 2 is odd; r: 2 is even.

1.	$p \; \square \rightarrow q$	A
2.	$p \; \square \rightarrow r$	A
3.	$\square \neg (q \; \& \; r)$	A
4.	Show $\square \neg p$	

5. $\quad\ulcorner\neg\Box\neg p$ AIP
6. $\quad\Diamond p$ Df$\Diamond$, 5
7. $\quad$ ~~Show~~ $p\ \Box\to\ \Diamond(q\ \&\ r)$
8. $\qquad\ulcorner\Box\to$
    $\quad p$ A$\Box\to$P
9. $\qquad q$ $\Box\to$R2, 1(8)
10. $\qquad r$ $\Box\to$R2, 2(8)
11. $\qquad q\ \&\ r$ &I, 9, 10
12. $\qquad\llcorner\Diamond(q\ \&\ r)$ $\Diamond$I, 11
13. $\quad\Diamond(q\ \&\ r)$ $\Diamond$E*, 6, 7
14. $\quad\neg\Box\neg(q\ \&\ r)$ Df$\Diamond$, 13
15. $\quad\llcorner\Box\neg(q\ \&\ r)$ R, 3

60. p: Kelly refuses the operation; q: Kelly regains full use of her arm; r: Kelly regrets not having the operation.

1. $p\ \Box\to\ \neg q$ A
2. $(p\ \&\ \neg q)\ \Box\to\ r$ A
3. ~~Show~~ $\neg p\lor r$
4. $\quad\ulcorner\neg(\neg p\lor r)$ AIP
5. $\quad p\ \&\ \neg r$ $\neg\lor$, 4
6. $\quad p$ &E, 5
7. $\quad\neg r$ &E, 5
8. $\quad\neg q$ $\Box\to$E, 1, 6
9. $\quad p\ \&\ \neg q$ &I, 6, 8
10. $\quad\llcorner r$ $\Box\to$E, 2, 9

10.3 System CS

5. 1. ~~Show~~ $p\to((p\ \Box\to\ q)\leftrightarrow q)$
2. $\quad\ulcorner p$ ACP
3. $\quad$ ~~Show~~ $(p\ \Box\to\ q)\to q$
4. $\qquad\ulcorner p\ \Box\to\ q$ ACP
5. $\qquad\llcorner q$ $\Box\to$E, 4, 2
6. $\quad$ ~~Show~~ $q\to(p\ \Box\to\ q)$
7. $\qquad\ulcorner q$ ACP
8. $\qquad$ ~~Show~~ $p\ \Box\to\ q$
9. $\qquad\quad\ulcorner\neg(p\ \Box\to\ q)$ AIP
10. $\qquad\quad p\ \Box\to\ \neg q$ S, 9
11. $\qquad\quad\neg q$ $\Box\to$E, 10, 2
12. $\qquad\quad\llcorner q$ R, 7
13. $\quad\llcorner(p\ \Box\to\ q)\leftrightarrow q$ $\leftrightarrow$I, 3, 6

10. 1. $p\ {\dashv}\ q$ A
2. $\neg(p\ \Box\to\ r)$ A
3. ~~Show~~ $(q\ \Box\to\ \neg p)\lor(q\ \Box\to\ \neg r)$
4. $\quad\ulcorner\neg((q\ \Box\to\ \neg p)\lor(q\ \Box\to\ \neg r))$ AIP
5. $\quad\neg(q\ \Box\to\ \neg p)\ \&\ \neg(q\ \Box\to\ \neg r)$ $\neg\lor$, 4
6. $\quad\neg(q\ \Box\to\ \neg p)$ &E, 5
7. $\quad\neg(q\ \Box\to\ \neg r)$ &E, 5
8. $\quad p\ \Box\to\ \neg r$ S, 2
9. $\quad q\ \Box\to\ \neg\neg p$ S, 6
10. $\quad q\ \Box\to\ \neg\neg r$ S, 7
11. $\quad q\ \Box\to\ p$ $\neg\neg$, 9
12. $\quad q\ \Box\to\ r$ $\neg\neg$, 10
13. $\quad p\ \Box\to\ q$ $\dashv\Box\to$, 1
14. $\quad$ ~~Show~~ $q\ \Box\to\ \neg r$
$\qquad\ulcorner\Box\to$
15. $\qquad q$ A$\Box\to$P
16. $\qquad\llcorner\neg r$ $\Box\to$R2, 8(11, 13, 15)

15.

1.	$p \square\rightarrow (q \vee r)$	A
2.	$(p \,\&\, q) \square\rightarrow s$	A
3.	$(p \,\&\, r) \square\rightarrow s$	A
4.	Show $p \square\rightarrow s$	
5.	$\neg(p \square\rightarrow s)$	AIP
6.	$p \square\rightarrow \neg s$	S, 5
7.	Show $p \square\rightarrow \neg q$	
	$\square\rightarrow$	
8.	p	A$\square\rightarrow$P
9.	Show $\neg q$	
10.	q	AIP
11.	s	$\square\rightarrow$R2, 2(8, 10)
12.	$\neg s$	$\square\rightarrow$R2, 6(8)
13.	Show $p \square\rightarrow (p \,\&\, r)$	
	$\square\rightarrow$	
14.	p	A$\square\rightarrow$P
15.	$q \vee r$	$\square\rightarrow$R2, 1(14)
16.	$\neg q$	$\square\rightarrow$R2, 7(14)
17.	r	$\vee$E*, 15, 16
18.	$p \,\&\, r$	&I, 14, 17
19.	Show $(p \,\&\, r) \square\rightarrow p$	
	$\square\rightarrow$	
20.	$p \,\&\, r$	A$\square\rightarrow$P
21.	p	&E, 20
22.	Show $p \square\rightarrow s$	
	$\square\rightarrow$	
23.	p	A$\square\rightarrow$P
24.	s	$\square\rightarrow$R2, 3(13, 19, 23)

10.4 System CL

5.

1.	Show $p \rightarrow ((p \square\rightarrow q) \leftrightarrow q)$	
2.	p	ACP
3.	Show $(p \square\rightarrow q) \rightarrow q$	
4.	$p \square\rightarrow q$	ACP
5.	q	$\square\rightarrow$E, 4, 2
6.	Show $q \rightarrow (p \square\rightarrow q)$	
7.	q	ACP
8.	Show $p \square\rightarrow q$	
	$\square\rightarrow$	
9.	p	A$\square\rightarrow$P
10.	q	$\square\rightarrow$R3, 5(2, 9)
11.	$(p \square\rightarrow q) \leftrightarrow q$	$\leftrightarrow$I, 3, 6

10.

1.	Show $((p \square\rightarrow q) \,\&\, \neg q) \rightarrow (\neg q \square\rightarrow \neg p)$	
2.	$(p \square\rightarrow q) \,\&\, \neg q$	ACP
3.	$p \square\rightarrow q$	&E, 2
4.	$\neg q$	&E, 2
5.	Show $\neg p$	
6.	p	AIP
7.	q	$\square\rightarrow$E, 3, 6
8.	$\neg q$	R, 4
9.	Show $\neg q \square\rightarrow \neg p$	
	$\square\rightarrow$	
10.	$\neg q$	A$\square\rightarrow$P
11.	$\neg p$	$\square\rightarrow$R3, 5(8, 10)

20.

1.	$p \diamondsuit \to q$	A
2.	$p \,\square \to r$	A
3.	$\neg r$	A
4.	~~Show~~ $\neg r \,\square \to \neg(p \,\&\, q)$	
5.	⌐~~Show~~ $\neg(p \,\&\, q)$	
6.	⌐ $p \,\&\, q$	AIP
7.	p	&E, 6
8.	r	$\square \to$E, 2, 7
9.	└$\neg r$	R, 3
10.	~~Show~~ $\neg r \,\square \to \neg(p \,\&\, q)$	
	⌐$\square \to$	
11.	$\neg r$	A$\square \to$P
12.	└└$\neg(p \,\&\, q)$	$\square \to$R3, 5(3, 11)

25.

1.	$r \dashv s$	A
2.	$(p \,\&\, q) \diamondsuit \to \neg s$	A
3.	$p \diamondsuit \to q$	A
4.	~~Show~~ $p \diamondsuit \to \neg r$	
5.	⌐$\neg(p \diamondsuit \to \neg r)$	AIP
6.	$p \,\square \to r$	CLDf$\diamondsuit \to$, 5
7.	~~Show~~ $\neg(p \,\square \to \neg(p \,\&\, q))$	
8.	⌐$p \,\square \to \neg(p \,\&\, q)$	AIP
9.	~~Show~~ $p \,\square \to \neg q$	
	⌐$\square \to$	
10.	p	A$\square \to$P
11.	$\neg(p \,\&\, q)$	$\square \to$R2, 8(10)
12.	$\neg p \lor \neg q$	$\neg\&$, 11
13.	└$\neg q$	$\lor$E*, 12, 10
14.	└$\neg(p \,\square \to \neg q)$	CLDf$\diamondsuit \to$, 3
15.	$p \diamondsuit \to (p \,\&\, q)$	CLDf$\diamondsuit \to$, 7
16.	~~Show~~ $(p \,\&\, q) \,\square \to s$	
	⌐$\square \to$	
17.	$p \,\&\, q$	A$\square \to$P
18.	r	$\square \to$R4, 6(15, 17)
19.	$r \dashv s$	$\square \to$R1, 1
20.	└s	$\dashv$E, 19, 18
21.	└$\neg((p \,\&\, q) \,\square \to s)$	CLDf$\diamondsuit \to$, 2

35.

1.	$p \,\&\, \neg r$	A
2.	$(p \,\&\, q) \,\square \to r$	A
3.	~~Show~~ $p \,\square \to \neg q$	
4.	⌐p	&E, 1
5.	$\neg r$	&E, 1
6.	~~Show~~ $p \,\square \to \neg q$	
	⌐$\square \to$	
7.	p	A$\square \to$P
8.	~~Show~~ $\neg q$	
9.	⌐q	AIP
10.	r	$\square \to$R2, 2(7, 9)
11.	└└$\neg r$	$\square \to$R3, 5(4, 7)

50.
1. $p \square \rightarrow q$ A
2. $q \square \rightarrow p$ A
3. Show $(p \diamond \rightarrow r) \leftrightarrow (q \diamond \rightarrow r)$
4. Show $(p \diamond \rightarrow r) \rightarrow (q \diamond \rightarrow r)$
5. $p \diamond \rightarrow r$ ACP
6. Show $q \diamond \rightarrow r$
7. $\neg(q \diamond \rightarrow r)$ AIP
8. $q \square \rightarrow \neg r$ CLDf$\diamond \rightarrow$, 7
9. Show $p \square \rightarrow \neg r$
     $\square \rightarrow$
10. p A$\square \rightarrow$P
11. $\neg r$ $\square \rightarrow$R2, 8(1, 2, 10)
12. $\neg(p \square \rightarrow \neg r)$ CLDf$\diamond \rightarrow$, 5
13. Show $(q \diamond \rightarrow r) \rightarrow (p \diamond \rightarrow r)$
14. $q \diamond \rightarrow r$ ACP
15. Show $p \diamond \rightarrow r$
16. $\neg(p \diamond \rightarrow r)$ AIP
17. $p \square \rightarrow \neg r$ CLDf$\diamond \rightarrow$, 16
18. Show $q \square \rightarrow \neg r$
     $\square \rightarrow$
19. q A$\square \rightarrow$P
20. $\neg r$ $\square \rightarrow$R2, 17(1, 2, 19)
21. $\neg(q \square \rightarrow \neg r)$ CLDf$\diamond \rightarrow$, 14
22. $(p \diamond \rightarrow r) \leftrightarrow (q \diamond \rightarrow r)$ $\leftrightarrow$I, 4, 13

70.
1. $\neg r \square \rightarrow \neg s$ A
2. $p \diamond \rightarrow s$ A
3. $\neg r \square \rightarrow p$ A
4. $(p \& q) \square \rightarrow \neg r$ A
5. Show $p \diamond \rightarrow \neg q$
6. $\neg(p \diamond \rightarrow \neg q)$ AIP
7. $p \square \rightarrow q$ CLDf$\diamond \rightarrow$, 6
8. Show $p \square \rightarrow (p \& q)$
     $\square \rightarrow$
9. p A$\square \rightarrow$P
10. q $\square \rightarrow$R2, 7(9)
11. $p \& q$ &I, 9, 10
12. Show $(p \& q) \square \rightarrow p$
     $\square \rightarrow$
13. $p \& q$ A$\square \rightarrow$P
14. p &E, 13
15. Show $p \square \rightarrow \neg r$
     $\square \rightarrow$
16. p A$\square \rightarrow$P
17. $\neg r$ $\square \rightarrow$R2, 4(8, 12)
18. Show $p \square \rightarrow \neg s$
     $\square \rightarrow$
19. p A$\square \rightarrow$P
20. $\neg s$ $\square \rightarrow$R2, 1(3, 15)
21. $\neg(p \square \rightarrow \neg s)$ CLDf$\diamond \rightarrow$, 2

Chapter 11
OBLIGATION

11.1 Deontic Connectives

5. *p*: Bertha left unannounced. O¬*p*.

10. *p*: Verne sends Harriet a copy; *q*: He agreed to keep the agreement secret. P*p* ∨ *q*.

15. *p*: Carl forgets; *q*: Penny forgets; *r*: The rules are fair. (O¬*p* & P*q*) ⊰ ¬*r*.

20. *p*: You do this; *q*: You do a good job. (*p* ⊰ O*p*) ⊰ ¬*q*.

25. *p*: You play it mean; *q*: They hate you on the field. O*p*. O*q*.

30. *p*: You execute every act of life as though it were your last. O*p*.

35. *p*: You promise rarely; *q*: It is lawful; *r*: You constantly perform. O*p* & (*q* ⊰ O*r*).

40. *p*: You win; *q*: You make the winner look great. ¬◇*p* ⊰ O*q*.

45. *p*: You are kind; *q*: Everyone you meet is fighting a hard battle. *q* ∴ O*p*.

50. *p*: You do; *q*: There is trying. O*p* ∨ O¬*p*. ¬*q*.

11.2 Deontic Tableaux

5. *p*: We elect Bob; *q*: We withdraw from the nuclear project. Not valid in D.

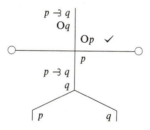

10. *p*: We increase incentives; *q*: Productivity improves. Valid in D.

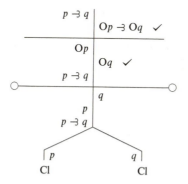

20. *p*: You are like me; *q*: You resist temptation. Valid in D.

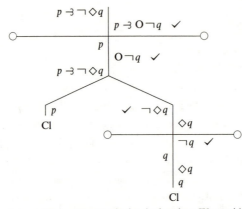

25. *p*: We've acted too late; *q*: The enemy attacks by daybreak; *r*: We avoid an all-out conflict. Valid in D.

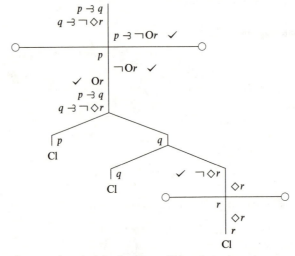

30. *p*: This act conforms to the principle of utility; *q*: This act's tendency to augment the happiness of the community is greater than its tendency to diminish it; *r*: You perform this act. Valid in D.

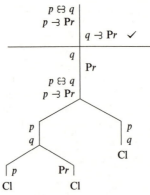

40. Not equivalent.

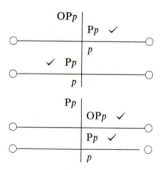

50. Not equivalent, though Pp implies $\Diamond$Pp.

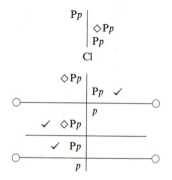

55. Not equivalent, though P$\Box p$ does imply $\Box$Pp.

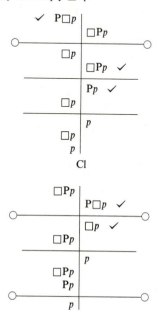

60. Valid in D.

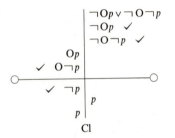

70. Valid in D.

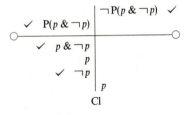

75. Valid in D.

80. Not valid in D.

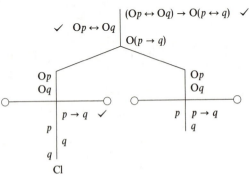

11.3 Deontic Deduction

5. 1. ¬Pp A
    2. ~~Show~~ P¬p

10. 1. ¬◇p A
     2. ~~Show~~ ¬Pp

3. ⌐¬P¬p   AIP
4. Op   P, 3
5. O¬p   P, 1
6. Show O(p & ¬p)
7. p   OR, 4
8. ¬p   OR, 5
9. p & ¬p   &I, 7, 8
10. ¬O(p & ¬p)   OC

15.
1. p ⊰ Oq   A
2. ¬◇q   A
3. Show ¬◇p
4. Show □¬p
5. Show ¬p
6. p   AIP
7. p ⊰ Oq   □R, 1
8. ¬◇q   □R, 2
9. Oq   ⊰E, 7, 6
10. ◇q   OE, 9
11. ¬◇p   Df◇, 4

23.
1. p → r   A
2. r → □¬q   A
3. Show p → ¬Pq
4. p   AIP
5. r   →E, 1, 4
6. □¬q   →E, 2, 5
7. Show O¬q
8. ¬q   □E, 6
9. ¬Pq   P, 7

3. Pp   AIP
4. ¬O¬p   P, 3
5. Show O¬p
6. Show ¬p
7. p   AIP
8. ◇p   ◇I, 7
9. ¬◇p   OR, 1

20.
1. p ⊰ Oq   A
2. ◇r ⊰ O¬q   A
3. Show p ⊰ □¬r
4. p   A□P
5. p ⊰ Oq   □R, 1
6. ◇r ⊰ O¬q   □R, 2
7. Oq   ⊰E, 5, 4
8. Show ¬◇r
9. ◇r   AIP
10. O¬q   ⊰E, 6, 9
11. Show O(q & ¬q)
12. q   OR, 7
13. ¬q   OR, 10
14. q & ¬q   &I, 12, 13
15. ¬O(q & ¬q)   OC
16. □¬r   Df◇, 8

31.
1. p ⊰ Oq   A
2. ◇q ⊰ Or   A
3. Pr ⊰ s   A
4. Show □¬s ⊰ □¬p
5. □¬s   A□P
6. Show □¬p
7. Show ¬p
8. p   AIP
9. p ⊰ Oq   □R, 1
10. ◇q ⊰ Or   □R, 2
11. Pr ⊰ s   □R, 3
12. Oq   ⊰E, 9, 8
13. ◇q   OE, 12
14. Or   ⊰E, 10, 13
15. □¬s   □R, 5
16. ¬s   □E, 15
17. ¬Pr   ⊰E*, 11, 16
18. O¬r   P, 17
19. Show O(r & ¬r)
20. r   OR, 14
21. ¬r   OR, 18
22. r & ¬r   &I, 20, 21
23. ¬O(r & ¬r)   OC

36. 1. Show (O*p* & O*q*) ↔ O(*p* & *q*)
    2. Show (O*p* & O*q*) → O(*p* & *q*)
      3. O*p* & O*q*   ACP
      4. O*p*   &E, 3
      5. O*q*   &E, 3
      6. *m* ⌐∃ *p*   Reduction, 4
      7. *m* ⌐∃ *q*   Reduction, 5
      8. Show *m* ⌐∃ (*p* & *q*)
        9. *m*   A□P
      10. *m* ⌐∃ *p*   □R, 6
      11. *m* ⌐∃ *q*   □R, 7
      12. *p*   ⌐∃E, 10, 9
      13. *q*   ⌐∃E, 11, 9
      14. *p* & *q*   &I, 12, 13
      15. O(*p* & *q*)   Reduction, 8
    16. Show O(*p* & *q*) → (O*p* & O*q*)
      17. O(*p* & *q*)   ACP
      18. *m* ⌐∃ (*p* & *q*)   Reduction, 17
      19. Show *m* ⌐∃ *p*
        20. *m*   A□P
      21. *m* ⌐∃ (*p* & *q*)   □R, 18
      22. *p* & *q*   ⌐∃E, 21, 20
      23. *p*   &E, 22
      24. Show *m* ⌐∃ *q*
        25. *m*   A□P
      26. *m* ⌐∃ (*p* & *q*)   □R, 18
      27. *p* & *q*   ⌐∃E, 26, 25
      28. *q*   &E, 27
      29. O*p*   Reduction, 19
      30. O*q*   Reduction, 24
      31. O*p* & O*q*   &I, 29, 30
    32. (O*p* & O*q*) ↔ O(*p* & *q*)   ↔I, 2, 16

11.4 Moral and Practical Reasoning

5. Valid in D.
  1. Show (*p* ⌐∃ *q*) → (O*p* □→ O*q*)
  2. *p* ⌐∃ *q*   ACP
  3. Show O*p* □→ O*q*
      □→
  4. O*p*   A□→P
  5. *p* ⌐∃ *q*   □→R1, 2
  6. Show O*q*
      O
  7. *p*   OR, 4
  8. *p* ⌐∃ *q*   OR, 5
  9. *q*   ⌐∃E, 8, 7

10. Not valid in D.
  1. *p* □→ *q*   A
  2. ¬O*q*   A
  3. Show P¬*p*
  4. Show ¬O*p*
  5. O*p*   AIP
  6. Show O*q*
      O
  7. *p*   OR, 5
  8. ????? (*p* □→ *q* can't reiterate.)

15. Valid in D.
  1. *p* □→ O*q*   A
  2. ¬◇*q*   A
  3. Show ¬◇*p*

    4. $\lceil \diamond p$ AIP

    5. $\;\;$ ~~Show~~ $p\,\square\!\rightarrow\,\diamond q$

         $\lceil \square\!\rightarrow$

    6. $\;\;p$ A$\square\!\rightarrow$P

    7. $\;\;Oq$ $\square\!\rightarrow$R2, 1(6)

    8. $\lfloor\diamond q$ OE, 7

    9. $\;\;\diamond q$ $\diamond$E*, 4, 5

  10. $\lfloor\neg\diamond q$ R, 2

20. Not valid in D.

25. Not valid in D.

    1. $p\,\square\!\rightarrow\,Oq$ A

    2. $\diamond r\,\square\!\rightarrow\,O\neg q$ A

    3. Show $p\,\square\!\rightarrow\,\square\neg r$

       $\square\!\rightarrow$

    4. p A$\square\!\rightarrow$P

    5. Show $\square\neg r$

    6. $\neg\square\neg r$ AIP

    7. $\diamond r$ Df$\diamond$, 6

    8. ????? (We can't use 2; conditions for $\square\!\rightarrow$R2 aren't fulfilled.)

30. Not valid in D.

Chapter 12
QUANTIFIERS AND MODALITY

12.1 System QS5

5. Px: x is paper; Fx: x is fire; Wxy: x wraps y. Ambiguous. (1) $\neg\diamond\exists x\exists y(Px \;\&\; Fy \;\&\; Wxy)$; (2) $\forall x(Px \rightarrow \neg\diamond\exists y(Fy \;\&\; Wxy))$; (3) $\forall x\forall y((Px \;\&\; Fy) \rightarrow \neg\diamond Wxy)$.

10. Px: x is a principle; Ix: x is important; Fx: x is flexible. Ambiguous. (1) $\diamond\forall x((Px \;\&\; Ix) \rightarrow Fx) \;\&\; \square\forall x((Px \;\&\; Ix) \rightarrow Fx)$; (2) $\forall x((Px \;\&\; Ix) \rightarrow (\diamond Fx \;\&\; \square Fx))$.

15. $Pxyzt$: x plans y by z at t; Tx: x is a time; Bx: x is a person; $f(t)$: The future from t; $g(t)$: The past from t. Ambiguous. (1) $\forall t\forall x((Tt \;\&\; Bx) \rightarrow \neg\diamond Pxf(t)g(t)t)$; (2) $\neg\diamond\exists t\exists x(Tt \;\&\; Bx \;\&\; Pxf(t)g(t)t)$.

25. Axt: x is an author at t; Pxt: x is a poet at t; Ixt: x is in its infancy at t; a: Society; Tt: t is a time. Ambiguous. (1) $\forall t(Iat \rightarrow \square\forall x(Axt \rightarrow Pxt))$; (2) $\forall t(Iat \rightarrow \forall x(Axt \rightarrow \square Pxt))$.

35. Px: x is a person; Wx: x works faithfully eight hours a day; Bx: x is boss; Tx: x works twelve hours a day. $\forall x(Px \rightarrow (Wx \multimap \diamond(Bx \;\&\; Tx)))$.

40. Px: x is a person; Cxy: x changes y; $f(x)$: x's fabric; $g(x)$: x's thought; a: Reality. $\forall x(Px \rightarrow (\neg\diamond Cxf(g(x)) \rightarrow \neg\diamond Cxa))$.

45. Sx: x is sincere; Px: x is a person; Tx: x is stupid. $\forall x(Px \rightarrow \diamond(Sx \;\&\; Tx))$.

55. Mx: x is a Maxim's; i: I; b: Peking; c: The moon; $Pxyz$: x puts y in (on) z. $\diamond\exists x(Mx \;\&\; Pixb) \multimap \diamond\exists x(Mx \;\&\; Pixc)$.

60. Mxy: x makes y understood; Sx: x speaks plain; Wx: x writes plain; Px: x is a person. $\forall x(Px \rightarrow (Mxx \multimap (Sx \;\&\; Px)))$.

64. $Sxyz$: x sells y to z; a: The public; b: Business; Pxy: x preserves y. $Pbb \multimap Sbba$.

70. Px: x is a person; Fxy: x finds time for y; a: Recreation; b: Illness. $\forall x(Px \rightarrow (\neg\diamond Fxa \rightarrow OFxb))$.

75. $Txyz$: x takes y away from z; a: Fortune; $g(x)$: x's goods; $h(x)$: x's heart; Px: x is a person. $\forall x(Px \rightarrow (\diamond Tag(x)x \;\&\; \neg\diamond Tah(x)x))$.

85. *Sxyz*: *x* sends *y* to *z*; *Mx*: *x* is an MX missile; *Rxyz*: *x* reports *y* to *z* for fraud; *a*: Mr. Newman; *b*: Me; c: The postal service. $\neg \exists x(Mx \ \& \ Saxb) \dashv 3 \ Rbac$.

90. *Fxy*: *x* forgives *y*; *Dxy*: *x* destroys *y*; *Bx*: *x* is a bridge; *Pxy*: *x* passes over *y*; *Qx*: *x* is a person. $\forall x(Px \to (\neg \Diamond \exists y(y \neq x \ \& \ Fxy) \to Dx \imath z(Bz \ \& \ \Box Pxz)))$ or $\forall x(Px \to (\neg \Diamond \exists y(y \neq x \ \& \ Fxy) \to (\tau x: (Bz \ \& \ \Box Pxz))Dxz))$.

95. *Dxyz*: *x* does *y* about *z*; *Px*: *x* is a person; *l*(*x*): *x*'s length; *f*(*x*): *x*'s life; *g*(*x*): *x*'s width; *h*(*x*): *x*'s depth. Ambiguous. (1) $\forall x(Px \to (\neg \Diamond \exists y Dxyl(f(x)) \ \& \ \Diamond \exists z(Dxzg(f(x)) \ \& \ Dxzh(f(x)))))$; (2) $\forall x(Px \to (\neg \exists y \Diamond Dxyl(f(x)) \ \& \ \exists z \Diamond(Dxzg(f(x)) \ \& \ Dxzh(f(x)))))$.

110. *Cx*: *x* is change; *Px*: *x* is progress; *Ex*: *x* is education. $\neg(\exists x Cx \dashv 3 \ \exists x Px)$; $\exists x Px \dashv 3 \ \exists x Cx$; $\exists x Cx \dashv 3 \ \exists x Ex$. This follows from Commager's statement.

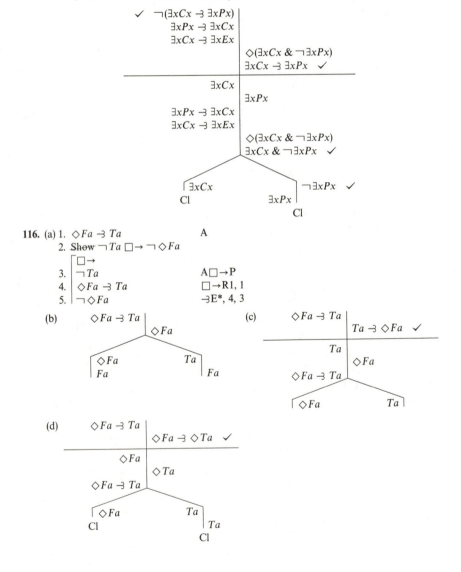

(e) $\quad \Diamond Fa \dashv 3\ Ta$

$\qquad\qquad\qquad$ | $\Box(\Diamond Fa \dashv 3\ Ta)$ ✓
$\underline{\qquad\qquad\qquad\qquad}$ | $\Diamond Fa \dashv 3\ Ta$
$\qquad \Diamond Fa \dashv 3\ Ta$ |
$\qquad\qquad\qquad$ Cl

Implies (a), (d), and (e).

140. Follows.

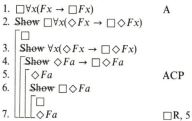

1. $\Box \forall x(Fx \to \Box Fx)$ $\qquad$ A
2. ~~Show~~ $\Box \forall x(\Diamond Fx \to \Box \Diamond Fx)$
$\qquad \Box$
3. $\quad$ ~~Show~~ $\forall x(\Diamond Fx \to \Box \Diamond Fx)$
4. $\qquad$ ~~Show~~ $\Diamond Fa \to \Box \Diamond Fa$
5. $\qquad\quad \Diamond Fa$ $\qquad$ ACP
6. $\qquad\quad$ ~~Show~~ $\Box \Diamond Fa$
$\qquad\qquad\quad \Box$
7. $\qquad\qquad \Diamond Fa$ $\qquad$ $\Box$R, 5

150. Does not follow.

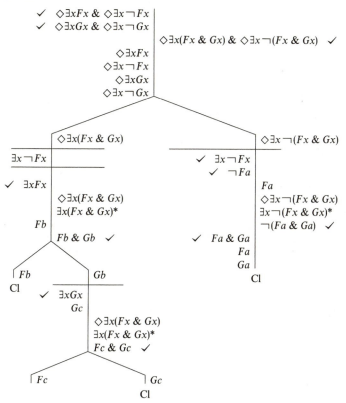

160. Follows.

1.	$\diamond\exists x(Fx \& Gx) \& \diamond\exists x\neg(Fx \& Gx)$	A
2.	Show $(\diamond\exists xFx \& \diamond\exists x\neg Fx) \vee (\diamond\exists xGx \& \diamond\exists x\neg Gx)$	
3.	$\neg((\diamond\exists xFx \& \diamond\exists x\neg Fx) \vee (\diamond\exists xGx \& \diamond\exists x\neg Gx))$	AIP
4.	$\neg(\diamond\exists xFx \& \diamond\exists x\neg Fx) \& \neg(\diamond\exists xGx \& \diamond\exists x\neg Gx)$	$\neg\vee$, 3
5.	$\neg(\diamond\exists xFx \& \diamond\exists x\neg Fx)$	&E, 4
6.	$\neg(\diamond\exists xGx \& \diamond\exists x\neg Gx)$	&E, 4
7.	$\neg\diamond\exists xFx \vee \neg\diamond\exists x\neg Fx$	$\neg\&$, 5
8.	$\neg\diamond\exists xGx \vee \neg\diamond\exists x\neg Gx$	$\neg\&$, 6
9.	$\diamond\exists x(Fx \& Gx)$	&E, 1
10.	$\diamond\exists x\neg(Fx \& Gx)$	&E, 1
11.	Show $\exists x(Fx \& Gx) \dashv3 \diamond\exists xFx$	
12.	$\exists x(Fx \& Gx)$	A□P
13.	$Fa \& Ga$	$\exists$E, 12
14.	Fa	&E, 13
15.	$\exists xFx$	$\exists$I, 14
16.	$\diamond\exists xFx$	$\diamond$I, 15
17.	$\diamond\exists xFx$	$\diamond$E, 9, 11
18.	Show $\exists x(Fx \& Gx) \dashv3 \diamond\exists xGx$	
19.	$\exists x(Fx \& Gx)$	A□P
20.	$Fa \& Ga$	$\exists$E, 19
21.	Ga	&E, 20
22.	$\exists xGx$	$\exists$I, 21
23.	$\diamond\exists xGx$	$\diamond$I, 22
24.	$\diamond\exists xGx$	$\diamond$E, 9, 18
25.	$\neg\diamond\exists x\neg Fx$	$\vee$E*, 7, 17
26.	$\neg\diamond\exists x\neg Gx$	$\vee$E*, 8, 24
27.	$\square\neg\exists x\neg Fx$	Df$\diamond$, 25
28.	$\square\neg\exists x\neg Gx$	Df$\diamond$, 26
29.	$\square\forall xFx$	QN, 27
30.	$\square\forall xGx$	QN, 28
31.	Show $\square\forall x(Fx \& Gx)$	
32.	Show $\forall x(Fx \& Gx)$	
33.	Show $Fb \& Gb$	
34.	$\square\forall xFx$	□R, 29
35.	$\square\forall xGx$	□R, 30
36.	$\forall xFx$	□E, 34
37.	$\forall xGx$	□E, 35
38.	Fb	$\forall$E, 36
39.	Gb	$\forall$E, 37
40.	$Fb \& Gb$	&I, 38, 39
41.	$\neg\diamond\neg\forall x(Fx \& Gx)$	Df$\diamond$, 31
42.	$\neg\diamond\exists x\neg(Fx \& Gx)$	QN, 41

12.2 Free Logic

5. Tx: x thinks; a: I. $Ta \therefore E!a$.

10. Valid in FS5.

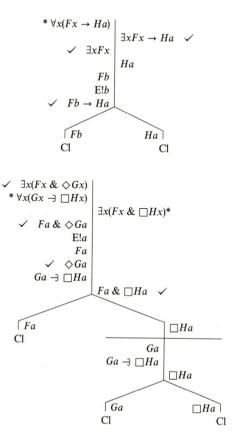

15. Valid in FS5.

20. Not valid in FS5.

25.
1. $\exists x(Fx \ \& \ \Box Gx)$ — A
2. $\forall x(Fx \rightarrow Hx)$ — A
3.  Show $\exists x(Hx \ \& \ \Box Gx)$
4. ⌈ $Fa \ \& \ \Box Ga \ \& \ E!a$ — F$\exists$E, 1
5. | $E!a$ — &E, 4
6. | Fa — &E, 4
7. | $\Box Ga$ — &E, 4
8. | $Fa \rightarrow Ha$ — F$\forall$E, 2, 5
9. | Ha — →E, 8, 6
10. | $Ha \ \& \ \Box Ga$ — &I, 9, 7
11. ⌊ $\exists x(Hx \ \& \ \Box Gx)$ — F$\exists$I, 10, 5

30.

1.	◇∃x(□Fx & Gx)	A
2.	□∀x(Gx ⊰ ◇∃yHxy)	A
3.	~~Show~~ ◇∃x◇∃y(Hxy & □Fx)	
4.	⌈~~Show~~ ∃x(□Fx & Gx) ⊰ ◇∃x◇∃y(Hxy & □Fx)	
	⌈□	
5.	│ ∃x(□Fx & Gx)	A□P
6.	│ □Fa & Ga & E!a	F∃E, 5
7.	│ □∀x(Gx ⊰ ◇∃yHxy)	□R, 2
8.	│ ∀x(Gx ⊰ ◇∃yHxy)	□E, 7
9.	│ E!a	&E, 6
10.	│ Ga ⊰ ◇∃yHay	F∀E, 8, 9
11.	│ Ga	&E, 6
12.	│ ◇∃yHay	⊰E, 10, 11
13.	│ □Fa	&E, 6
14.	│ ~~Show~~ ∃yHay ⊰ ◇∃y(Hay & □Fa)	
	│ ⌈□	
15.	│ │ ∃yHay	A□P
16.	│ │ Hab & E!b	F∃E, 15
17.	│ │ Hab	&E, 16
18.	│ │ E!b	&E, 16
19.	│ │ □Fa	□R, 13
20.	│ │ Hab & □Fa	&I, 17, 19
21.	│ │ ∃y(Hay & □Fa)	F∃I, 20, 18
22.	│ └◇∃y(Hay & □Fa)	◇I, 21
23.	│ ◇∃y(Hay & □Fa)	◇E, 12, 14
24.	│ ∃x◇∃y(Hxy & □Fx)	F∃I, 9, 23
25.	└◇∃x◇∃y(Hxy & □Fx)	◇I, 24
26.	└◇∃x◇∃y(Hxy & □Fx)	◇E, 1, 4

12.3 Identity and Descriptions

5. c: Cicero; d: Tully; Sxy: x speaks more effectively than y; Rx: x is a Roman; a: Cataline. Not valid in QS5 (because 'Cataline' and 'Cicero' could name the same person); not valid in FS5 (because, in addition, Cataline might not have existed). QS5 tableau:

```
                              * c = d │
           * ∀x((Rx & x ≠ c) → ◇Scx) │
                                   Ra │
                                      │ ◇Sda
                  ✓ (Ra & a ≠ c) → ◇Sca │
                               _____
                              │                   │
                   ┌ Ra & a ≠ c  ✓         ◇Sca │
                   │                        ◇Sda │
             _____│_____                   Cl
            │             │
          ┌ Ra       a ≠ c  ✓
          Cl      * a = c
                     Rc
                  * a = d
                     Rd
                        │ ◇Sdc
                        │ ◇Sac
                        │ ◇Sad
                        │ ◇Sca
                        │ ◇Scd
                        │ ◇Saa
                        │ ◇Scc
                        │ ◇Sdd
```

10. *a*: Pegasus; *Wx*: *x* has wings; *Hx*: *x* is a horse. Valid in QS5; not valid in FS5 (because Pegasus could exist without being a winged horse). QS5 tableau:

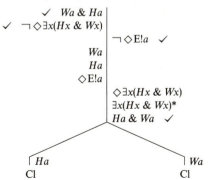

FS5 tableau:

15. Follows in QS5; does not follow in FS5 (because *t* and *t'* might not denote). QS5 tableau:

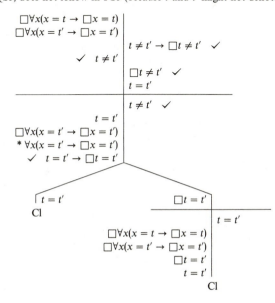

20. Follows in QS5 (whether or not t is rigid). Does not follow in FS5 (because t and t' might not denote). QS5 tableau:

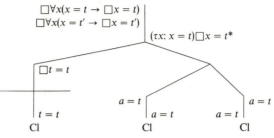

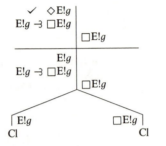

25. g: God. Valid in QS5 and FS5.

12.4 Quantification and Obligation

5. Px: x is a person; $f(x)$: The color of x; a: Life; Axy: x absorbs y; $Rxyz$: x remembers y at time z; Tx: x is a time; $g(x)$: x's details. $\forall x(Px \to (ORxf(a)\ \&\ O\neg\exists z(Tz\ \&\ Rxg(a)z)))$.

10. Wx: x will work; Ex: x eats. $\forall x(\neg Wx \to O\neg Ex)$.

15. Px: x is a person; Cx: x is a conviction; Hx: x is honest; Bxy: x has y; Sx: x is strong; Fxy: x follows y. $\forall x(Px \to O(\forall y((Cy\ \&\ Hy\ \&\ Bxy) \to Fxy)\ \&\ Sx))$.

20. Px: x is a person; $Cxyz$: x cuts y according to z; $f(x)$: x's coat; $g(x)$: x's cloth; $Axyz$: x adapts y to z; Hx: x is changing; Sx: x is a circumstance. $\forall x(Px \to O(Cxf(x)g(x)\ \&\ \forall y((Sy\ \&\ Hy) \to Axxy)))$.

25. Px: x is a person; Gxy: x goes confidently in the direction of y; $f(x)$: x's dreams; Lxy: x lives y; Ixy: x has imagined y; Fx: x is a life. $\forall x(Px \to (OGxf(x)\ \&\ OLx\imath y(Fy\ \&\ Ixy)))$ or $\forall x(Px \to (OGxf(x)\ \&\ O(\tau y\colon Fy\ \&\ Ixy)Lxy))$.

30. Px: x is a person; $f(x)$: x's best; Dxy: x has done y; Fx: x fails; $g(x)$: x's worst; $h(x)$: x's error; Cx: x succeeds; Sxy: x sees y. $\forall x(Px \to (((Dxf(x)\ \&\ Fx)\ \multimap ODxg(x))\ \&\ \Diamond(Sxh(x)\ \&\ Cx)))$.

35. Px: x is a person; Hx: x heard; $Sxyz$: x sees y at z; a: Them; c: There; Cxy: x cries y; b: 'Beware! Beware!' $\forall x(Px \to ((Hx \to OSxac)\ \&\ Oxb))$.

40. a: Failure; $Cxyz$: x challenges y to z; Px: x is a person; Hxy: x is a new height of y; c: Accomplishment; Dxy: x is a new depth of y; d: Despair; $Pxyz$: x pulls y to z. $\forall x(Px \to (O\exists y(Caxy\ \&\ Hyc)\ \&\ \neg O\exists z(Paxz\ \&\ Dzd)))$ or $\forall x(Px \to (O\exists y(Caxy\ \&\ Hyc)\ \&\ O\neg\exists z(Paxz\ \&\ Dzd)))$.

45. Px: x is a person; Ex: x is an evil; $Sxyzw$: x speaks y of z at time w; $Kxyz$: x knows y for a certainty at time z; $Axyzwt$: x asks y, at time t, why z should tell w; Tx: x is a time. $\forall x(Px \to \forall y\forall z\forall w((Ey\ \&\ Pz\ \&\ Tw) \to ((O\neg Sxyzw \lor Kxyw)\ \&\ (Kxyw \to OAxxxyw))))$.

50. Px: x is a person; Wx: x wins; Mxy: x makes y look great. $\forall x(Px \to (\neg\Diamond Wx \multimap OMx\imath yWy))$ or $\forall x(Px \to (\neg\Diamond Wx \multimap O(\tau y\colon Wy)Mxy))$.

55. *Px*: *x* is a person; *Vx*: *x* is perverted; *Mx*: *x* is a mind; *Hxyz*: *x* hears *y* in defense of *z*; *Sx*: *x* is silent. $\forall x(Mx \to (\Diamond Px \,\rJ\, \forall y((Py \,\&\, \exists z OHzyx) \to Sy)))$.

70. Follows in both QS5 and FS5. FS5 tableau:

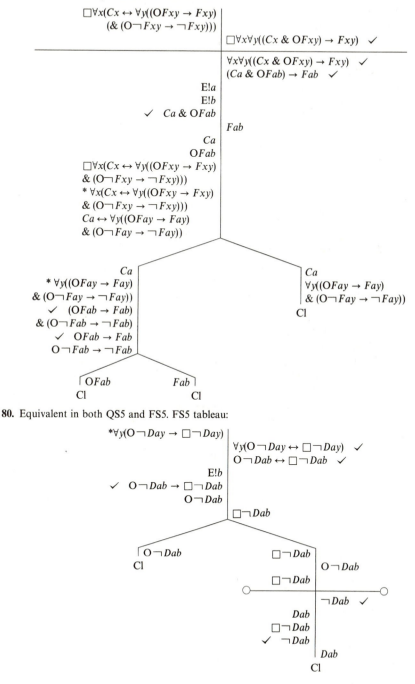

80. Equivalent in both QS5 and FS5. FS5 tableau:

90. QS5 deduction:

1. $\forall x \forall y (PDxy \lor \exists z(Lz \ \& \ O_z \neg Dxy))$ A
2. ~~Show~~ $\neg \exists xLx \rightarrow \forall x \forall y PDxy$
3. $\lceil \neg \exists xLx$ ACP
4. $\mid$ ~~Show~~ $\forall x \forall y PDxy$
5. $\mid$ $\lceil$ ~~Show~~ $PDab$
6. $\mid$ $\mid$ $\lceil PDab \lor \exists z(Lz \ \& \ O_z \neg Dxy)$ $\forall E^2$, 1
7. $\mid$ $\mid$ $\mid$ ~~Show~~ $\neg \exists z(Lz \ \& \ O_z \neg Dxy)$
8. $\mid$ $\mid$ $\mid$ $\lceil \exists z(Lz \ \& \ O_z \neg Dxy)$ AIP
9. $\mid$ $\mid$ $\mid$ $\mid$ $Lc \ \& \ O_c \neg Dxy$ $\exists E$, 8
10. $\mid$ $\mid$ $\mid$ $\mid$ Lc &E, 9
11. $\mid$ $\mid$ $\mid$ $\mid$ $\exists xLx$ $\exists I$, 10
12. $\mid$ $\mid$ $\mid$ $\llcorner \neg \exists xLx$ R, 3
13. $\llcorner\llcorner\llcorner PDab$ $\lor E^*$, 6, 7

FS5 deduction:

1. $\forall x \forall y (PDxy \lor \exists z(Lz \ \& \ O_z \neg Dxy))$ A
2. ~~Show~~ $\neg \exists xLx \rightarrow \forall x \forall y PDxy$
3. $\lceil \neg \exists xLx$ ACP
4. $\mid$ ~~Show~~ $\forall x \forall y PDxy$
5. $\mid$ $\lceil$ ~~Show~~ $E!a \rightarrow (E!b \rightarrow PDab)$
6. $\mid$ $\mid$ $\lceil E!a$ ACP
7. $\mid$ $\mid$ ~~Show~~ $E!b \rightarrow PDab$
8. $\mid$ $\mid$ $\lceil E!b$ ACP
9. $\mid$ $\mid$ $\mid$ $PDab \lor \exists z(Lz \ \& \ O_z \neg Dab)$ F$\forall E^2$, 1, 6, 8
10. $\mid$ $\mid$ $\mid$ ~~Show~~ $\neg \exists z(Lz \ \& \ O_z \neg Dab)$
11. $\mid$ $\mid$ $\mid$ $\lceil \exists z(Lz \ \& \ O_z \neg Dxy)$ AIP
12. $\mid$ $\mid$ $\mid$ $\mid$ $Lc \ \& \ O_c \neg Dxy \ \& \ E!c$ $\exists E$, 11
13. $\mid$ $\mid$ $\mid$ $\mid$ Lc &E, 12
14. $\mid$ $\mid$ $\mid$ $\mid$ $E!c$ &E, 12
15. $\mid$ $\mid$ $\mid$ $\mid$ $\exists xLx$ F$\exists I$, 13, 14
16. $\mid$ $\mid$ $\mid$ $\llcorner \neg \exists xLx$ R, 3
17. $\llcorner\llcorner\llcorner\llcorner PDab$ $\lor E^*$, 9, 10

INDEX

DEDUCTION RULES

&E

n. $\underline{\mathscr{A} \ \& \ \mathscr{B}}$

n + m. $\mathscr{A}$ (or $\mathscr{B}$)   &E, n

&I

n. $\mathscr{A}$

m. $\underline{\mathscr{B}}$

p. $\mathscr{A} \ \& \ \mathscr{B}$ & I, n, m

∨E

k. $\mathscr{A} \lor \mathscr{B}$

l. $\mathscr{A} \to \mathscr{C}$

m. $\underline{\mathscr{B} \to \mathscr{C}}$

n. $\mathscr{C}$ ∨E, k, l, m

∨I

n. $\underline{\mathscr{A}}$

n + m. $\mathscr{A} \lor \mathscr{B}$ (or $\mathscr{B} \lor \mathscr{A}$)   ∨I, n

↔E

n. $\mathscr{A} \leftrightarrow \mathscr{B}$

m. $\underline{\mathscr{A}}$ (or $\mathscr{B}$)

p. $\mathscr{B}$ (or $\mathscr{A}$)   ↔E, n, m

↔I

n. $\mathscr{A} \to \mathscr{B}$

m. $\underline{\mathscr{B} \to \mathscr{A}}$

p. $\mathscr{A} \leftrightarrow \mathscr{B}$ ↔I, n, m

→E

n. $\mathscr{A} \to \mathscr{B}$

m. $\underline{\mathscr{A}}$

p. $\mathscr{B}$ →E, n, m

Conditional Proof

n.   Show $\mathscr{A} \to \mathscr{B}$

n + 1. $\lceil \mathscr{A}$ ACP

       $\vdots$

m. $\lfloor \mathscr{B}$